Selected Papers from the 5th International Symposium on Mycotoxins and Toxigenic Moulds: Challenges and Perspectives

Special Issue Editors

Sarah De Saeger

Siska Croubels

Kris Audenaert

MDPI • Basel • Beijing • Wuhan • Barcelona • Belgrade

Special Issue Editors
Sarah De Saeger
Ghent University
Belgium

Siska Croubels
Ghent University
Belgium

Kris Audenaert
Ghent University
Belgium

Editorial Office
MDPI AG
St. Alban-Anlage 66
Basel, Switzerland

This edition is a reprint of the Special Issue published online in the open access journal *Toxins* (ISSN 2072-6651) from 2016–2017 (available at: http://www.mdpi.com/journal/toxins/special_issues/papers_5MYTOX).

For citation purposes, cite each article independently as indicated on the article page online and as indicated below:

Author 1; Author 2. Article title. *Journal Name* **Year**, *Article number*, page range.

First Edition 2017

ISBN 978-3-03842-484-0 (Pbk)
ISBN 978-3-03842-485-7 (PDF)

Articles in this volume are Open Access and distributed under the Creative Commons Attribution license (CC BY), which allows users to download, copy and build upon published articles even for commercial purposes, as long as the author and publisher are properly credited, which ensures maximum dissemination and a wider impact of our publications. The book taken as a whole is © 2017 MDPI, Basel, Switzerland, distributed under the terms and conditions of the Creative Commons license CC BY-NC-ND (http://creativecommons.org/licenses/by-nc-nd/4.0/).

Table of Contents

About the Special Issue Editors v

Preface to "Selected Papers from the 5th International Symposium on Mycotoxins and Toxigenic Moulds: Challenges and Perspectives" vii

Ilse Vanhoutte, Laura De Mets, Marthe De Boevre, Valdet Uka, José Diana Di Mavungu, Sarah De Saeger, Leen De Gelder and Kris Audenaert
Microbial Detoxification of Deoxynivalenol (DON), Assessed via a *Lemna minor* L. Bioassay, through Biotransformation to 3-epi-DON and 3-epi-DOM-1
Reprinted from: *Toxins* **2017**, *9*(2), 63; doi: 10.3390/toxins9020063 1

Valdet Uka, Geromy G. Moore, Natalia Arroyo-Manzanares, Dashnor Nebija, Sarah De Saeger and José Diana Di Mavungu
Unravelling the Diversity of the Cyclopiazonic Acid Family of Mycotoxins in *Aspergillus flavus* by UHPLC Triple-TOF HRMS
Reprinted from: *Toxins* **2017**, *9*(1), 35; doi: 10.3390/toxins9010035 19

Wen Shi, Yanglan Tan, Shuangxia Wang, Donald M. Gardiner, Sarah De Saeger, Yucai Liao, Cheng Wang, Yingying Fan, Zhouping Wang and Aibo Wu
Mycotoxigenic Potentials of *Fusarium* Species in Various Culture Matrices Revealed by Mycotoxin Profiling
Reprinted from: *Toxins* **2017**, *9*(1), 6; doi: 10.3390/toxins9010006 40

Cynthia Adaku Chilaka, Marthe De Boevre, Olusegun Oladimeji Atanda and Sarah De Saeger
Occurrence of *Fusarium* Mycotoxins in Cereal Crops and Processed Products (*Ogi*) from Nigeria
Reprinted from: *Toxins* **2016**, *8*(11), 342; doi: 10.3390/toxins8110342 55

Mohamed F. Abdallah, Rudolf Krska and Michael Sulyok
Mycotoxin Contamination in Sugarcane Grass and Juice: First Report on Detection of Multiple Mycotoxins and Exposure Assessment for Aflatoxins B_1 and G_1 in Humans
Reprinted from: *Toxins* **2016**, *8*(11), 343; doi: 10.3390/toxins8110343 73

Silvio Uhlig, Ana Stanic, Ingerd S. Hofgaard, Bernhard Kluger, Rainer Schuhmacher and Christopher O. Miles
Glutathione-Conjugates of Deoxynivalenol in Naturally Contaminated Grain Are Primarily Linked via the Epoxide Group
Reprinted from: *Toxins* **2016**, *8*(11), 329; doi: 10.3390/toxins8110329 85

Vitaly G. Dzhavakhiya, Tatiana M. Voinova, Sofya B. Popletaeva, Natalia V. Statsyuk, Lyudmila A. Limantseva and Larisa A. Shcherbakova
Effect of Various Compounds Blocking the Colony Pigmentation on the Aflatoxin B1 Production by *Aspergillus flavus*
Reprinted from: *Toxins* **2016**, *8*(11), 313; doi: 10.3390/toxins8110313 97

H. J. van der Fels-Klerx and Louise Camenzuli
Effects of Milk Yield, Feed Composition, and Feed Contamination with Aflatoxin B1 on the Aflatoxin M1 Concentration in Dairy Cows' Milk Investigated Using Monte Carlo Simulation Modelling
Reprinted from: *Toxins* **2016**, *8*(10), 290; doi: 10.3390/toxins8100290 108

Martina Loi, Francesca Fanelli, Paolo Zucca, Vania C. Liuzzi, Laura Quintieri, Maria T. Cimmarusti, Linda Monaci, Miriam Haidukowski, Antonio F. Logrieco, Enrico Sanjust and Giuseppina Mulè
Aflatoxin B_1 and M_1 Degradation by Lac2 from *Pleurotus pulmonarius* and Redox Mediators
Reprinted from: *Toxins* **2016**, *8*(9), 245; doi: 10.3390/toxins8090245 119

Rachelle El Khoury, Ali Atoui, Carol Verheecke, Richard Maroun, Andre El Khoury and Florence Mathieu
Essential Oils Modulate Gene Expression and Ochratoxin A Production in *Aspergillus carbonarius*
Reprinted from: *Toxins* **2016**, *8*(8), 242; doi: 10.3390/toxins8080242 135

Sarah De Saeger, Kris Audenaert and Siska Croubels
Report from the 5th International Symposium on Mycotoxins and Toxigenic Moulds: Challenges and Perspectives (MYTOX) Held in Ghent, Belgium, May 2016
Reprinted from: *Toxins* **2016**, *8*(5), 146; doi: 10.3390/toxins8050146 149

About the Special Issue Editors

Sarah De Saeger is full professor and head of a research group of about 15 PhD students and five full-time postdoctoral researchers. Her expertise relates to chemical analysis of food and feed, chemical food safety and mycotoxins. The Laboratory of Food Analysis performs, in particular, research on the issue of mycotoxins on both a national and an international level (including developing countries). In the last three years, 27 research projects were granted with funding from EU, FAO, EFSA, as well as national funding (HERCULES, FWO, FOD, BELSPO, IWT, BOF, VLIR-UOS). The laboratory focuses on four research lines: mycotoxins and human health; detection methods; fungal metabolomics and untargeted analysis, and mycotoxin occurrence. The research covers the characterization (e.g., modified mycotoxins), exposure and screening through biomarkers, as well as the development of innovative detection methods such as 'molecularly imprinted polymers' and biosensors, next to chromatographic and immuno-based techniques.

Sarah De Saeger is coordinator of the MYTOX research association platform (www.mytox.be). She has been an expert in EFSA CONTAM working groups since 2011 and a member of the Scientific Committee (SciCom) of the Federal Agency for Food Chain Safety since 2015. In June 2015, she established the Joint Laboratory of Mycotoxin Research of the Ghent University–Shanghai Jiao Tong University–Chinese Academy of Sciences (Shanghai Institutes of Biological Sciences). In 2015 she was awarded the Ghent University Prometheus Award for research and the International DBN Science and Technology Award, registered by the Ministry of Science and Technology, China.

Sarah De Saeger teaches all food-related courses in the Faculty of Pharmaceutical Sciences: bromatology (3rd Ba), bioanalytical practical (3rd Ba), food safety (1st Ma) and special nutrition (1st Ma).
Sarah De Saeger is author and co-author of more than 200 peer-reviewed international publications and she has been the (co)promoter of many doctoral theses. A full bibliographic list can be found on: https://biblio.ugent.be/person/801000957125.

Siska Croubels is full professor and head of the Department of Pharmacology, Toxicology and Biochemistry at the Faculty of Veterinary Medicine of Ghent University in Belgium. She is an associate member of the European College of Veterinary Pharmacology and Toxicology. Her research in the field of veterinary toxicology, focuses on the toxic effects and toxicokinetics of (emerging and modified) mycotoxins; on the interactions between mycotoxins and pathogens in several animal species (e.g., pigs as model for humans), and on the development of in vitro and in vivo models for efficacy and safety testing of mycotoxin detoxifiers. Her research group in veterinary toxicology is a member of the MYTOX association research platform (www.mytox.be), dealing with the effects of mycotoxins on human and animal health.

Her research in veterinary pharmacology focuses on pharmacokinetics, pharmacodynamics, bioanalysis and residues of veterinary drugs in several animal species; and on the development of a suitable animal pig(let) model for (paediatric) drug research (www.safepedrug.eu).
In order to support these two main research goals, the laboratory has more than 20 years of experience in the development of analytical methodologies for the detection and quantification of drugs and toxic agents (mainly mycotoxins) and relevant metabolites in different biological samples such as plasma, urine, tissues and cell media (mainly (U)HPLC–MS/MS and UHPLC–HRMS methodologies).

Siska Croubels teaches veterinary toxicology (1st Ma), general pharmacology (3rd Ba), pharmacotherapy of food-producing animals (2nd Ma) and horses (2nd Ma), veterinary legislation i.c. drug prescription (1st & 3rd Ma), and toxicology and risk assessment of residues in food for the Institute for Continuing Education in Veterinary Medicine.

Prof. Dr. Siska Croubels is author and co-author of more than 200 peer-reviewed international publications and she was a promoter of 17 doctoral theses in the field of veterinary pharmacology and toxicology. She currently (co)-supervises 14 PhD students and two post-doctoral researchers. A full bibliographic list can be found on: https://biblio.ugent.be/person/801000884373

Kris Audenaert is head of the Laboratory of Applied Mycology and Phenomics. This research group was recently established (2015) and is part of the Department of Applied Biosciences of the Faculty of Bioscience Engineering, Ghent University. He is a molecular plant pathologist who focuses on innovative agronomic tools such as endophytic fungi and plant volatiles to control plant pathogens. In addition, his research group focuses on toxigenic and secondary metabolites of plant pathogenic fungi within the framework of the MYTOX research association platform (www.mytox.be). Kris Audenaert has a keen interest in the role of mycotoxins in the interaction of fungi with their environment. In addition, a new research topic on biodegradation of mycotoxins using bacteria and fungi has recently been initiated. For this research topic, in-house developed bio-assays are combined with HRMS analyses to decipher detoxification- and degradation pathways. Finally, the research group uses phenotyping platforms to characterize the interaction of fungi with their environment (plant, food matrix, etc.) with multispectral imaging.

Kris Audenaert teaches genetics (1st Ba), cell biology (1st Ba), experimental design and applied data analysis (3rd Ba and linking course Ma), plant physiology (2nd Ba), and is a member of the Cell Statistical Consultancy, offeringstatistical assistance for Ma-students with their theses.

He has 56 international publications in peer reviewed journals and his SCI is >1200. He has participated in more than 130 international conferences, workshops and meetings. Currently, he is involved as a supervisor or co-supervisor in nine national and international projects on mycotoxins. A full bibliographic list can be found on: https://biblio.ugent.be/person/801001189016.

Preface to "Selected Papers from the 5th International Symposium on Mycotoxins and Toxigenic Moulds: Challenges and Perspectives"

Mycotoxins—toxic fungal secondary metabolites—play a significant role in food and feed safety. They have shown to be the number one threat regarding chronic toxicity related to food and feed contaminants. Economic losses are due to effects on livestock productivity and to direct losses in crop yield and stored agricultural products. Legislative limits for a range of mycotoxins worldwide, as well as the presence of emerging mycotoxins result in an increased number of official controls derived from national food safety plans. The challenges in mycotoxin and toxigenic mould research are still enormous due to the frequency, complexity and variability in occurrence.

The MYTOX association research platform MYTOX "Mycotoxins and Toxigenic Moulds" was established in 2007 and consists of more than 50 researchers from 12 research laboratories in the Ghent University Association. MYTOX deals with mycotoxin research in a multi-disciplinary way, based on four main themes: (1) mycotoxins; (2) toxigenic fungi; (3) mycotoxins and animal health; and (4) mycotoxins and human health. In this way, MYTOX tackles the mycotoxin issue along the production chain from the field to the end consumer, within the 'One Health' concept.

This Special Issue aimed to bring together active researchers to present their current work in mycotoxins and mycotoxigenic fungi.

For additional links regarding the Symposium, please follow: http://en.mytox.be/conferences/.

Sarah De Saeger, Siska Croubels and Kris Audenaert
Special Issue Editors

Article

Microbial Detoxification of Deoxynivalenol (DON), Assessed via a *Lemna minor* L. Bioassay, through Biotransformation to 3-epi-DON and 3-epi-DOM-1

Ilse Vanhoutte [1], Laura De Mets [1], Marthe De Boevre [2], Valdet Uka [2], José Diana Di Mavungu [2], Sarah De Saeger [2], Leen De Gelder [1,*,†] and Kris Audenaert [3,†]

1. Laboratory of Environmental Biotechnology, Department of Applied Biosciences, Faculty of Bioscience Engineering, Ghent University, 9000 Ghent, Belgium; Ilse.Vanhoutte@UGent.be (I.V.); Laura.DeMets@UGent.be (L.D.M.)
2. Laboratory of Food Analysis, Department of Bioanalysis, Faculty of Pharmaceutical Sciences, Ghent University, 9000 Ghent, Belgium; Marthe.DeBoevre@UGent.be (M.D.B.); Valdet.Uka@UGent.be (V.U.); Jose.DianaDiMavungu@UGent.be (J.D.D.M.); Sarah.DeSaeger@UGent.be (S.D.S.)
3. Laboratory of Applied Mycology and Phenomics, Department of Applied Biosciences, Faculty of Bioscience Engineering, Ghent University, 9000 Ghent, Belgium; Kris.Audenaert@UGent.be

* Correspondence: Leen.DeGelder@UGent.be; Tel.: +32-9-243-24-75

† These authors contributed equally to this work.

Academic Editor: Massimo Reverberi

Received: 31 August 2016; Accepted: 4 February 2017; Published: 13 February 2017

Abstract: Mycotoxins are toxic metabolites produced by fungi. To mitigate mycotoxins in food or feed, biotransformation is an emerging technology in which microorganisms degrade toxins into non-toxic metabolites. To monitor deoxynivalenol (DON) biotransformation, analytical tools such as ELISA and liquid chromatography coupled to tandem mass spectrometry (LC-MS/MS) are typically used. However, these techniques do not give a decisive answer about the remaining toxicity of possible biotransformation products. Hence, a bioassay using *Lemna minor* L. was developed. A dose–response analysis revealed significant inhibition in the growth of *L. minor* exposed to DON concentrations of 0.25 mg/L and higher. Concentrations above 1 mg/L were lethal for the plant. This bioassay is far more sensitive than previously described systems. The bioassay was implemented to screen microbial enrichment cultures, originating from rumen fluid, soil, digestate and activated sludge, on their biotransformation and detoxification capability of DON. The enrichment cultures originating from soil and activated sludge were capable of detoxifying and degrading 5 and 50 mg/L DON. In addition, the metabolites 3-epi-DON and the epimer of de-epoxy-DON (3-epi-DOM-1) were found as biotransformation products of both consortia. Our work provides a new valuable tool to screen microbial cultures for their detoxification capacity.

Keywords: deoxynivalenol (DON); *Lemna minor*; bioassay; biotransformation; detoxification; 3-epi-DON; 3-epi-de-epoxy-DON (3-epi-DOM-1)

---

1. Introduction

Mycotoxins are secondary metabolites produced by fungi, posing serious risks to health and economy when present in food or feed products. In order to reduce these risks, pre-harvest crop management strategies have been introduced. Fungicides, well-considered crop rotation, turning tillage techniques, resistant varieties and biocontrol all contribute to reducing mycotoxins in the crop [1,2]. In addition, methods of downstream post-harvest processing such as sorting, dehulling and milling amongst others help to reduce the mycotoxin level in agricultural commodities [3]. Although implementing good agricultural and processing practices may diminish fungal infestation and mycotoxin production, full prevention of mycotoxin contamination is impossible to achieve. Therefore, detoxification strategies have been introduced as remediation tools for contaminated food and feed batches. Mycotoxin binders can be used. However, these can interact with other molecules in the gastro-intestinal tract of animals (e.g., medicines and antibiotics) [4]. Microbial and enzymatic degradation in which the mycotoxin molecules are effectively altered and thereby detoxified, pose a more attractive alternative [2,5].

The mycotoxin deoxynivalenol (DON), occurring worldwide, is produced by *Fusarium culmorum* and *Fusarium graminearum* in cereals (e.g., maize, wheat and barley) [6,7]. DON is a sesquiterpenoid trichothecene containing a 12,13-epoxide group, which is responsible for its toxicity by inhibiting protein synthesis [8,9]. Acute exposure can cause vomiting, nausea and diarrhea. Effects of chronic low-dose exposure are decreased weight gain, anorexia, decreased nutritional efficiency and altered immune function [10,11]. The European Commission has set a maximum level of DON for humans in unprocessed grains at 1.25 mg/kg; the guidance level for animals in feed is in general 8 mg/kg (dependent of type of feed or animal) [12,13]. DON is a recalcitrant molecule, resisting most downstream processing operations [14], and is not effectively removed from the matrix by binders [15]. Therefore, elimination of DON and other mycotoxins from contaminated matrices via microbial biotransformation might be a valuable emerging technology [16].

Biotransformation of DON by mixed cultures or isolates originating from different environmental sources has been reported. A large number of DON-degrading bacteria are found in rumen fluid or intestines, where DON is anaerobically transformed into de-epoxy-deoxynivalenol (DOM-1) [17–23]. Soil is also a promising source of DON-degrading organisms [24–29]. Bacterium E3-39, classified in the *Agrobacterium-Rhizobium* group, can transform DON into 3-keto-DON [29]. *Nocardioides* WSN05-2 and *Devosia mutans* 17-2-E-8 convert DON into 3-epi-DON in aerobic conditions [24,25], whereas *Citrobacter freundii* degrades DON into DOM-1 aerobically and anaerobically [26]. Besides soil, other microbial communities capable of DON biotransformation have been reported [28,30].

However, modification of a compound does not automatically entail detoxification [16,31], which is of course the ultimate goal. Metabolites can have residual or even heightened toxicity, which is often overlooked in biotransformation studies when using solely analytical tools [16]. For DON, some known derivatives have been tested on toxicity. In acetylated trichothecenes, loss of side groups on C4, C15 or C8 resulted in reduced protein synthesis inhibition [32]. However, 15-acetyl-deoxynivalenol (15-ADON) had a similar toxicity as the parent toxin DON, whereas 3-acetyl-deoxynivalenol (3-ADON) was less toxic than DON [33]. Toxicity tests have also been performed on metabolites of DON. It seems that 3-keto-DON is three to ten times less toxic than DON [29,34] and DOM-1 and 3-epi-DON are at least 50 times less toxic than DON [33–35]. When studying detoxification of mycotoxins, toxicity assays are crucial to assess toxicology. Animal trials can be performed but are expensive, time consuming and hampered by ethical issues. Cell culture-based systems [10,19,32,33,36–38] provide information about the metabolism of mycotoxins, but require a high workload (e.g., acquirement of cell lines, sterile work environment, long preparations and expensive reagents). Therefore, researchers have looked for alternative assays to estimate toxicity. Mycotoxins are known to induce adverse effects in many other organisms including birds, amphibians, arthropods, crustaceans, unicellular organisms, microorganisms and plants [39]. DON has been tested on (phyto)toxicity and relative toxicity towards other mycotoxins with several bioassays (e.g., *Arabidopsis thaliana*, wheat, *Lemna pausicostata*, *Chlamydomonas reinhardtii*, *Artemia salina* L (brine shrimp larvae), *Tetrahymena pyriformis* (ciliated protozoa) and (engineered) yeasts [40–47]. These bioassays are inexpensive, fast and require a lower workload.

In this work, a highly sensitive DON bioassay was developed and implemented to screen bacterial cultures for their detoxification capacity of DON and other trichothecenes using the aquatic macrophyte *Lemna minor* L. as an indicator organism. The goal was to develop a robust, inexpensive, highly sensitive and readily applicable high-throughput method to screen bacterial strains or enrichment cultures for their ability to biotransform and detoxify DON. The trichothecene DON is, in addition to being a mycotoxin, also a phytotoxin, and due to this feature, a plant-based bioassay can be used. *Lemna minor* L. is a well-known plant for use in bioassays and has been previously used to evaluate the biodegradation of herbicides [48] and to determine the toxicity of fumonisins [49–51]. To our knowledge, it was never used as an indicator organism for the toxicity of DON or trichothecenes. After development of the bioassay, a diverse set of matrices comprising rumen fluid, soil, digestate from an anaerobic digester and activated sludge from a water treatment plant, were used as inoculum for DON-degrading enrichment cultures. These cultures were analyzed on their detoxification capabilities with the bioassay and on their biotransformation capabilities with analytical tools.

2. Results

2.1. Developing the Bioassay Using Lemna Minor

2.1.1. Linearity

In order to assess the sensitivity of *Lemna minor* to DON, a wide concentration range was tested (0 (control)–0.1–0.5–1–5–10–50–100 mg/L DON). After 7 days, DON caused a reduction of 41% ± 12% in growth of *Lemna minor* at a concentration of 0.5 mg/L DON. Growth was completely inhibited at concentrations of 1 to 100 mg/L DON. In addition, an increase of bleached fronds was observed at 0.1 to 10 mg/L DON, while at concentrations of 10, 50 and 100 mg/L, 100% of the fronds were bleached. A correlation between the presence of DON and growth inhibition was further investigated for lower concentrations between 0 and 1 mg/L DON.

A calibration curve was set up with concentrations of 0, 0.0625, 0.125, 0.25, 0.5 and 1 mg/L DON starting from six fronds to assess the sensitivity of the bioassay (Figure 1e). The number of fronds (Figure 1a,b) and frond area (Figure 1c,d) are observed as growth parameters. For each parameter, a sigmoid correlation was found between the frond growth (Figure 1a,c) and the concentration of DON, confirming the conventional response of *Lemna minor* to growth-inhibiting compounds [52]. A log/logit transformation was carried out to obtain a linear relationship (Figure 1b,d). This transformation resulted in a good correlation between log(DON), and logit($growth_{number\ of\ fronds}$) and logit($growth_{frond\ area}$) with an R^2 of 0.996 (p-value < 0.001) and 0.947 (p-value = 0.005) respectively.

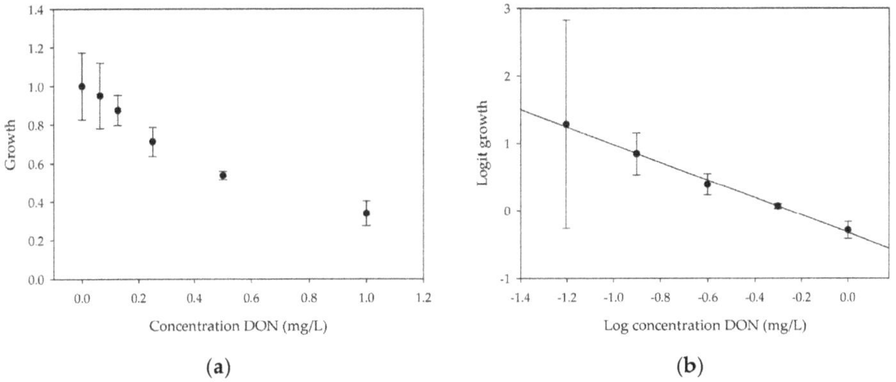

Figure 1. *Cont.*

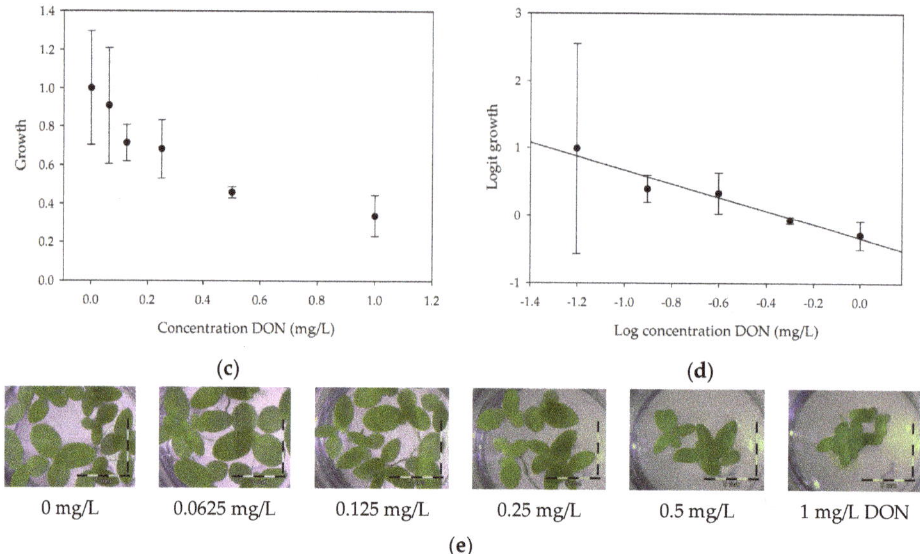

Figure 1. Growth of *Lemna minor* in response to deoxynivalenol (DON): (**a**) Data calculated based on number of fronds: correlation of growth and concentration DON (0–1 mg/L DON); (**b**) Data calculated based on number of fronds: correlation of logit growth and log concentration DON (0.0625–1 mg/L DON); (**c**) Data calculated based on frond area: correlation of growth and concentration DON (0–1 mg/L DON); (**d**) Data calculated based on frond area: correlation of logit growth and log concentration DON (0.0625–1 mg/L DON); (**e**) Response of *Lemna minor* after 7 days to increasing DON concentrations. Legend: black-white bar = 5 mm.

From this tight correlation between *Lemna* growth reduction and DON concentration, it can be concluded that this bioassay is suitable as a tool to assess the toxicity mediated by DON. Next to frond growth and frond area, pulse amplitude-modulated chlorophyll fluorescence was also used to evaluate the impact of the toxin on the plant. After 12 h, a decrease in the quantum yield of Photosystem II (ϕ_{PSII}) was observed, indicating that DON interferes with photosynthesis (Figure S1).

2.1.2. Repeatability and Sensitivity

To determine the variability of the impact of DON on plant growth, calibration curves (ranging from 0, 0.125, 0.25, 0.5 to 1 mg/L DON) were tested in triplicate on different days, performed by two people. The 95% confidence intervals were calculated for the main calibration curve. The results are shown in Figure 2.

As seen in Figure 2, there is a high variation in response of the concentration at 0.125 mg/L DON. Some data points are located outside the 95% confidence interval. However, at higher concentrations, the variation within all three independent experiments is lower. The plant is significantly sensitive to DON at 0.250 mg/L: from a non-parametric Kruskal–Wallis test followed by a one-sided post-hoc Dunn's test (α: 0.05) it could be concluded that a concentration of 0.250 mg/L DON ($n = 9$) significantly differs from the control. This result is in concordance with the chlorophyll fluorescence measurements carried out at 24 h after DON application (Figure S1).

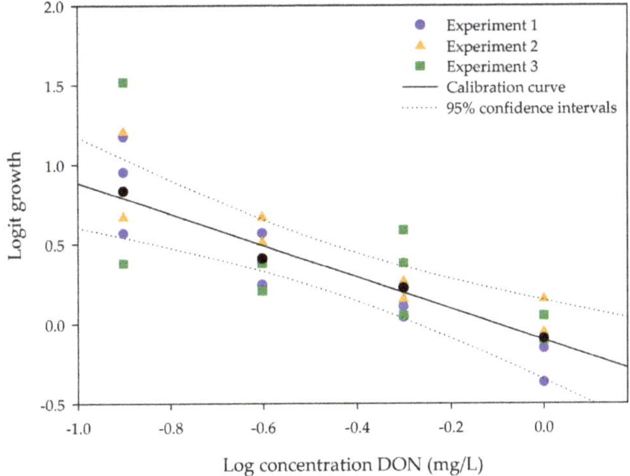

Figure 2. The calibration curve is expressed as the logit growth of number of fronds in function of the log concentration DON (mg/L). Data is shown from 0.125 to 1 mg/L DON. Triplicates of each experiment are illustrated in the same color.

2.1.3. Sensitivity of *Lemna Minor* towards Other *Fusarium* Mycotoxins

In order to assess the applicability of this bioassay to other trichothecenes and to zearalenone (ZEN), the bioassay was performed in triplicate on diacetoxyscirpenol (DAS), fusarenon X (FUS-X), T-2 toxin, HT-2 toxin, and nivalenol (NIV), which are trichothecene mycotoxins, and on the estrogenic mycotoxin ZEN at concentration level 1 mg/L. The plant *Lemna minor* was sensitive to DAS, FUS-X, T-2 and HT-2, but not to NIV and ZEN (Table 1).

Table 1. Sensitivity of *Lemna minor* towards other *Fusarium* mycotoxins.

Mycotoxin	% Growth at 1 mg/L Mycotoxin
DAS	26 ± 4 [b]
DON	34 ± 6 [b]
FUS-X	44 ± 12 [b]
T-2 toxin	52 ± 12 [b]
HT-2 toxin	54 ± 15 [b]
ZEN	90 ± 42 [a]
NIV	92 ± 11 [a]
Control	100 ± 7 [a]

[a,b] Statistically analyzed via a non-parametric Kruskal–Wallis test followed by a one sided post-hoc Dunnett's test (α: 0.05). DAS: diacetoxyscirpenol; FUS-X: fusarenon X; ZEN: zearalenone; NIV: nivalenol.

2.2. *Implementing the Bioassay to Screen for DON Detoxification by the Enrichment Cultures*

The obtained enrichment cultures (originating from a 6-week enrichment of rumen fluid, soil, digestate and activated sludge) were inoculated in "minimal incubation medium" (MMO) with 5 and 50 mg/L DON and incubated for four weeks at 30 °C and 100 rpm. The samples taken at four weeks of incubation were analyzed for the detoxification potential of the enrichment cultures (Figure S2). Sterilized culture filtrates were analyzed with the bioassay with a previous dilution step to an estimated 1 mg/L based on the amended DON-concentrations (5 and 50 mg/L) (Table 2). If a microbial community present in an enrichment culture was able to detoxify DON, we would expect the *Lemna* plants to show a better growth compared to the *Lemna* plants in control wells effectively exposed to 1 mg/L of DON.

Table 2. Screening of enrichment cultures (rumen fluid, soil, digestate and activated sludge) after four weeks of incubation at 0, 5 and 50 mg/L DON, analyzed with the bioassay.

Concentration DON (mg/L)	Rumen fluid	Soil	Digestate	Activated sludge
0 mg/L DON	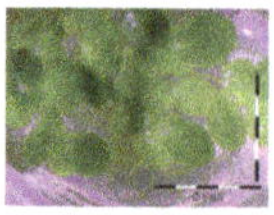 101 ± 9% [a]	 93 ± 6% [a]	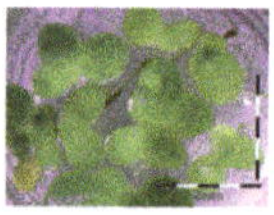 96 ± 4% [a]	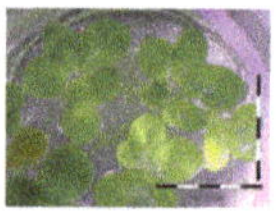 104 ± 10% [a]
5 mg/L DON [1]	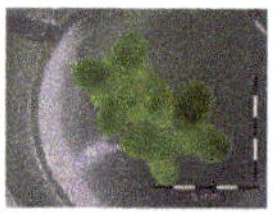 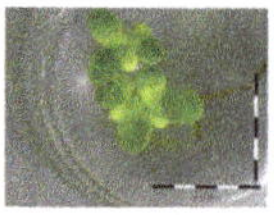 28 ± 0% [b]	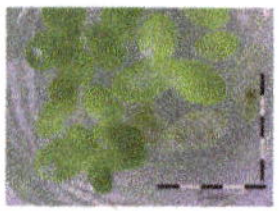 83 ± 6% [a]	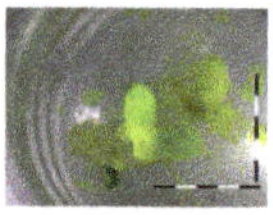 34 ± 5% [b]	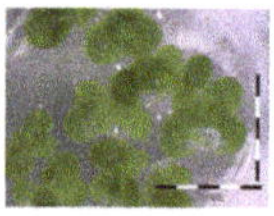 99 ± 13% [a]
50 mg/L DON [1]	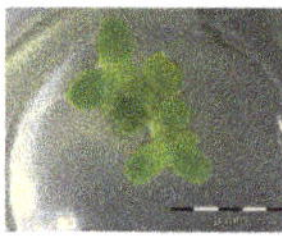 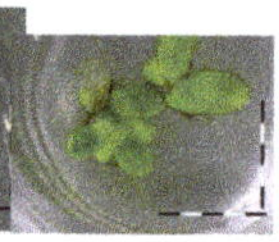 36 ± 2% [b]	 87 ± 10% [a]	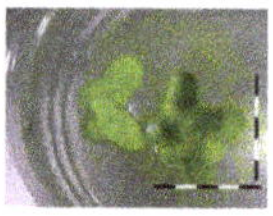 43 ± 4% [b]	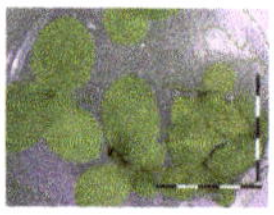 83 ± 4% [a]

Growth (%) is mentioned below each figure. Legend: black-white bar = 5 mm. [a,b] Statistically analyzed via a non-parametric Kruskal–Wallis test followed by a one sided post-hoc Dunn's test (α: 0.05). [1] Controls 5 and 50 mg/L DON diluted to 1 mg/L DON (5 and 50 times respectively), as well as the samples.

A high growth reduction is observed when *Lemna* plants are incubated with culture filtrate originating from enrichment cultures of rumen fluid and digestate. The growth reduction is similar to *Lemna* plants incubated with 1 mg/L of DON, indicating that no detoxification occurred. *Lemna* plants incubated with culture filtrate of the enrichment cultures of soil or activated sludge showed a similar growth compared to the control where no DON was added. These findings indicate that the enrichment cultures originating from soil and activated sludge contain some promising microorganisms capable of detoxifying DON aerobically within four weeks.

These data were also confirmed with LC-MS/MS. No modification of DON occurred with digestate and rumen fluid cultures within four weeks, whereas soil and activated sludge cultures fully converted 50 mg/L DON. The same trend is observed for 5 mg/L. The cultures originated from soil and activated sludge could convert 100% ± 0% of 5 mg/L DON, whereas the cultures of digestate and rumen fluid only converted 11% ± 3% and 9% ± 9% of 5 mg/L DON, respectively. These findings have been statistically evaluated using a one-way ANOVA test followed by a Dunnett T3 test for pairwise multiple comparisons (α: 0.05). The samples of the control, digestate and rumen fluid are significantly different from the samples of activated sludge and soil. In addition, a quantitative determination was performed of known derivatives or metabolites of DON: 3-ADON, 15-ADON, deoxynivalenol-3-glucoside (DON-3G) and DOM-1. However, none of these molecules were found.

Subsequently, a more detailed investigation of the samples at 50 mg/L DON was performed by LC-high resolution MS (LC-HRMS) (Figure 3). The samples of digestate and rumen fluid showed a metabolite profile that was similar to the control (medium and DON). DON was still present and no novel entities could be detected. In contrast, investigation of the chromatograms of soil and activated sludge revealed two additional compounds, which were eluted before DON, i.e., at retention time (RT) 2.4 min and at RT 3.9 min. The compound at RT 2.4 min, with $[M + H^+] = 297.1333$, was putatively assigned as 3-epi-DON ($C_{15}H_{21}O_6$, mass error = −1.68 ppm), while the other biotransformation compound (RT 3.9 min, $[M + H^+] = 281.1392$) was tentatively identified as 3-epi-DOM-1 ($C_{15}H_{21}O_5$, mass error = 1.06 ppm). The identification of these compounds was confirmed by close examination of their chromatographic retention and fragmentation pattern (Figures S3 and S4) in light of data acquired for authentic standards of DON and DOM-1. The same fragment ions can be seen in the MS/MS spectra of each pair of the parent- and the epi-compounds. Interestingly, the intensities of the ions are different, as commonly observed with these types of isomers [53]. Our data corroborate previous studies that identified 3-epi-DON as biotransformation product of DON, and which showed that the former is eluted before the latter in reversed-phase chromatography [53–55]. Similarly, those studies also support our identification of 3-epi-DOM-1, in that this epimer was also eluted before the main compound DOM-1 (authentic standard).

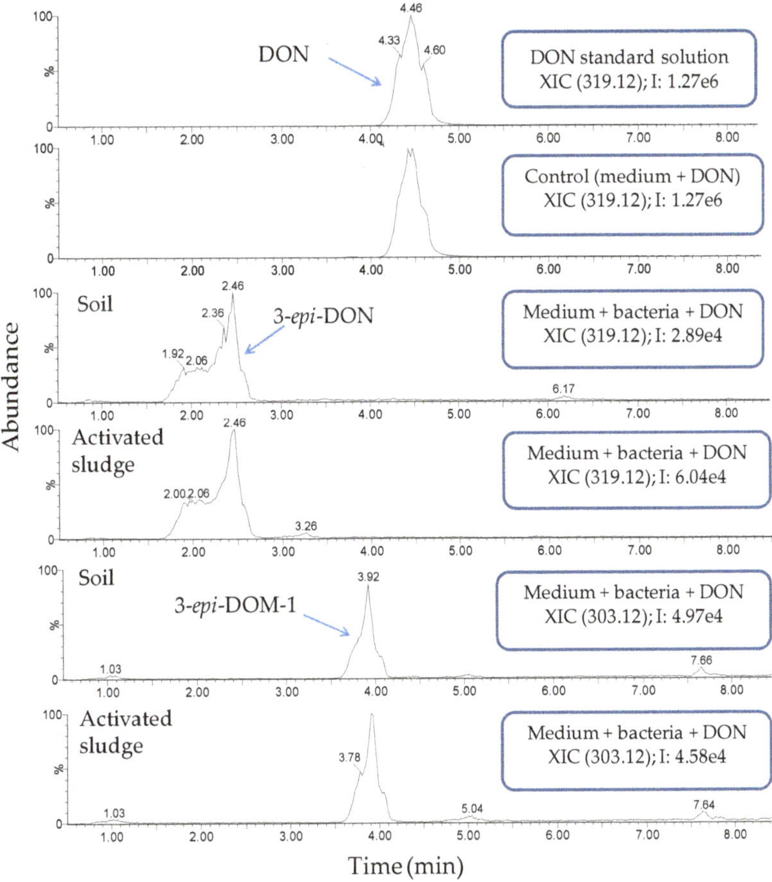

Figure 3. Extracted ion chromatograms (XICs) of the samples at 50 mg/L DON soil and activated sludge after four weeks, including the control (medium and DON). DON and 3-epi-DON were detected as both the protonated molecules and sodium adducts. Only the XICs for the sodium adducts are shown.

2.3. Monitoring Detoxification (Bioassay) and Biotransformation (LC-MS/MS) by the Enrichment Cultures Soil and Activated Sludge through Time

In order to get a sense of DON detoxification and biotransformation kinetics exhibited by enrichment cultures of soil and activated sludge, we analyzed the supernatant of the intermediate samples taken at weeks 1, 2 and 3 of incubation (Figure S2). For DON detoxification, results from the bioassay are shown in Table 3, expressed as % growth, directly related to % detoxification. For DON biotransformation, results from the LC-MS/MS are shown in Table 4, expressed as % biotransformation.

Table 3. Detoxification kinetics of 5 and 50 mg/L DON by enrichment cultures (soil and activated sludge) assessed through the *Lemna minor* bioassay.

Matrix	Growth of *Lemna minor* (%) after exposure to culture supernatant from the DON detoxification experiment after x weeks assessed by the bioassay		
	After 1 week of detoxification	After 2 weeks of detoxification	After 3 weeks of detoxification
0 mg/L DON			
Sterile control			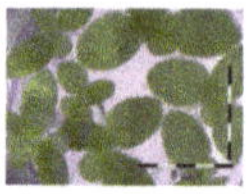 100 ± 4 [a]
5 mg/L DON			
Sterile control [1]	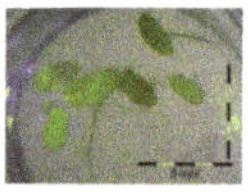 30 ± 2 [b]	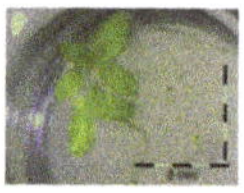 32 ± 1 [b]	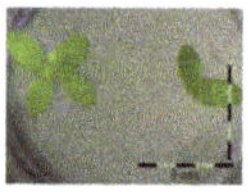 30 ± 2 [b]
Soil	 32 ± 6 [b]	 93 ± 3 [a]	 86 ± 21 [a]
Activated sludge	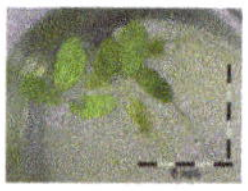 33 [2]	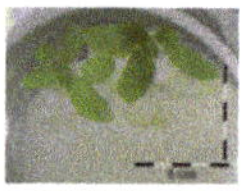 53 ± 4 [b]	 69 ± 4 [b]
50 mg/L DON			
Sterile control [1]	 35 ± 5 [b]	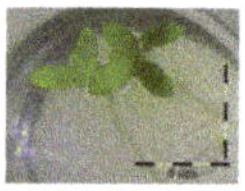 31 ± 4 [b]	 32 ± 3 [b]
Soil	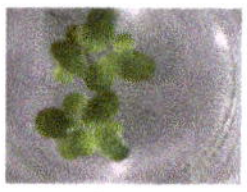 47 ± 1 [b]	 88 ± 4 [a]	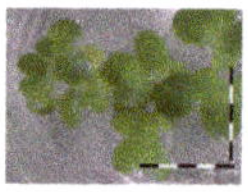 90 ± 3 [a]
Activated sludge	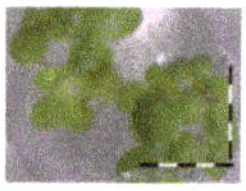 99 ± 1 [a]	 78 ± 15 [a]	 83 ± 3 [a]

Legend: black-white bar = 5 mm. [a,b] Statistically analyzed via a non-parametric Kruskal–Wallis test followed by a one sided post-hoc Dunn's test (α: 0.05). [1] Controls 5 and 50 mg/L DON diluted to 1 mg/L DON (5 and 50 times respectively), as well as the samples. [2] Sample activated sludge at 5 mg/L DON in week one exceptionally not in triplicate.

A stable concentration of DON is observed over time for the controls 5 mg/L (respectively 5 ± 1 mg/L, 5 ± 3 mg/L and 4 ± 1 mg/L for weeks 1, 2 and 3) and 50 mg/L DON (respectively 51 ± 11 mg/L, 48 ± 16 mg/L and 45 ± 6 mg/L for weeks 1, 2 and 3), indicating no abiotic influences on DON during the experiment. Analysis with the bioassay shows that exposure of each control supernatant (originating from the 5 mg/L and 50 mg/L control experiments, diluted to 1 mg/L) resulted in approximately 30% growth of *Lemna minor* (Table 3), which is in agreement with other experiments at 1 mg/L DON (Figure 1).

Table 4. Biotransformation kinetics of 5 and 50 mg/L DON by soil and activated sludge enrichment cultures assessed by LC-MS/MS.

Matrix	DON Biotransformation (%) in the Culture Supernatant from the DON Biotransformation Experiment after x Weeks Assessed by LC-MS/MS Analysis		
	After 1 Week of Biotransformation	After 2 Weeks of Biotransformation	After 3 Weeks of Biotransformation
5 mg/L DON			
Sterile control	0 ± 27 [b]	0 ± 63 [1,b]	0 ± 20 [b]
Soil	56 ± 15 [b]	100 ± 0 [a]	100 ± 0 [a]
Activated sludge	72 ± 2 [a]	100 ± 0 [a]	100 ± 0 [a]
50 mg/L DON			
Sterile control	0 ± 22 [b]	0 ± 32 [b]	0 ± 14 [b]
Soil	28 ± 6 [b]	100 ± 0 [a]	100 ± 0 [a]
Activated sludge	68 ± 14 [a]	100 ± 0 [a]	100 ± 0 [a]

[a,b] Statistically analyzed via a non-parametric Kruskal–Wallis test followed by a one-sided post-hoc Dunn's test (α: 0.05). [1] Sample control 5 mg/L DON in week two exceptionally in duplicate.

Supernatant for soil enrichment culture starting at 5 mg/L DON results after one week in the same growth reduction as in the sterile DON control. However, after two and three weeks, 93% ± 3% and 86% ± 21% growth was detected, respectively (compared to the control without DON). This is in agreement with the LC-MS/MS analysis where DON was no longer detected after two weeks. Similar results were obtained starting from 50 mg/L DON. The activated sludge enrichment cultures displays other kinetics. At low concentrations (5 mg/L DON), DON was no longer detected with LC-MS/MS after week two and three. However, in week two and three, a slightly phytotoxic effect is still observed. At higher concentrations (50 mg/L DON), DON was again fully biotransformed after week two and three, whereas detoxification already occurred in week one.

3. Discussion

In this study, we have developed and implemented a highly sensitive bioassay using *Lemna minor* L. for screening bacterial cultures on their DON detoxification capacities. To assess the sensitivity of *Lemna minor* L. to DON, a dose–response curve was established and a linear relation was found between logit (*Lemna* growth) and log (DON concentration). It should be mentioned that for the low concentrations 0.0625 and 0.125 mg/L DON, the variation in replicates was fairly high. Although we do not have any evidence, this might be due to small differences in the physiological fitness of the *Lemna* plants where some plants are inhibited by such low DON doses while others are not. This is in contrast to higher DON concentrations, for which plants are equally sensitive, resulting in very reproducible results. Therefore, the bioassay should be used to detect loss of toxicity starting from 0.25 mg/L DON. Growth of *Lemna minor* L. in response to DON was assessed by frond area and number of fronds. In the future, other response parameters might be included in the bioassay. We provide evidence that chlorophyll fluorescence might be a fast alternative and a sensitive parameter to implement in detoxification studies as previously shown for herbicides [48]. Measuring the conductivity of the medium might be a fourth parameter to determine the electrolyte release [40]. Subsequently, the usability of the *Lemna* bioassay was tested for other trichothecenes and for the estrogenic mycotoxin ZEN. ZEN and NIV seemed to be non-toxic for *Lemna minor*, in contrast to DON, T-2 toxin, HT-2 toxin, DAS and FUS-X. Therefore, the bioassay can be used as a cheap bio-tool for screening the phytotoxicity of DON, other tricothecenes and DON derivatives.

The sensitivity of the bioassay was statistically evaluated, where a concentration of 0.25 mg/L DON was found to be significantly different from the control. Between 1 and 100 mg/L DON, complete growth inhibition and leaf necrosis occurred, whereas at 0.25 until 1 mg/L growth reduction occurred. In our bioassay, 34% ± 6% growth was observed at 1 mg/L DON after 7 days resulting in a highly sensitive bioassay. In Table 5, our assay is compared to other existing bioassays, including information about advantages, drawbacks and limitations, as well as the workload, time requirements and applicability of the bioassays. The various bioassays are ranged from lowest to highest sensitivity for DON.

Table 5. Comparison of bioassays used for assessing DON toxicity.

Organism [Reference]	Application	Characteristics		
		Sensitivity	**Time**	**Workload**
Brine shrimp larvae *Artemia salina* L. [47,56]	General screening for trichothecenes in grains	600–1200 ng/disc (~30–60 mg/L) DON → 50% of mortality	30 h of preparation, 30 h of incubation	Preparation of larvae (including separation eggs) * Disc screening method: addition of 20 μL/disc toxin in a well with addition of 2 drops (±100 μL) of a suspension of larvae * Measuring mortality (counting immobile larvae under microscope, killing larvae, counting total number)
Unicellular algae *Chlamydomonas reinhardtii* [42]	* Testing cloned genes (resistance to trichothecenes) * Comparing trichothecenes with C3-OH group with their acetylated derivatives	25 mg/L DON → clear toxic effect	8 days	Preparation of preculture Measuring growth (haemocytometer), cell viability (plating), calculating number of culture doublings
Yeast *Klyveromyces marxianus* [44]	Screening DON-degrading organisms	23 mg/L DON → 50% growth inhibition 300 mg/L DOM-1 → no growth inhibition	Overnight preparation, 22 h of incubation	Preparation of preculture Incubation of sterile supernatants Measuring optical density at 650 nm
Engineered baker's yeast [41]	* Use as a bioassay indicator organism * Screening for DON-detoxifying bacteria	5 mg/L DON → 50% growth inhibition	16 h	Preparation yeast culture * Agar diffusion test * Measuring optical density at 620 nm
Arabidopsis thaliana [45,46]	Studying the phytotoxic action of trichothecenes	* 10 μM (~3 mg/L) DON → no inhibition of seed germination, inhibition of root growth * 23.0 ± 6.8 μM (~6,8 ± 2 mg/L) DON → mortality of 50% of leaves	* 3 days of preparation, 3 days of incubation * 3 weeks of preparation, 7 days of incubation	* Preparation of seeds Investigation of growth and morphology * Preparation of 3-week-old plants Leave protocol: investigation of shriveling, chlorosis, death
Wheat plants [46]	Studying the phytotoxic action of trichothecenes	>15 μM (~4.5 mg/L) DON → inhibition of root elongation of wheat plants	3 days of preparation, 3 days of incubation	Preparation of seeds Investigation of growth and morphology
Lemna pausicostata [40]	Screening trichothecenes for bioherbicides	10 μM (~3 mg/L DON) → 56.0% ± 5.7% growth	72 h	No need for preparation Measuring electrolyte release (conductivity), growth inhibition, chlorophyll reduction
Ciliated protozoa *Tetrahymena pyriformis* [43]	* Testing toxicity of mycotoxins * Screening of cereals	0.6 mg/L DON → minimum active dose	Preparation of heat shocks, 150 min incubation	Preparation heat shocks for good division synchrony, requires good heating and cooling devices Measuring delay between start of division in control and toxin-treated culture, counting number of cells
Lemna minor L.	* Screening for phytotoxicity of DON and other trichothecenes * Screening for DON-degrading organisms	1 mg/L DON → 34% ± 6% growth 0.250 mg/L → minimum active dose (significantly different)	* 7 days * 12 h with chlorophyll fluorescence	No need for preparation Measuring number of fronds, frond area, chlorophyll fluorescence

As seen in Table 5, the sensitivity of most bioassays ranged from a concentration of 30 down to 3 mg/L DON. Only the ciliated protozoa *Tetrahymena pyriformis* was also reported as a very sensitive organism for DON with 0.6 mg/L DON as minimum active dose. However, our bioassay can detect analyte concentrations as low as 0.25 mg/L DON, which to our knowledge has not been reported before.

Implementing bioassays in biotransformation experiments is of great importance because the toxicity of metabolites is often overlooked in screening assays for new promising microbial strains. Several examples in literature are available showing that mycotoxin modification does not always result in mycotoxin detoxification, which illustrates the necessity and benefits of bioassays as developed in this study. For aflatoxin B1, the metabolite aflatoxin M1 is a commonly known metabolite which is categorized as possibly carcinogenic to humans (Group 2B) by the International Agency for Research on Cancer [57]. The same accounts for ZEN, for which several metabolites have been found with similar or even more estrogenic activity [16,58]. Furthermore, this problem is nicely illustrated by the following 'detoxification' reaction of 3-ADON which is first converted to DON as an initial step which leads to an increased toxicity, subsequently followed by de-epoxidation into the non-toxic metabolite DOM-1 [44].

Enrichment cultures were evaluated for their DON detoxification capacity through the *Lemna minor* bioassay. The microbial communities originating from digestate and rumen fluid did not show this capability, which is not entirely surprising as they are adapted to an anaerobic environment, whereas the enrichment and the biotransformation experiment were performed in the presence of oxygen. The cultures obtained from soil and activated sludge showed a clear detoxification through time with the total disappearance of 5 and 50 mg/L DON after two weeks, as assessed by LC-MS/MS. Soil has already been reported as a source for DON-transforming microorganisms [24–29], but to our knowledge, this is not the case for activated sludge from a water treatment plant.

LC-HRMS analysis of the soil and activated sludge cultures after 4 weeks of biotransformation revealed two DON metabolites, namely 3-epi-DON and 3-epi-DOM-1. It has been shown that 3-epi-DON is less toxic than DON [34,35] and has been reported to be formed by several bacteria belonging to the genera *Devosia* and *Nocardioides* [25,28,59]. It has been shown that de-epoxidation of DON to DOM-1 already lowers toxicity [33,35], so although toxicological studies for 3-epi-DOM-1 are not available yet, one can assume that this compound is also less toxic than the parent mycotoxin DON as it is its de-epoxidized and epimerized form. To date, 3-epi-DOM-1 has never been reported as a microbial transformation product of DON.

The observation that the disappearance of DON, as detected by LC-MS/MS, after two weeks in soil and activated sludge cultures, does not completely match the detoxification data obtained via the *Lemna minor* bioassay, points to differences in kinetics of DON transformation between soil and activated sludge cultures and/or small differences in toxicity between the detoxification products 3-epi-DON and 3-epi-DOM-1. Therefore, an in-depth study on the kinetics of DON transformation by these enrichment cultures will be a very interesting step in further research. Thereby, the bioassay presented here will be an indispensable tool in assessing detoxification and therefore complement purely analytical tools in this and other biotransformation studies.

4. Materials and Methods

4.1. Sources

Rumen fluid was sampled via a rumen fistula of three sheep at the experimental farm of Ghent University. Soil originated from a monoculture corn field in Nazareth, Belgium. Digestate came from an anaerobic digester provided by Innolab (Ghent, Belgium). Activated sludge originated from a municipal water treatment plant (Gavere, Belgium). The aquatic plant *Lemna minor* was maintained in an aquarium at room temperature under a light regime of 16 h/8 h (light/dark) illuminated with a Solar Lux 26 Watt lamp (Aquatic nature, Ghent, Belgium). The aquarium contained 40 L of tap water each week supplemented with 1 mL of macro elements stock solution containing 60 g/L KNO_3, 12 g/L KH_2PO_4 and 32 g/L K_2SO_4.

4.2. Source Material Preparation

Before use as inoculants for the enrichment experiments, the microbial community sources were prepared. Rumen fluid was directly used as inoculum. A soil sample weighing 20 g was suspended in 25 mL physiological water (8.5 g/L NaCl in distilled water) and added to a stomacher for 1 minute at low speed in order to homogenize the sample and separate the liquid phase. Digestate (30 mL) was transferred into a stomacher for 2 min at normal speed to homogenize the sample and separate the liquid phase. An aliquot of activated sludge sample was mixed homogeneously and added to a 1.5-mL recipient. The floc structure of activated sludge was broken down through pipetting up and down with a sterile syringe and needle releasing the bacteria into suspension.

4.3. Standards

The individual mycotoxin solid calibration standards (1 mg) of DON, 3-ADON, 15-ADON, DOM-1, NIV and zearalanone (ZAN, internal standard) were purchased from Sigma-Aldrich (Bornem, Belgium). DON-3G (50.2 ng/μL, in acetonitrile) was purchased from Biopure Referenzsubstanzen (Tulln, Austria).

All mycotoxin solid standards for LC-MS/MS analysis were dissolved in methanol (1 mg/mL), and were storable for a minimum of 1 year at −18 °C [60]. DON-3G was kept at 4 °C. The working solutions of DON, 3-ADON, 15-ADON, DOM-1 and NIV (10 ng/μL) were prepared in methanol, and stored at −18 °C, while DON-3G (10 ng/μL) was dissolved in acetonitrile (4 °C, monthly renewed).

DON standard for the bioassay and the enrichment protocol was dissolved in dimethyl sulfoxide (DMSO) at a concentration of 1 mg/mL and stored at −18 °C. Working solutions of 100 and 10 mg/L DON were prepared in "*Lemna* medium" or "minimal incubation medium" (MMO) and stored at −18 °C.

4.4. Enrichment for DON-Degrading Microorganisms

Screening of microorganisms with respect to DON biotransformation capacity was performed by adding the four inoculants at 0.1% in 50 mL sterile MMO. This medium is based on Stanier medium [61] and contained 1.4 g/L Na_2HPO_4, 1.4 g/L KH_2PO_4, 0.3 g/L $(NH_4)_2SO_4$, 98.5 mg/L $MgSO_4$, 5.9 mg/L $CaCl_2.H_2O$, 3.2 mg/L Na_2EDTA, 2.8 mg/L $FeSO_4.7H_2O$, 1.2 mg/L $ZnSO_4.7H_2O$, 1.7 mg/L $MnSO_4.H_20$, 0.4 mg/L $CuSO_4.5H_2O$, 0.2 mg/L $CoCl_2.6H_2O$ and 0.1 mg/L $(NH_4)_6Mo_{24}.4H_2O$. The liquid medium was supplemented with 50 mg/L DON as the only carbon source. The enrichment was incubated for 6 weeks at 100 rpm and 30 °C. Controls without bacteria and without DON were included. After the incubation period, enrichment cultures were obtained and archived at −80 °C in 20% *v*/*v* glycerol.

4.5. Biotransformation Experiment

For the biotransformation experiment, precultures were made in MMO medium. The obtained enrichment cultures (stored at −80 °C) were added into 10 mL MMO medium at 1% with 5 or 50 mg/L DON. The aliquots were incubated at 30 °C and 100 rpm. Microorganism growth of the precultures was monitored with a spectrophotometer, measuring the optical density at 600 nm. The precultures were used as an inoculum for the biotransformation experiment at an optical density value of 1. Subsequently, 2% of each preculture inoculum was added with 5 or 50 mg/L DON into 10 mL MMO medium. The experiment was performed in triplicate. Controls are included (MMO medium with 5 or 50 mg/L DON without addition of microbial source and MMO medium with microbial sources without DON). The enrichment cultures were incubated for four weeks at 30 °C and 100 rpm. Samples were taken weekly and stored at −18 °C.

4.6. Bioassay Protocol

A mineral growth medium for *Lemna minor* was prepared according to Megateli et al. [62]. A sterile 24-well plate was used to grow *Lemna minor* in "*Lemna* medium", in a total volume of 2 mL per well. Starting from six fronds, the plants were incubated in triplicate for 7 days in a growth chamber (16 h light exposure, 22 °C). For each analysis of samples, a calibration curve was included (0, 0.0625, 0.125, 0.25, 0.5 and 1 mg/L DON). Samples were filter sterilized using a sterile filter and syringe. Each sample was diluted to the range of calibration curves. Samples starting from 5 mg/L DON were diluted 5 times; samples starting from 50 mg/L DON were diluted 50 times resulting in a final concentration of 1 mg/L in the *Lemna* bioassay. After 7 days of incubation, the plates were analyzed based on frond growth. The number of fronds are counted with a microscope (Stereo IX). Photos are made of each well with the microscope and frond area is calculated with the program APS Assess.

4.7. Detection of DON and Its Possible Metabolites by LC-MS/MS

4.7.1. Sample Preparation

One milliliter of MMO medium was weighed in a glass tube (10 mL). A calibration curve was set up by spiking three blank medium samples with DON, NIV, 3-ADON, 15-ADON, DOM-1 and DON-3G (10 ng/µL, range 100 µg/L–200 µg/L–400 µg/L). The stock solution (25 µL) of the internal standard ZAN (10 ng/µL, 250 ng on sample) was added to both calibrators as unknowns. Samples were vortexed (Labinco BV, Breda, The Netherlands) for 2 min. The medium samples were evaporated until dryness under nitrogen at 60 °C using the TurboVap® LV (Biotage, Dusseldorf, Germany), and redissolved in 100 µL of injection solvent (methanol/water/acetic acid (41.8/57.2/1, *v*/*v*/*v*) with 5 mM of ammonium acetate). Finally, the redissolved sample was vortexed for 3 min. Prior to LC-MS/MS analysis, the samples were collected in an Ultrafree-MC centrifugal device (0.22 µm, Millipore, Bedford, MA, USA) and centrifuged for 10 min at 10,000 g.

4.7.2. LC-MS/MS Analysis

LC-MS/MS analysis was performed using a Waters Acquity HPLC system coupled to a Quattro Micro mass spectrometer (Waters, Milford, MA, USA) equipped with an electrospray interface (ESI) by injecting a volume of 20 µL. Chromatographic separation was performed using an Acquity HPLC Symmetry C18 column (5 µm, 150 mm × 2.1 mm) and a guard column of the same material (10 mm × 2.1 mm i.d.) (Waters, Milford, MA, USA). The column was kept at 60 °C, and the temperature of the autosampler was set at 10 °C. A mobile phase consisting of variable mixtures of mobile phase A (water/methanol/acetic acid, 94/5/1 (*v*/*v*/*v*), 5 mM ammonium acetate) and mobile phase B (methanol/water/acetic acid, 97/2/1 (*v*/*v*/*v*), 5 mM ammonium acetate) was used at a flow rate of 0.3 mL/min with a gradient elution program. The gradient elution started at 99% mobile phase A with a linear decrease to 35% in 7 min. The next 4-min mobile phase A decreased to 25%. An isocratic period of 100% mobile phase B started at 11 min for 2 min. Initial column conditions were reached at 23 min using a linear decrease of mobile phase B, and over 5 min mobile phase A was used to recondition the column. The duration of each HPLC run was 28 min. The mass spectrometer was operated in the positive electrospray ionization (ESI^+) mode. MS parameters were as follows: ESI source block and desolvation temperatures: 150 and 350 °C, respectively; capillary voltage: 3.2 kV; argon collision gas: 1.15×10^{-2} mbar; cone nitrogen and gas flow: 50 L/h and 800 L/h, respectively. The data acquisition was performed using selected reaction monitoring (SRM, Table 6). Masslynx™ version 4.1 and Quanlynx® version 4.1. software (Waters, Milford, MA, USA) were used for data acquisition and processing.

Table 6. Selected reaction monitoring (SRM) transitions for the analyzed mycotoxins in liquid medium.

Analyte	Precursor Ion (m/z)	Molecular Ion	Cone Voltage (V)	Product Ions (m/z)	Collision Energy (eV)
NIV	313.1	$[M + H]^+$	26	125.0 * 205.0	13 12
DON	297.1	$[M + H]^+$	26	231.2 * 249.2	15 10
DOM-1	281.0	$[M + H]^+$	26	109.1 * 137.0	19 15
3-ADON	339.0	$[M + H]^+$	28	203.2 * 231.2	12 13
15-ADON	339.0	$[M + H]^+$	26	261.0 * 321.2	10 10
DON-3G	476.1	$[M + NH_4]^+$	15	248.6 * 296.9	18 12
ZAN (IS)	321.0	$[M + H]^+$	26	189.2 * 303.3	19 13

* quantifier ion. DOM-1: de-epoxy-deoxynivalenol; 3-ADON: 3-acetyl-deoxynivalenol; 15-ADON: 15-acetyl-deoxynivalenol; DON-3G: deoxynivalenol-3-glucoside; ZAN: zearalanone.

The analytical method was validated according to Commission Decision 2002/657/EC [63], and all validation parameters met the criteria mentioned. Validation data for DON is shown in Table S1. Limit of detection (LOD) and limit of quantification (LOQ) ranged from 30 µg/L to 45 µg/L, and 61 µg/L to 89 µg/L, respectively.

4.7.3. LC-HRMS Analysis

Chromatographic separation was achieved on an ACQUITY UPLC I-class FTN system (Waters, Manchester, UK), using a ZORBAX RRHD Eclipse Plus C18 (1.8 µm, 2.1 mm × 100 mm). The mobile phase consisted of H_2O:MeOH (99:1, v/v) containing 0.05% HCOOH and 5 mM $HCOONH_4$ (solvent A) and MeOH (solvent B). A gradient elution program was applied as follows: 0–0.5 min: 5% B, 0.5–20 min: 5%–95% B, 20–21 min: 95% B, 21–24 min: 95%–5% B, 24–28 min: 5% B. The flow rate was 0.3 mL/min. The column temperature was set at 40 °C and the temperature of the autosampler was 10 °C. Five microliters of the sample was injected.

HRMS analyses were performed using a hybrid quadrupole (Q) orthogonal acceleration time-of-flight (TOF) high-definition mass spectrometer, the Synapt G2-Si HDMS (Waters), equipped with an electrospray ionization (ESI) source. Data were acquired as positive ion (ESI^+) polarity runs in resolution mode (> 20000 FWHM). The MS parameters were as follows: capillary voltage 2.8 kV; sample cone voltage 30 V; source temperature 150 °C; desolvation gas flow 800 L/h at a temperature of 550 °C and cone gas flow 50 L/h. Nitrogen was used as the desolvation and cone gases. Argon was employed as the collision gas at a pressure of 9.28×10^{-3} mbar. The instrument was calibrated using sodium formate clusters. During the MS analysis, a leucine-enkephalin solution (200 pg/µL) was continuously infused into the mass spectrometer at a flow rate of 20 µL/min via the lockspray interface, generating the reference ion ($[M + H]^+$ = 556.2771) used for mass correction. Mass spectra were collected in continuum mode from m/z 50 to 1200 with a scan time of 0.1 s, an inter-scan delay of 0.01 s and a lockspray frequency of 20 s. A data-dependent acquisition (DDA) mode was implemented to obtain the simultaneous acquisition of exact mass data for the precursor and fragment ions. The top five ions were selected for MS/MS from a single MS survey scan. The scan time for MS/MS was 0.2 s. The collision energy in the trap cell was ramped from 10/15 V (low mass, start/end) up to 60/150 V (high mass, start/end). Instrument control and data processing were carried out using Masslynx 4.1 software (Waters).

4.8. Data Processing

For developing the bioassay, data was processed for the number of fronds, as well as for frond area. For each calibration curve, the growth was calculated with formula (1) and (2). Growth is plotted as function of concentration DON (mg/L). Subsequently, logit of growth was calculated (3) and plotted in function of logarithm of concentration DON (mg/L) to obtain a linear correlation. In every step, standard deviations were determined. These data were plotted with Sigmaplot 13. The regression coefficients and corresponding *p*-values were calculated with Sigmaplot 13 with linear regression.

$$Growth_{number\ of\ fronds} = \frac{number\ of\ fronds}{number\ of\ fronds_{control}}, \quad (1)$$

$$Growth_{frond\ area} = \frac{frond\ area}{frond\ area_{control}}, \quad (2)$$

$$Logit\ growth = Log\frac{growth}{1 - growth}, \quad (3)$$

The three calibration curves used for evaluating the repeatability were also plotted in Sigmaplot 13, together with the 95% confidence intervals.

For processing data of the bioassay samples, growth was calculated and expressed in % relative to the control. For processing data of LC-MS/MS, % biotransformation of treatments was calculated relative to the corresponding control at the same time point when the sample was taken. Data was statistically analyzed via a one-way ANOVA test or a non-parametric Kruskal–Wallis test followed by a Dunnett T3 test or Dunn's test for pairwise multiple comparisons (α: 0.05) in SPSS Statistics 23. Standard errors were given as ±.

Supplementary Materials: The following are available online at www.mdpi.com/2072-6651/9/2/63/s1, Figure S1: A dose–response curve of chlorophyll fluorescence in *Lemna minor* at 12 h after exposure to different DON concentrations, Figure S2: Scheme of experimental setup of enrichment and biotransformation experiment, Figure S3: MS/MS spectra of DON and 3-epi-DON, Figure S4: MS/MS spectra of DOM-1 and 3-epi-DOM-1, Table S1: Validation parameters for DON.

Acknowledgments: We would like to acknowledge the financial support received for this study from MYCOKEY, a project funded within the Horizon2020 Research and Innovation program of the European Commission (project number 678781) and from the Agency for Innovation by Science and Technology (IWT-LA, project 110776) (Brussels, Belgium). The authors also acknowledge the Hercules infrastructure funding AUGE/13/13.

Author Contributions: K.A. and L.D.G. conceived the experiments. K.A., L.D.G., M.D.B. and J.D.D.M. designed the experiments. I.V., L.D.M. and V.U. performed the experiments. I.V., V.U. and J.D.D.M. analyzed the data. S.D.S. contributed reagents, materials and analysis for the LC-MS/MS analysis. I.V., K.A., L.D.G., V.U., J.D.D.M. and M.D.B. wrote the paper.

Conflicts of Interest: The authors declare no conflict of interest.

References

1. Jouany, J.P. Methods for preventing, decontaminating and minimizing the toxicity of mycotoxins in feeds. *Anim. Feed Sci. Technol.* **2007**, *137*, 342–362. [CrossRef]
2. Wambacq, E.; Vanhoutte, I.; Audenaert, K.; De Gelder, L.; Haesaert, G. Occurrence, prevention and remediation of toxigenic fungi and mycotoxins in silage: A review. *J. Sci. Food Agric.* **2016**, *96*, 2284–2302. [CrossRef] [PubMed]
3. Grenier, B.; Loureiro-Bracarense, A.-P.; Leslie, J.F.; Oswald, I.P. Physical and chemical methods for mycotoxin decontamination in maize. In *Mycotoxin Reduction in Grain Chains*; Leslie, J.F., Logrieco, A., Eds.; John Wiley & Sons, Inc.: Hoboken, NJ, USA, 2014.
4. Goossens, J.; Vandenbroucke, V.; Pasmans, F.; De Baere, S.; Devreese, M.; Osselaere, A.; Verbrugghe, E.; Haesebrouck, F.; De Saeger, S.; Eeckhout, M.; et al. Influence of mycotoxins and a mycotoxin adsorbing agent on the oral bioavailability of commonly used antibiotics in pigs. *Toxins* **2012**, *4*, 281–295. [CrossRef] [PubMed]

5. Schatzmayr, G.; Zehner, F.; Taubel, M.; Schatzmayr, D.; Klimitsch, A.; Loibner, A.P.; Binder, E.M. Microbiologicals for deactivating mycotoxins. *Mol. Nutr. Food Res.* **2006**, *50*, 543–551. [CrossRef] [PubMed]
6. Hope, R.; Aldred, D.; Magan, N. Comparison of environmental profiles for growth and deoxynivalenol production by fusarium culmorum and f-graminearum on wheat grain. *Lett. Appl. Microbiol.* **2005**, *40*, 295–300. [CrossRef] [PubMed]
7. Magan, N.; Hope, R.; Colleate, A.; Baxter, E.S. Relationship between growth and mycotoxin production by fusarium species, biocides and environment. *Eur. J. Plant Pathol.* **2002**, *108*, 685–690. [CrossRef]
8. Ehrlich, K.C.; Daigle, K.W. Protein-synthesis inhibition by 8-oxo-12,13-epoxytrichothecenes. *Biochim. Biophys. Acta* **1987**, *923*, 206–213. [CrossRef]
9. Sobrova, P.; Adam, V.; Vasatkova, A.; Beklova, M.; Zeman, L.; Kizek, R. Deoxynivalenol and its toxicity. *Interdiscip. Toxicol.* **2010**, *3*, 94–99. [CrossRef] [PubMed]
10. Pestka, J.J. Deoxynivalenol: Toxicity, mechanisms and animal health risks. *Anim. Feed Sci. Technol.* **2007**, *137*, 283–298. [CrossRef]
11. Pestka, J.J. Deoxynivalenol: Mechanisms of action, human exposure, and toxicological relevance. *Arch. Toxicol.* **2010**, *84*, 663–679. [CrossRef] [PubMed]
12. European Commission. Commission recommendation of 17 august 2006 on the presence of deoxynivalenol, zearalenone, ochratoxin a, t-2 and ht-2 and fumonisins in products intended for animal feeding. *Off. J. Eur. Union* **2006**, *L 229*, 7–9.
13. European Commission. Commission regulation no 1881/2006 of 19 december 2006 setting maximum levels for certain contaminants in foodstuffs. *Off. J. Eur. Union* **2006**, *L 364*, 5–24.
14. Lauren, D.R.; Smith, W.A. Stability of the fusarium mycotoxins nivalenol, deoxynivalenol and zearalenone in ground maize under typical cooking environments. *Food Addit. Contam.* **2001**, *18*, 1011–1016. [CrossRef] [PubMed]
15. De Mil, T. Safety of Mycotoxin Binders Regarding Their Use with Veterinary Medicinal Products in Poultry and pigs: An In Vitro and Pharmacokinetic Approach. Ph.D. Thesis, Ghent University, Ghent, Belgium, 2016.
16. Vanhoutte, I.; Audenaert, K.; De Gelder, L. Biodegradation of mycotoxins: Tales from known and unexplored worlds. *Front. Microbiol.* **2016**, *7*. [CrossRef] [PubMed]
17. Binder, E.M.; Binder, J. Strain of Eubacterium That Detoxyfies Trichothenes. U.S. Patent No. 6,794,175, 21 September 2004.
18. King, R.R.; Mcqueen, R.E.; Levesque, D.; Greenhalgh, R. Transformation of deoxynivalenol (vomitoxin) by rumen microorganisms. *J. Agric. Food Chem.* **1984**, *32*, 1181–1183. [CrossRef]
19. Kollarczik, B.; Gareis, M.; Hanelt, M. In vitro transformation of the fusarium mycotoxins deoxynivalenol and zearalenone by the normal gut microflora of pigs. *Nat. Toxins* **1994**, *2*, 105–110. [CrossRef] [PubMed]
20. Yu, H.; Zhou, T.; Gong, J.H.; Young, C.; Su, X.J.; Li, X.Z.; Zhu, H.H.; Tsao, R.; Yang, R. Isolation of deoxynivalenol-transforming bacteria from the chicken intestines using the approach of PCR-DGGE guided microbial selection. *BMC Microbiol.* **2010**, *10*, 182. [CrossRef] [PubMed]
21. Guan, S.; He, J.W.; Young, J.C.; Zhu, H.H.; Li, X.Z.; Ji, C.; Zhou, T. Transformation of trichothecene mycotoxins by microorganisms from fish digesta. *Aquaculture* **2009**, *290*, 290–295. [CrossRef]
22. Young, J.C.; Zhou, T.; Yu, H.; Zhu, H.H.; Gong, J.H. Degradation of trichothecene mycotoxins by chicken intestinal microbes. *Food Chem. Toxicol.* **2007**, *45*, 136–143. [CrossRef] [PubMed]
23. Zhou, T.; Gong, J.; Young, J.C.; Yu, H.; Li, X.Z.; Zhu, H.; Hill, A.; Yang, R.; de Lange, C.F.M.; Du, W.; et al. Microorganisms isolated from chicken gut can effectively detoxify don and other trichothecene mycotoxins. Available online: http://www.centraliaswineresearch.ca/proceedings/2007/csru2007zhou.pdf (accessed on 8 February 2017).
24. He, J.W.; Hassan, Y.I.; Perilla, N.; Li, X.Z.; Boland, G.J.; Zhou, T. Bacterial epimerization as a route for deoxynivalenol detoxification: The influence of growth and environmental conditions. *Front. Microbiol.* **2016**, *7*. [CrossRef] [PubMed]
25. Ikunaga, Y.; Sato, I.; Grond, S.; Numaziri, N.; Yoshida, S.; Yamaya, H.; Hiradate, S.; Hasegawa, M.; Toshima, H.; Koitabashi, M.; et al. *Nocardioides* sp. Strain wsn05-2, isolated from a wheat field, degrades deoxynivalenol, producing the novel intermediate 3-epi-deoxynivalenol. *Appl. Microbiol. Biot.* **2011**, *89*, 419–427. [CrossRef] [PubMed]

26. Islam, R. Isolation, Characterization and Genome Sequencing of a Soil-Borne Citrobacter Freundii Strain Capable of Detoxifying Trichothecene Mycotoxins. Ph.D. Thesis, University of Guelph, Guelph, ON, Canada, 2012.
27. Islam, R.; Zhou, T.; Young, J.C.; Goodwin, P.H.; Pauls, K.P. Aerobic and anaerobic de-epoxydation of mycotoxin deoxynivalenol by bacteria originating from agricultural soil. *World J. Microb. Biot.* **2012**, *28*, 7–13. [CrossRef] [PubMed]
28. Sato, I.; Ito, M.; Ishizaka, M.; Ikunaga, Y.; Sato, Y.; Yoshida, S.; Koitabashi, M.; Tsushima, S. Thirteen novel deoxynivalenol-degrading bacteria are classified within two genera with distinct degradation mechanisms. *FEMS Microbiol. Lett.* **2012**, *327*, 110–117. [CrossRef] [PubMed]
29. Shima, J.; Takase, S.; Takahashi, Y.; Iwai, Y.; Fujimoto, H.; Yamazaki, M.; Ochi, K. Novel detoxification of the trichothecene mycotoxin deoxynivalenol by a soil bacterium isolated by enrichment culture. *Appl. Environ. Microb.* **1997**, *63*, 3825–3830.
30. Volkl, A.; Vogler, B.; Schollenberger, M.; Karlovsky, P. Microbial detoxification of mycotoxin deoxynivalenol. *J. Basic Microb.* **2004**, *44*, 147–156. [CrossRef] [PubMed]
31. Sinclair, C.J.; Boxall, A.B.A. Ecotoxicity of transformation products. In *Transformation Products of Synthetic Chemicals in the Environment*; Boxall, A.B.A., Ed.; Springer: Heidelberg, Germany, 2009; Volume 2.
32. Thompson, W.L.; Wannemacher, R.W. Structure-function-relationships of 12,13-epoxytrichothecene mycotoxins in cell-culture—Comparison to whole animal lethality. *Toxicon* **1986**, *24*, 985–994. [CrossRef]
33. Eriksen, G.S.; Pettersson, H.; Lundh, T. Comparative cytotoxicity of deoxynivalenol, nivalenol, their acetylated derivatives and de-epoxy metabolites. *Food Chem. Toxicol.* **2004**, *42*, 619–624. [CrossRef] [PubMed]
34. He, J.W.; Bondy, G.S.; Zhou, T.; Caldwell, D.; Boland, G.J.; Scott, P.M. Toxicology of 3-epi-deoxynivalenol, a deoxynivalenol-transformation product by devosia mutans 17-2-e-8. *Food Chem Toxicol.* **2015**, *84*, 250–259. [CrossRef] [PubMed]
35. Pierron, A.; Mimoun, S.; Murate, L.S.; Loiseau, N.; Lippi, Y.; Bracarense, A.P.F.L.; Schatzmayr, G.; He, J.W.; Zhou, T.; Moll, W.D.; et al. Microbial biotransformation of don: Molecular basis for reduced toxicity. *Sci. Rep.-UK* **2016**, *6*. [CrossRef] [PubMed]
36. Hanelt, M.; Gareis, M.; Kollarczik, B. Cytotoxicity of mycotoxins evaluated by the mtt-cell culture assay. *Mycopathologia* **1994**, *128*, 167–174. [CrossRef] [PubMed]
37. Kouadio, J.H.; Mobio, T.A.; Baudrimont, I.; Moukha, S.; Dano, S.D.; Creppy, E.E. Comparative study of cytotoxicity and oxidative stress induced by deoxynivalenol, zearalenone or fumonisin b1 in human intestinal cell line caco-2. *Toxicology* **2005**, *213*, 56–65. [CrossRef] [PubMed]
38. Rocha, O.; Ansari, K.; Doohan, F.M. Effects of trichothecene mycotoxins on eukaryotic cells: A review. *Food Addit. Contam. A* **2005**, *22*, 369–378. [CrossRef] [PubMed]
39. Panigrahi, S. Bioassay of mycotoxins using terrestrial and aquatic, animal and plant-species. *Food Chem. Toxicol.* **1993**, *31*, 767–790. [CrossRef]
40. Abbas, H.K.; Yoshizawa, T.; Shier, W.T. Cytotoxicity and phytotoxicity of trichothecene mycotoxins produced by *fusarium* spp. *Toxicon* **2013**, *74*, 68–75. [CrossRef] [PubMed]
41. Abolmaali, S.; Mitterbauer, R.; Spadiut, O.; Peruci, M.; Weindorfer, H.; Lucyshyn, D.; Ellersdorfer, G.; LemmenS, M.; Moll, W.D.; Adarn, G. Engineered bakers yeast as a sensitive bioassay indicator organism for the trichothecene toxin deoxynivalenol. *J. Microbiol. Meth.* **2008**, *72*, 306–312. [CrossRef] [PubMed]
42. Alexander, N.J.; McCormick, S.P.; Ziegenhorn, S.L. Phytotoxicity of selected trichothecenes using chlamydomonas reinhardtii as a model system. *Nat. Toxins* **1999**, *7*, 265–269. [CrossRef]
43. Bijl, J.P.; Rousseau, D.M.; Dive, D.G.; Vanpeteghem, C.H. Potentials of a synchronized culture of tetrahymena-pyriformis for toxicity studies of mycotoxins. *J. Assoc. Off. Anal. Chem.* **1988**, *71*, 282–285.
44. Binder, J.; Horvath, E.M.; Heidegger, J.; Ellend, N.; Danner, H.; Krska, R.; Braun, R. A bioassay for comparison of the toxicity of trichothecenes and their microbial metabolites. *Cereal Res. Commun.* **1997**, *25*, 489–491.
45. Desjardins, A.E.; McCormick, S.P.; Appell, M. Structure-activity relationships of trichothecene toxins in an arabidopsis thaliana leaf assay. *J. Agric. Food Chem.* **2007**, *55*, 6487–6492. [CrossRef] [PubMed]
46. Masuda, D.; Ishida, M.; Yamaguchi, K.; Yamaguchi, I.; Kimura, M.; Nishiuchi, T. Phytotoxic effects of trichothecenes on the growth and morphology of arabidopsis thaliana. *J. Exp. Bot.* **2007**, *58*, 1617–1626. [CrossRef] [PubMed]
47. Scott, P.M.; Harwig, J.; Blanchfield, B.J. Screening fusarium strains isolated from overwintered canadian grains for trichothecenes. *Mycopathologia* **1980**, *72*, 175–180. [CrossRef] [PubMed]

48. Hulsen, K.; Minne, V.; Lootens, P.; Vandecasteele, P.; Hofte, M. A chlorophyll a fluorescence-based lemna minor bioassay to monitor microbial degradation of nanomolar to micromolar concentrations of linuron. *Environ. Microbiol.* **2002**, *4*, 327–337. [CrossRef] [PubMed]
49. Burgess, K.M.; Renaud, J.B.; McDowell, T.; Sumarah, M.W. Mechanistic insight into the biosynthesis and detoxification of fumonisin mycotoxins. *ACS Chem. Biol.* **2016**, *9*, 2618–2625. [CrossRef] [PubMed]
50. Tanaka, T.; Abbas, H.K.; Duke, S.O. Structure-dependent phytotoxicity of fumonisins and related-compounds in a duckweed bioassay. *Phytochemistry* **1993**, *33*, 779–785. [CrossRef]
51. Vesonder, R.F.; Wu, W.; Weisleder, D.; Gordon, S.H.; Krick, T.; Xie, W.; Abbas, H.K.; McAlpin, C.E. Toxigenic strains of fusarium moniliforme and fusarium proliferatum isolated from dairy cattle feed produce fumonisins, moniliformin and a new c21h38n2o6 metabolite phytotoxic to lemna minor l. *J. Nat. Toxins* **2000**, *9*, 103–112. [PubMed]
52. Wang, W.C. Toxicity tests of aquatic pollutants by using common duckweed. *Environ. Pollut. B* **1986**, *11*, 1–14. [CrossRef]
53. Diana Di Mavungu, J.; Malysheva, S.V.; Sanders, M.; Larionova, D.; Robbens, J.; Dubruel, P.; Van Peteghem, C.; De Saeger, S. Development and validation of a new LC-MS/MS method for the simultaneous determination of six major ergot alkaloids and their corresponding epimers. Application to some food and feed commodities. *Food Chem.* **2012**, *135*, 292–303. [CrossRef]
54. Hassan, Y.I.; Zhu, H.; Zhu, Y.; Zhou, T. Beyond ribosomal binding: The increased polarity and aberrant molecular interactions of 3-epi-deoxynivalenol. *Toxins (Basel)* **2016**. [CrossRef] [PubMed]
55. He, J.W.; Yang, R.; Zhou, T.; Boland, G.J.; Scott, P.M.; Bondy, G.S. An epimer of deoxynivalenol: Purification and structure identification of 3-epi-deoxynivalenol. *Food Addit. Contam. Part A Chem. Anal. Control Expo. Risk Assess.* **2015**, *32*, 1523–1530. [CrossRef] [PubMed]
56. Harwig, J.; Scott, P.M. Brine shrimp (artemia salina L.) larvae as a screening system for fungal toxins. *Appl. Microbiol.* **1971**, *21*, 1011–1016. [PubMed]
57. IARC. *Iarc Monographs on the Evaluation of Carcinogenic Risks to Humans: Aflatoxins*; IARC Press: Lyon, France, 2002; Volume 82.
58. Shier, W.T.; Shier, A.C.; Xie, W.; Mirocha, C.J. Structure-activity relationships for human estrogenic activity in zearalenone mycotoxins. *Toxicon* **2001**, *39*, 1435–1438. [CrossRef]
59. He, J.W. Detoxification of Deoxynivalenol by a Soil Bacterium Devosia Mutans 17-2-e-8. Ph.D. Thesis, The University of Guelph, Guelph, ON, Canada, 2015.
60. Spanjer, M.C.; Rensen, P.M.; Scholten, J.M. Lc-ms/ms multi-method for mycotoxins after single extraction, with validation data for peanut, pistachio, wheat, maize, cornflakes, raisins and figs. *Food Addit. Contam. A* **2008**, *25*, 472–489. [CrossRef] [PubMed]
61. Stanier, R.Y.; Palleroni, N.J.; Doudoroff, M. The aerobic pseudomonads: A taxonomic study. *J. Gen. Microbiol.* **1966**, *43*, 159–271. [CrossRef] [PubMed]
62. Megateli, S.; Dosnon-Olette, R.; Trotel-Aziz, P.; Geffard, A.; Semsari, S.; Couderchet, M. Simultaneous effects of two fungicides (copper and dimethomorph) on their phytoremediation using lemna minor. *Ecotoxicology* **2013**, *22*, 683–692. [CrossRef] [PubMed]
63. European Commission. Commission decision 2002/657/ec implementing council directive 96/23/ec concerning the performance of analytical methods and the interpretation of results. *Off. J. Eur. Communities* **2002**, *L 221*, 8–36.

© 2017 by the authors. Licensee MDPI, Basel, Switzerland. This article is an open access article distributed under the terms and conditions of the Creative Commons Attribution (CC BY) license (http://creativecommons.org/licenses/by/4.0/).

 *toxins*

Article

Unravelling the Diversity of the Cyclopiazonic Acid Family of Mycotoxins in *Aspergillus flavus* by UHPLC Triple-TOF HRMS

Valdet Uka [1,2], Geromy G. Moore [3], Natalia Arroyo-Manzanares [1,4], Dashnor Nebija [2], Sarah De Saeger [1] and José Diana Di Mavungu [1,*]

1 Laboratory of Food Analysis, Faculty of Pharmaceutical Sciences, Ghent University, Ottergemsesteenweg 460, 9000 Ghent, Belgium; valdet.uka@ugent.be (V.U.); natalia.arroyomanzanares@ugent.be (N.A.-M.); sarah.desaeger@ugent.be (S.D.S.)

2 Department of Pharmacy, Faculty of Medicine, University of Prishtina, Rrethi i Spitalit p.n, 10000 Prishtina, Kosovo; dashnor.nebija@uni-pr.edu

3 Southern Regional Research Center, Agricultural Research Service, United States Department of Agriculture (ARS-USDA), New Orleans, 70124 LA, USA; geromy.moore@ars.usda.gov

4 Department of Analytical Chemistry, Faculty of Sciences, University of Granada, Campus Fuentenueva s/n, E-18071 Granada, Spain

* Correspondence: jose.dianadimavungu@ugent.be; Tel.: +32-9-264-8134

Academic Editor: Massimo Reverberi
Received: 9 November 2016; Accepted: 10 January 2017; Published: 13 January 2017

Abstract: Cyclopiazonic acid (α-cyclopiazonic acid, α-CPA) is an indole-hydrindane-tetramic acid neurotoxin produced by various fungal species, including the notorious food and feed contaminant *Aspergillus flavus*. Despite its discovery in *A. flavus* cultures approximately 40 years ago, its contribution to the *A. flavus* mycotoxin burden is consistently minimized by our focus on the more potent carcinogenic aflatoxins also produced by this fungus. Here, we report the screening and identification of several CPA-type alkaloids not previously found in *A. flavus* cultures. Our identifications of these CPA-type alkaloids are based on a dereplication strategy involving accurate mass high resolution mass spectrometry data and a careful study of the α-CPA fragmentation pattern. In total, 22 CPA-type alkaloids were identified in extracts from the *A. flavus* strains examined. Of these metabolites, 13 have been previously reported in other fungi, though this is the first report of their existence in *A. flavus*. Two of our metabolite discoveries, 11,12-dehydro α-CPA and 3-hydroxy-2-oxo CPA, have never been reported for any organism. The conspicuous presence of CPA and its numerous derivatives in *A. flavus* cultures raises concerns about the long-term and cumulative toxicological effects of these fungal secondary metabolites and their contributions to the entire *A. flavus* mycotoxin problem.

Keywords: cyclopiazonic acid; ergot-like alkaloid; dereplication; HRMS

---

1. Introduction

The ergot-like alkaloid cyclopiazonic acid (α-cyclopiazonic acid, α-CPA) is an indole-hydrindane-tetramic acid mycotoxin produced by many fungal species in the Ascomycete genera *Penicillium* and *Aspergillus*. α-CPA was first isolated from a liquid culture of *Penicillium cyclopium* Westling in 1968, as the main toxic compound of this microorganism [1]. Afterwards, in 1973, Ohmomo et al. [2] reported its production by a strain of *Aspergillus versicolor*. In 1977, it was also demonstrated that α-CPA can be produced by *Aspergillus flavus*, a prolific food and feed contaminant [3]. Since then, CPA-producing strains have been identified in other fungal species such as *Penicillium griseofulvum*, *Penicillium commune*, *Penicillium chrysogenum*, *Aspergillus oryzae*, *Aspergillus fumigatus* and *Aspergillus tamarii* [4–7].

A gene cluster for the biosynthesis of α-CPA, containing three essential genes, was identified in the genome of *A. flavus* and *A. oryzae*, situated adjacent to the aflatoxin gene cluster [8,9]. α-CPA is biosynthesized from three precursors including a tryptophan residue, two units of acetic acid and an isoprenoid moiety (dimethylallyl diphosphate—DMAPP) in a three-enzyme biochemical pathway. Through this short metabolic pathway, two biosynthetic intermediates are generated, *cyclo*-acetoacetyl-L-tryptophan (cAATrp) and β-cyclopiazonic acid (β-CPA), by consecutive action of three enzymes, CpaS, CpaD and CpaO. The hybrid two-module polyketide synthase-nonribosomal peptide synthetase (PKS-NRPS), i.e., CpaS (also known as *CpaA*), is responsible for the formation of the tetramic acid *cyclo*-acetoacetyl-L-tryptophan (cAATrp). cAATrp is then prenylated by the prenyltransferase CpaD (also known as *dmaT*), which leads to the generation of β-CPA, the ultimate tricyclic precursor of α-CPA. The final conversion of β- to α-CPA is catalyzed by the putative monoamine cyclo-oxidase CpaO (also known as *maoA*) in a redox reaction forming two rings (ring C and D) [10,11].

The biochemical mechanism beyond the toxicological profile of α-CPA is proved to be its specific ability to inhibit sarco/endoplasmic reticulum Ca^{2+}-ATPase (SERCA) in different tissues and cell types [4,12]. Hence, α-CPA is considered one of the few potent, selective and reversible SERCA inhibitors, another example being thapsigargin [13]. SERCA is an active membrane pump responsible for the transfer of Ca^{2+} ions from the cytosol of the cell to the lumen of the sarco/endoplasmic reticulum, thus maintaining a low concentration of free calcium ions in the cytosol. Physiologically, these levels of intracellular calcium are essential for housekeeping activities of each cellular entity, like cell signaling, proliferation, differentiation and muscle contraction-relaxation. For this reason, SERCA blockage by α-CPA disrupts the normal intracellular calcium gradient and eventually leads to cell damage and death. Although the main target organs of CPA poisoning seem to be skeletal muscle, hepatic tissues and spleen, several animal studies have also reported pathological lesions in the kidney, pancreas, heart and gastro-intestinal tract [14,15]. Moreover, it has been demonstrated that CPA exposure can be associated with several neurological symptoms like hypokinesia, catalepsy, hypothermia, tremors, convulsions, cessation of food intake and resulting cachexia [16]. In general, α-CPA is not considered an acute mycotoxin due to its relatively high LD_{50} ranges in rats (30–70 mg/kg) and because of the benign nature of the intoxication [17]. Although confirmed CPA-mycotoxicoses have not been reported in humans, it was speculated that CPA exposure may be linked with "kodo poisoning," a toxic syndrome characterized by nausea, vomiting, depression and unconsciousness after intake of CPA-contaminated Kodo millet [18].

The risk of human exposure to α-CPA could arise from the consumption of contaminated food commodities. α-CPA has been found to contaminate various grains and seeds, as well as different food matrices such as cheese, nuts and meat products [19–22]. Furthermore, this mycotoxin has been reported in milk and eggs, most likely due to animal consumption of contaminated feeds [23]. The importance of α-CPA as a toxic contaminant of different foodstuffs has been neglected and masked by other concurrent mycotoxins such as aflatoxins, especially when co-occurrence of CPA and aflatoxins has been observed [24–28].

Based on our literature review, since 1968 when α-CPA was first isolated, approximately 30 cyclopiazonic acid (CPA)-type alkaloids have been reported in different fungal extracts of aspergilli and penicillium (Figure 1; Table 1). These naturally occurring CPA analogues, containing the tetramate moiety as key structural motif, are characterized by some minor structural variations, and all belong to indole or oxindole (indolinone) subclasses of alkaloids. Very soon after the α-CPA discovery, Holzapfel et al. [29] reported the identification of two other CPA derivatives: β-CPA (also termed *bis*-secodehydrocyclopiazonic acid) and α-CPA imine. Later on, *iso*-α-CPA, the isomer of α-CPA, was structurally characterized in *A. flavus* [30]. Very recently, a new CPA derivative, pseuboydone E, has been isolated in *Pseudallescheria boydii* [31].The above-mentioned metabolites, namely α-CPA, *iso*-α-CPA, β-CPA, α-CPA imine, cAATrp and pseuboydone E, are the only CPA derivatives belonging to the indole subclass of CPA-type alkaloids (Figure 1A). All the remaining analogues, i.e., 2-oxoCPA, speradines, cyclopiamides and aspergillines, are classified in the oxindole subclass of CPA-type alkaloids. The first member of oxindoles possessing a keto-group in the C2-position of the indole nucleus is 2-oxoCPA; its presence has been reported in *A. oryzae* [32,33]. The first N-methylated pentacyclic oxindole analogues of α-CPA, speradine A and 3-hydroxyl-speradine A were isolated in fungal cultures of *A. tamarii* [34,35]. Four other tetracyclic oxindole alkaloids, named speradine B, C, D and E, were identified from *A. oryzae* [36]. A rare hexacyclic oxindole alkaloid, speradine F (also termed penicamedine A), together with two novel tetracyclic oxindoles, speradine G and H, were characterized in *A. oryzae* isolated from river sediments in China [37,38] (Figure 1B). In the literature, the nomenclature of oxindoles has been utilized incorrectly. For example, Ma et al. [39] reported the identification of speradine B, C and D from a sponge-derived strain of *A. flavus*, but these molecules do not correspond with metabolites previously described by Hu and coworkers [36]. Speradine B and D from Ma et al. [39] actually correspond with speradine F and C, respectively, as described by Hu et al. [37]. Speradine C reported in Ma et al. [39] was actually a new compound possessing an unprecedented 6/5/6/5/5/6 hexacyclic system with a unique 4-oxo-1,3-oxazinane ring. In order to differentiate the metabolite previously isolated by Hu et al. [36], we decided to name this compound speradine I. Afterwards, five highly oxygenated CPA-related alkaloids, aspergillines A–E (Figure 1C), all having a rigid and sterically congested hexacyclic 6/5/6/5/5/5 indole-tetrahydrofuran-tetramate scaffold, were isolated from *Aspergillus versicolor* [40]. Aspergillines B and E possess a butanoic acid methyl ester moiety, whereas aspergilline C contains an extra isoprenoid moiety attached to the indole nucleus (Figure 1C). In addition, another group of CPA-related oxindoles, named cyclopiamides A–J, were isolated from a deep-sea-derived strain of *Penicillium commune* [41,42] (Figure 1D). Cyclopiamides H and I isolated in *P. commune* prove to be the same chemical entities with speradine B and aspergilline D, respectively. To avoid future confusion regarding the nomenclature of the CPA-related alkaloids, we suggest they are named as they were discovered chronologically (Figure 1; Table 1).

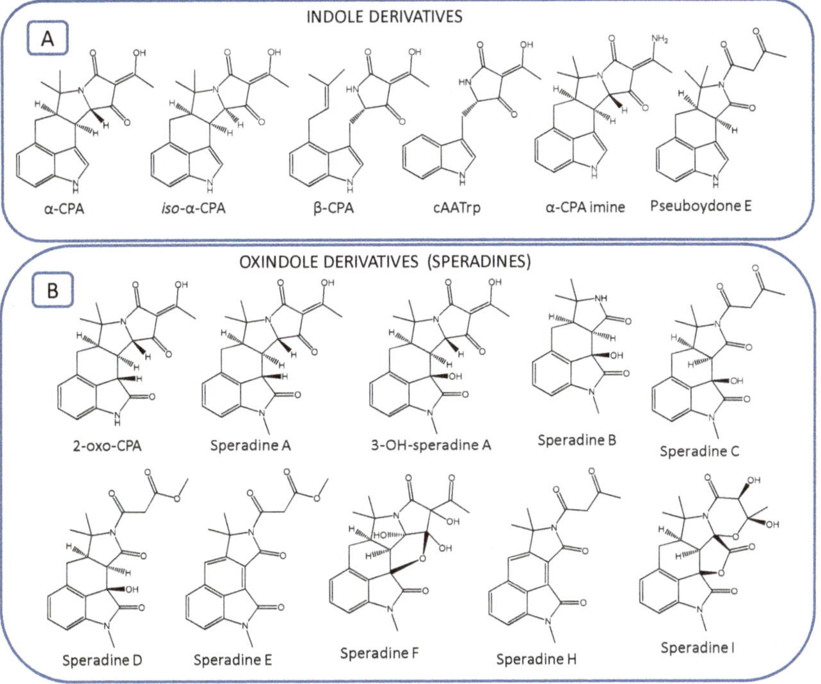

Figure 1. *Cont.*

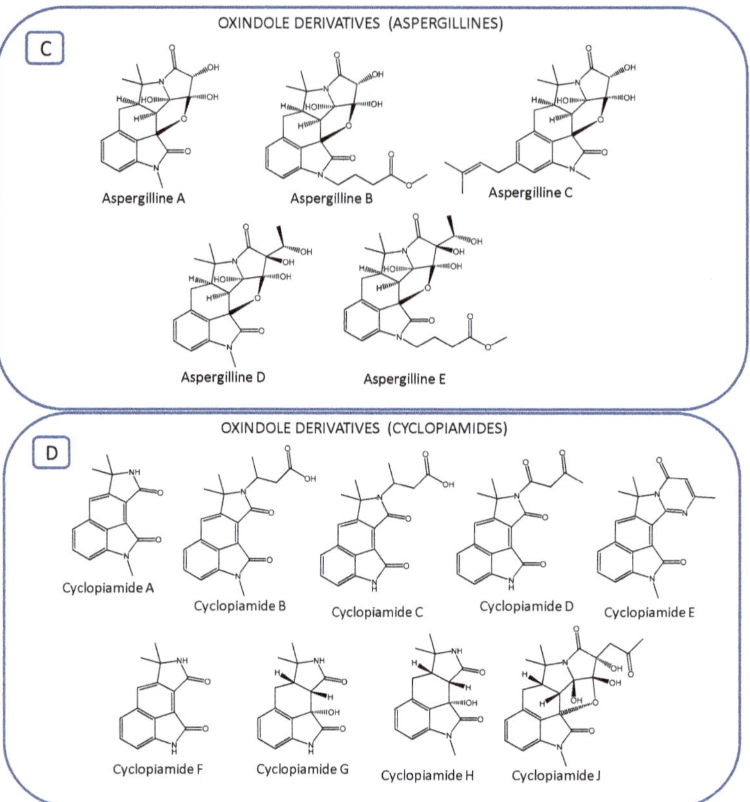

Figure 1. Structures of cyclopiazonic acid (CPA)-type alkaloids: (**A**) Indole derivatives; (**B**) Speradines; (**C**) Aspergillines; (**D**) Cyclopiamides.

Table 1. CPA-type alkaloids identified in different fungal sources.

Compound	Name	Formula	Exact Mass	Source	Reference
1	α-CPA	$C_{20}H_{20}N_2O_3$	336.1474	Various *Aspergillus* and *Penicillium* species	[1,3,29]
2	Iso-α-CPA	$C_{20}H_{20}N_2O_3$	336.1474	*Aspergillus flavus*	[30]
3	β-CPA	$C_{20}H_{22}N_2O_3$	338.1630	Various *Aspergillus* and *Penicillium* species	[8,9,29]
4	cAATrp	$C_{15}H_{14}N_2O_3$	270.1004	Various *Aspergillus* and *Penicillium* species	[8–11]
5	α-CPA imine	$C_{20}H_{21}N_3O_2$	335.1634	*Penicillium cyclopium*	[29]
6	Pseuboydone E	$C_{19}H2_{0}N_2O_3$	324.1474	*Pseudallescheria boydii*	[31]
7	2-oxoCPA	$C_{20}H_{20}N_2O_4$	352.1423	*Aspergillus oryzae*	[32,33]
8	Speradine A	$C_{21}H_{22}N_2O_4$	366.1580	*A. tamarii, A. oryzae*	[33,34]
9	3-OH-speradine A	$C_{21}H_{22}N_2O_5$	382.1529	*A. tamarii*	[35]
10	Speradine B	$C_{16}H_{18}N_2O_3$	286.1317	*A. oryzae, A. flavus*	[36,39]
11	Speradine C	$C_{20}H_{22}N_2O_5$	370.1529	*A. oryzae, A. flavus*	[36,39]
12	Speradine D	$C_{20}H_{22}N_2O_6$	386.1478	*A. oryzae*	[36]
13	Speradine E	$C_{20}H_{18}N_2O_5$	366.1216	*A. oryzae*	[36]
14	Speradine F (Pencamedine A)	$C_{21}H_{22}N_2O_7$	414.1427	*A. oryzae, P. commune, A. flavus, P. camemberti*	[37–39,41]
15	Speradine H	$C_{20}H_{18}N_2O_4$	350.1267	*A. oryzae, P. commune*	[37,41]
16	Speradine I	$C_{21}H_{22}N_2O_7$	414.1427	*A. flavus*	[39]
17	Aspergilline A	$C_{19}H_{20}N_2O_6$	372.1321	*A. versicolor*	[40]
18	Aspergilline B	$C_{23}H_{26}N_2O_8$	458.1689	*A. versicolor*	[40]
19	Aspergilline C	$C_{24}H_{28}N_2O_6$	440.1947	*A. versicolor*	[40]
20	Aspergilline D (Cyclopiamide I)	$C_{21}H_{24}N_2O_7$	416.1584	*A. versicolor, P. commune*	[41,42]
21	Aspergilline E	$C_{25}H_{30}N_2O_9$	502.1951	*A. versicolor*	[40]
22	Cyclopiamide A	$C_{16}H_{14}N_2O_2$	266.1055	*P. cyclopium, P. commune, A. flavus*	[39,41,42]
23	Cyclopiamide B	$C_{20}H_{20}N_2O_4$	352.1423	*P. commune*	[41]
24	Cyclopiamide C	$C_{19}H_{18}N_2O_4$	338.1267	*P. commune*	[41]
25	Cyclopiamide D	$C_{19}H_{16}N_2O_4$	336.1110	*P. commune*	[41]
26	Cyclopiamide E	$C_{20}H_{17}N_3O_2$	331.1321	*P. commune*	[41]
27	Cyclopiamide F	$C_{15}H_{12}N_2O_2$	252.0899	*P. commune*	[41]
28	Cyclopiamide G	$C_{15}H_{16}N_2O_3$	272.1161	*P. commune*	[41]
29	Cyclopiamide H	$C_{16}H_{18}N_2O_3$	286.1317	*A. oryzae, P. commune, A. flavus*	[36,39,41]
30	Cyclopiamide J	$C_{22}H_{24}N_2O_7$	428.1584	*P. commune*	[41]

Since *Aspergillus flavus* is an important mycotoxigenic mold, and a very frequent food and feed contaminant with ubiquitous nature, the probability of human and animal exposure to CPA, as well as its associated health hazard, is higher compared to other fungal species. On the other hand, one method for preventing aflatoxin contamination of crops is by introducing a non-aflatoxigenic competitor strain of *A. flavus* to compete with natural aflatoxin-producing fungi. Although this approach may reduce aflatoxin levels in food and feed commodities, the accumulation of other mycotoxins such as CPA has been observed [43]. In this regard, it is of utmost importance to thoroughly investigate this fungus for its capability to produce known and yet unknown CPA-type alkaloids. This can be achieved through a dereplication strategy based on accurate mass high resolution mass spectrometry (HRMS) and fragmentation data [44]. Nowadays, accurate mass measurements, isotope-model fitting, tandem mass spectrometry (MS/MS) spectra and chemical databases are integrated in single software packages, thus allowing a fast and aggressive dereplication of known metabolites (Figure 2). A careful study of the fragmentation pattern of known compounds can be used to help detect and identify novel and previously unreported analogues. Hence, the main aim of this work was to investigate the diversity of the CPA family of alkaloids in different strains of *A. flavus* by accurate mass HRMS, thereby building knowledge towards a better assessment of the global *A. flavus* mycotoxin burden.

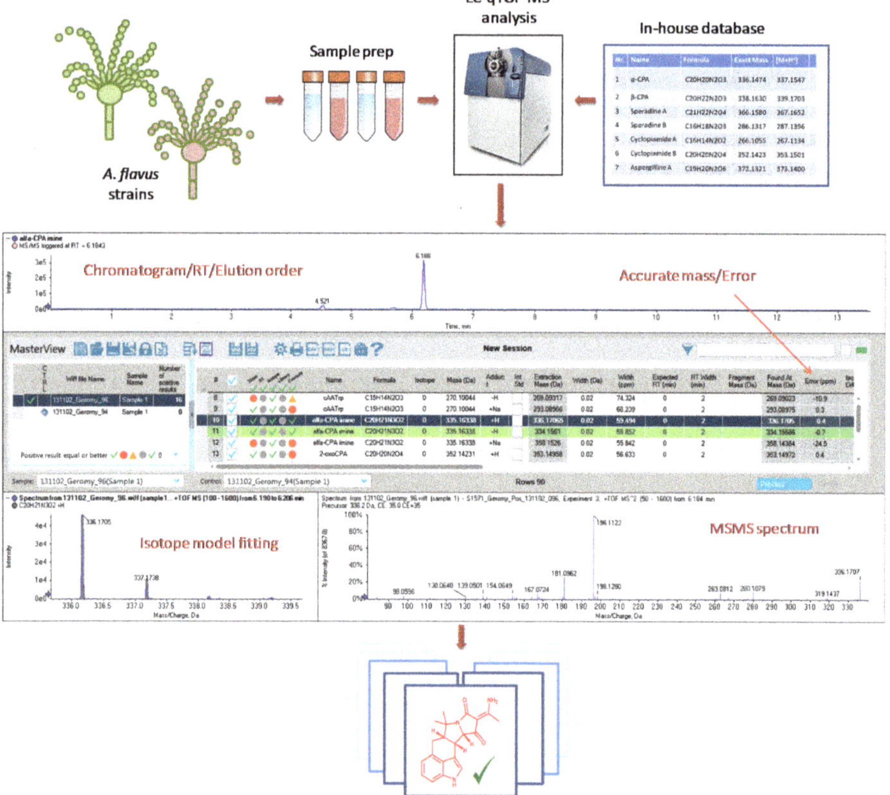

Figure 2. Dereplication workflow of already reported CPA-related alkaloids by using *MasterView*^{TM} *software* as an integrated package displaying: Extracted chromatogram (RT—retention time), mass accuracy (error—ppm), relative quantity (area—arbitrary units), isotope-model fitting and tandem mass spectrometry (MS/MS) spectra.

2. Results

2.1. Identification of Indole Cyclopiazonic Acid (CPA)-Type Derivatives

A dereplication approach based on accurate mass HRMS data, combined with a careful examination of fragmentation spectra, was applied to ascertain the presence of previously identified CPA-type alkaloids and to establish an unambiguous identification strategy for further screening work. The employed analytical methodology involves an untargeted data acquisition (consisting of full scan time-of-flight (TOF) HRMS survey and information-dependent acquisition, (IDA) MS/MS scans) and the processing of data using both targeted and untargeted approaches. An α-CPA reference standard was available, and therefore, this compound was identified in the fungal extracts by comparison of retention time, accurate mass HRMS and HRMS/MS data with the reference standard. A careful investigation and interpretation of α-CPA fragmentation data was the basis for the identification of other CPA-type alkaloids as described below.

A perfect match was observed between the α-CPA MS/MS spectrum of the reference standard and that of the putative CPA in *A. flavus* extracts. A typical MS/MS spectrum of the protonated α-CPA ion ($[m + H]^+/z$ 337.1536; $\Delta = -0.9$ ppm) is shown in Figure 3A. The observed fragmentation pattern is also in accordance with reports from previous studies [1,45,46]. This spectrum, apart from the parent ion ($[m + H]^+/z$ 337), also shows some prominent fragment ions at m/z 196, 182, 167, 154 and 140 together with their "sibling species" at m/z 197, 181, 168, 155 and 141, thereby corroborating the findings from Holzapfel [1]. Generation of these metastable species most likely is achieved by ejection of a hydrogen atom from the main original fragment or vice versa [1]. For example, at least a part of the m/z 154 signal is yielded by proton ejection from m/z 155. High resolution mass measurements revealed that fragments at mass 154 and 155 correspond to the chemical formula of $C_{11}H_8N^+$ and $C_{11}H_9N^+$, respectively. Based on their chemical composition, these fragments must contain the indole system and three additional carbon atoms, resulting from the cleavage of the C4-C5 and C10-C11 bonds of ring D. The fragment ion at m/z 196 has the chemical composition of $C_{14}H_{14}N^+$ and corresponds to the fragment ion 154 with three additional carbon atoms. This ion (m/z 196) arises by cleavage of the C4-C5 and C9-C10 bonds of ring D. Accurate mass measurements showed that the ion at mass 182 corresponds to the chemical formula $C_9H_{12}NO_3{}^+$. This fragment ion contains all the oxygen atoms of the parent ion, and therefore represents the tetramic acid moiety of the molecule as depicted in Figure 3A. The ion with m/z 167 is most likely formed by further ejection of a methyl group ($-CH_3$) from the ion at m/z 182, whilst the mass 130 represents the indole nucleus of the molecule. On the other hand, the ion at m/z 140 represents ring E of α-CPA with the hydroxy-ethyl moiety ($=C(OH)CH_3$) attached. Moreover, an ion with 18 Da difference (m/z 319) from the parent ion can be seen in the spectrum, which is attributed to the loss of a water molecule. β-CPA ($[m + H]^+/z$ 339.1700; $\Delta = -0.9$ ppm), also known as bissecodehydrocyclopiazonic acid, is a biosynthetic precursor of α-CPA with opened rings C and D. The MS/MS spectrum of this compound was similar to that previously reported by Holzapfel et al. [29], showing a peak at m/z 283 ($C_{16}H_{15}N_2O_3{}^+$) which corresponds to a loss of the $(CH_3)C{=}CH_2$- group from the parent ion (Figure 3B). Another prominent peak, with m/z 198, can be seen in the spectrum, representing the indole nucleus together with the dimethylallyl moiety. This ion was generated by the cleavage of the C4-C5 bond. Moreover, accurate mass measurements of the product ion 198 revealed a chemical composition of $C_{14}H_{16}N^+$, excluding oxygen functionalities which further supports its structural formula as depicted in Figure 3B. In the same spectrum, three other fragment ions can be seen at m/z 156 ($C_{11}H_{10}N^+$), 155 ($C_{11}H_9N^+$) and 154 ($C_{11}H_8N^+$), which were also observed in the mass spectrum of α-CPA but with different abundances. The presence of these tricyclic ions in the fragmentation pattern of the seco-molecule β-CPA can be explained by the fact that initial fragments of the precursor ion subsequently undergo a cyclization step of the two side chains of the indole system. More precisely this cyclization process is carried out by formation of the C4-C11 bond in the fragment ion at m/z198 and closure of ring C (Figure S1). The ion at m/z 144 ($C_{10}H_{10}N^+$), most likely arises from the ion of mass 198 by loss of the isoprenoid moiety ($(CH_3)C{=}CH_2$-).

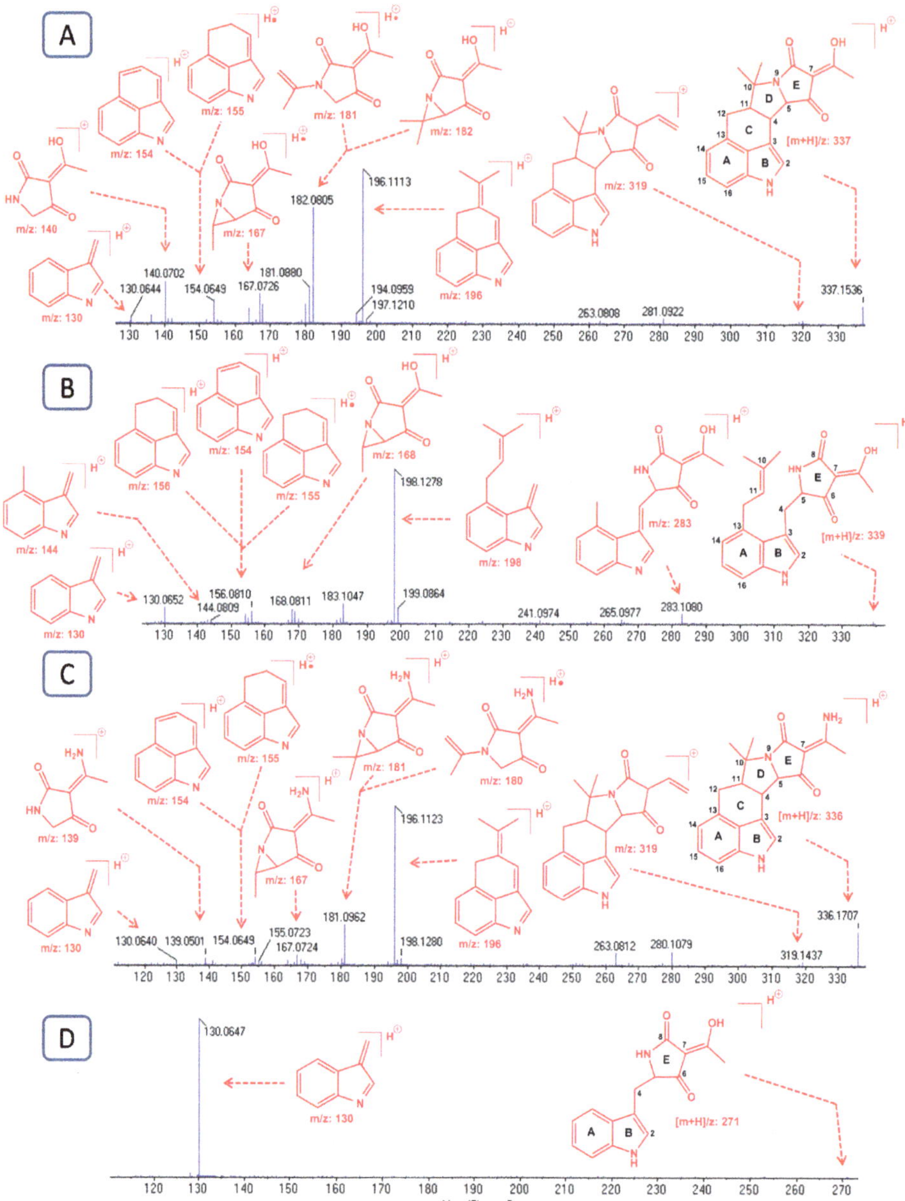

Figure 3. MS/MS spectra and putative structural fragments of: (**A**) α-CPA; (**B**) β-CPA; (**C**) α-CPA imine; (**D**) *cyclo*-acetoacetyl-L-tryptophan (cAATrp). The MS/MS spectra were acquired in IDA (information dependent acquisition) mode using a CE (collision energy) of 35 V with a collision energy spread (CES) of 15 V.

α-CPA imine is a very similar chemical entity to α-CPA, in which the hydroxyl functionality is substituted with an amino group. The fact that these metabolites are closely related to each other is also demonstrated by their shared fragmentation behavior. The MS/MS spectrum of α-CPA imine

([m + H]$^+$/z 336.1707; $\Delta < 0.1$ ppm) showed ions with m/z 319, 196, 154, 155 and 130 identical with those observed for α-CPA (Figure 3C). The fragment ions containing the amino group in their structure, namely m/z 181, 180 and 139, showed 1 Da difference with their corresponding fragments from the α-CPA molecule, which is consistent with the mass difference between -OH and -NH_2 groups. The biosynthetic intermediate of α-CPA, *cyclo*-acetoacetyl-L-tryptophan (cAATrp), showed only one prominent fragment peak at m/z 130, corresponding to the indole nucleus of this metabolite (Figure 3D). Pseuboydone E was not detected in our samples.

2.2. Identification of Oxindole CPA-Type Derivatives

2-oxo-CPA is an oxygenated derivative of α-CPA. More precisely, it is an oxygen atom in the form of a keto-group that is added at the C2 position of the indole nucleus of the α-CPA. The MS/MS spectrum of 2-oxo-CPA ([m + H]$^+$/z 353.1504; Δ = 0.8 ppm) showed, besides the precursor ion at [m + H]$^+$/z 353, characteristic fragments at m/z 335, 212, 182, 170, 154, 146 and 140 (Figure S2A). Since 2-oxo-CPA has an extra oxygen atom in its structure compared to its precursor, α-CPA, there is a mass difference of 16 Da between the two compounds. In this regard, all fragment ions in which the oxygenated indole system is incorporated show this 16 Da difference with their corresponding α-CPA fragments. Hence, fragment ions at m/z 335, 212, 170 and 146 are the oxygenated analogues of the fragment ions at m/z 319, 196, 154 and 130 in the MS/MS spectrum of α-CPA. On the other hand, peaks at m/z 182 and 140, which contain the tetramate moiety are identical to their corresponding fragments in the α-CPA MS/MS spectrum. The presence of the tricyclic fragment ion at m/z 154 may be justified by a possible cleavage of the hydroxyl group at C2 position of the ion at m/z 170.

Speradine A also known as 1-*N*-methyl-2-oxo-CPA, is a methylated derivative of 2-oxo-CPA first reported in a marine-derived isolate of *A. tamarii*. The fragmentation pattern of speradine A ([m + H]$^+$/z 367.1648; $\Delta = -1.0$ ppm) was similar to that of 2-oxo-CPA and α-CPA (Figure S2B). Due to the extra *N*-methyl group attached in the indole nucleus, speradine A has a 14 Da mass difference with 2-oxo-CPA. This 14 Da mass difference can be observed in all the typical fragments of speradine A (m/z 349, 325, 226 and 160) as compared to the MS/MS spectrum of 2-oxo-CPA. It is worth noting that HRMS and MS/MS data observed in our study for speradine A are in accordance with previous reports [33,34]. 3-Hydroxy-speradine A ([m + H]$^+$/z 383.1610; Δ = 2.0 ppm) is a hydroxylated derivative of speradine A, in which a hydroxyl functionality is added in the C3 position of the speradine A. The extraction of the mass corresponding to 3-hydroxy-speradine A, i.e., [m + H]/z 383.1610 resulted in three different peaks at RT 5.03, 5.76 and 5.97 min (Figure 4). The fragmentation pattern of the peaks at RT 5.76 and 5.97 min corresponded to the 3-OH-speradine A, whereas the metabolite eluting at RT 5.03 min had a slightly different fragmentation pattern, suggesting that the hydroxyl group may be attached at a different position. MS/MS spectrum of 3-OH-speradine A is depicted in Figure S2C. The two peaks with the mass of precursor ion and fragmentation pattern corresponding to 3-OH-speradine A can be explained by existence of different diastereoisomers.

Speradine B ([m + H]/z 287.1391; $\Delta = -0.4$ ppm) eluted at two different retention times, 4.15 and 4.99 min, both displaying identical MS/MS spectra with a typical fragmentation pattern for cyclopiazonic acid derivatives missing ring E (Figure S3A; Figure 4). The elution of speradine B at two different peaks can be justified by the existence of the same chemical compound in several stereoisomeric forms.

Speradine C has a very similar core structure with speradine B, except that in the N-6 position of speradine C a diketide moiety is attached. Based on this structural analogy, speradine C ([m + H]$^+$/z 371.1600; $\Delta = -0.2$ ppm) underwent a very similar mode of fragmentation with that of speradine B, as depicted in Figure S3B. MS/MS spectrum of speradine C showed a peak at m/z 353, which corresponds to the loss of a water molecule (18 Da) from the parent ion. The fragment ion at m/z 287 represents the molecular ion of speradine B, which is generated after the loss of the diketide moiety from the N-6 position of speradine C. The rest of the fragments' ions (i.e., m/z 226, 184, 169, 156 and 129) are identical with those in the fragmentation pattern of speradine B.

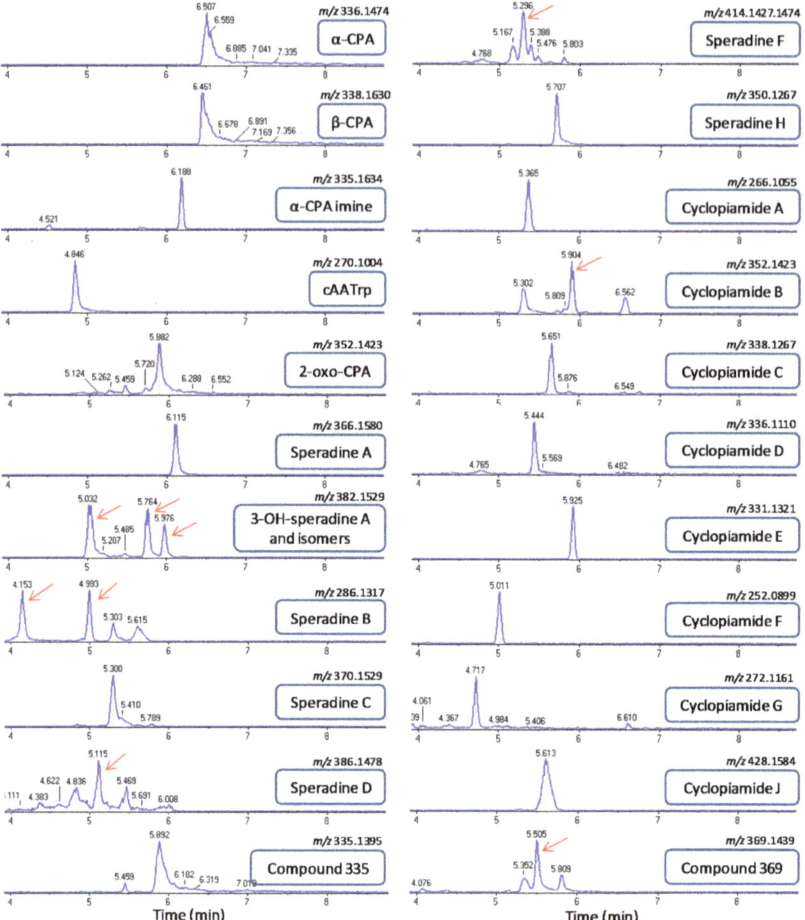

Figure 4. Extracted ion chromatograms (XIC) and elution order of CPA-type alkaloids detected in fungal extracts of different strains of *Aspergillus flavus*. The extraction mass window was set at 25 mDa. Red arrows indicate the correct retention time.

Speradine D is a closely related metabolite to speradine C showing a typical pathway of fragmentation for cyclopiazonic acids missing ring E and possessing a saturated ring C (Figure S3C). In the same fashion as speradine C, initially the OH group speradine D ([m + H]/z 387.1554; $\Delta = 1.0$ ppm) in the C3 position was lost through the elimination of a water molecule (-18 Da); subsequently, a cleavage of the N6-linked side chain occurred. The downstream fragmentation pathway of speradine D was identical with that of speradine B and C.

Speradine F is a highly oxygenated hexacyclic oxindole-tetrahidrofuran-tetramate metabolite having a very similar structural scaffold as the aspergillines isolated in fungal extracts of *A. versicolor*. Since speradine F is an *N*-methyl-2-oxo-indole with a saturated ring C, in the MS/MS spectrum of this compound ([m + H]$^+$/z 415.1497; $\Delta = -1.9$ ppm), we could see the typical mass fragments for this subgroup of oxindoles like m/z 269, 226, 184 and 169, as were observed in the fragmentation pattern of speradine B, C and D (Figure S3D). Besides these typical ions, two other ions could be assigned in this MS/MS spectrum, namely the fragment ions at m/z 397 and 379. The fragment ion at m/z 397 is attributed to the disruption of ring F (tetrahydrofuran ring) as a water molecule, whilst the 379 ion is generated from the subsequent loss of a water molecule from the C7-linked side chain. On the other hand, speradine H ($[m + H]^+/z$ 351.1341; $\Delta = -1.1$ ppm) is a tetracyclic *N*-methyl-2-oxo-indole with an unsaturated ring C, thus exerting some typical fragments for this subgroup of oxindoles such as, *m/z* 267, 250, 222, 207 and 194 (Figure S3E).

Cyclopiamide A ($[m + H]^+/z$ 267.1132; $\Delta = -0.7$ ppm) and cyclopiamide B ($[m + H]^+/z$ 353.1498; $\Delta = -0.8$ ppm), both tetracyclic *N*-methyl-2-oxo-indoles with an unsaturated ring C, shared the same fragmentation pathway as speradine H (Figure S4A,B). Cyclopiamide C ($[m + H]^+/z$ 339.1339; $\Delta = -0.2$ ppm), cyclopiamide D ($[m + H]^+/z$ 337.1178; $\Delta = -2.9$ ppm) and cyclopiamide F ($[m + H]^+/z$ 253.0966; $\Delta = -3.0$ ppm), all belonging to the subgroup of 2-oxo-indoles with an unsaturated ring C, shared the same fragmentation pathway, with the most prominent fragments at *m/z* 253, 236, 208, 180 and 165 (Figures S4C,D and S5B). Cyclopiamide E is actually a pentacyclic *N*-methyl-2-oxo-indole with a very characteristic ring E, a 4-oxo-1,3-diazine, which seems to influence the fragmentation pattern of the parent ion. Hence, although it belongs to *N*-methyl-2-oxo-indoles with unsaturated ring C, cyclopiamide E ($[m + H]^+/z$ 332.1409; $\Delta = 2.7$ ppm) showed a unique pattern of fragmentation with the most prominent fragment peaks at *m/z* 317, 288, 274 and 249 (Figure S5A).

Cyclopiamide G is a 2-oxo-indole with a saturated ring C, thus sharing some key structural fragments (*m/z* 212 and 170) with 2-oxo-CPA. Furthermore, in the MS/MS spectrum of cyclopiamide G ($[m + H]^+/z$ 273.1233; $\Delta = -1.5$ ppm), an ion with *m/z* 255 demonstrates the loss of a hydroxyl group from the C3 position of the parent molecule (Figure S5C). Cyclopiamide J ($[m + H]^+/z$ 429.1656; $\Delta = -0.9$ ppm) underwent a similar fashion of fragmentation as speradine F, because they share almost the same chemical scaffold with a small difference in the C7-linked side chain. Hence, the typical fragments at *m/z* 397, 379, 269, 226, 184 and 169 can be observed in the MS/MS spectrum of cyclopiamide J (Figure S5D). None of the aspergillines could be detected in the different strains of *A. flavus*.

2.3. Identification of Previously Unreported CPA-Type Derivatives

To investigate other derivatives that were not included in our list of target CPA-type alkaloids, the MS/MS data were checked for the presence of diagnostic ions of this class of compounds (Table S2). Besides the above known compounds, this untargeted analysis uncovered two other metabolites, *m/z 335.1395* and *m/z 369.1439*, which shared the same fragmentation pattern and whose MS/MS spectra encompassed diagnostic ions of the CPA-type alkaloids. Accurate mass measurements revealed that the metabolite at $[m + H]^+/z$ 335.1395 ($\Delta = 1.4$ ppm) corresponds to the chemical formula $C_{20}H_{19}N_2O_3$, possessing two protons less than the original molecule of α-CPA. These two protons' (2 Da) difference implies an extra double bond in the chemical scaffold of α-CPA. The location of this double bond is most likely in ring C of the molecule because fragment ions at *m/z* 130 (ring A and B), *m/z* 168 and 182 (ring D and E) are the same as described in the MS/MS spectrum of α-CPA, whilst the fragments at *m/z* 194, 223, 251, 317 containing ring C in their structure exhibit that 2 Da difference with their corresponding fragments from the α-CPA MS/MS spectrum (Figure S6A). On the other hand, the metabolite eluting at RT 5.5 min had the chemical formula $C_{20}H_{21}N_2O_5$ ($[m + H]^+/z$ 369.1439; $\Delta = -1.6$ ppm). Chemical composition as well as fragmentation pattern of this metabolite provided enough evidence to be assigned as a C3-hydroxylated analogue of 2-oxo CPA (Figure S6B).

2.4. Screening of CPA-Type Alkaloids in Different A. flavus Strains

Based on the data described above, a screening of CPA-type alkaloids was performed on a set of 55 *A. flavus* strains. In total, 22 CPA-type alkaloids were identified in extracts of the strains investigated, demonstrating the great potential of this ubiquitous fungus in producing secondary metabolites (Table 2). Of these metabolites, 13 have been previously reported in other fungi, but here they are reported for the first time in *A. flavus*. We also report the occurrence of two novel CPA-related metabolites in these samples.

Table 2. CPA-type alkaloids detected in our study.

A. flavus Strains (SRRC)	CPA-Type Alkaloids																					
	α-CPA	β-CPA	α-CPA Imine	cAATrp	2-oxoCPA	Speradine A	3-OH-Speradine A	Speradine B	Speradine C	Speradine D	Speradine F	Speradine H	Cyclopiamide A	Cyclopiamide B	Cyclopiamide C	Cyclopiamide D	Cyclopiamide E	Cyclopiamide F	Cyclopiamide G	Cyclopiamide J	Compound 335	Compound 369
0038	++	++	++	+	++	+	-	-	-	-	++	-	-	-	++	+	-	+	+	-	++	++
0141	+	-	-	-	+	++	+	-	+	-	-	-	-	+	-	-	-	-	-	-	-	-
0144	+++	++	-	++	++	+	+	-	-	-	-	-	-	++	++	+	-	+	+	-	++	++
0150	+++	++	++	++	++	+	+	-	-	-	-	-	-	-	++	-	-	++	+	-	++	++
0151	-	-	-	-	-	-	-	-	-	-	-	-	-	-	-	-	-	-	-	-	-	-
0167	+++	++	+	++	++	-	-	-	-	-	-	-	-	-	-	-	-	-	-	-	++	++
0283	++	++	-	++	++	++	++	++	++	+	++	-	++	++	++	+	-	+	+	++	++	++
0295	++	++	-	++	++	++	+++	++	++	+	++	++	++	++	+	+	++	+	+	++	++	++
1000F	+++	++	-	++	++	-	-	-	-	-	-	-	-	-	-	-	-	+	-	-	++	++
1006	++	++	++	++	+	++	++	-	-	-	-	-	-	+	+	+	-	+	-	-	++	++
1020	+++	++	++	++	+	-	-	-	-	-	-	-	-	-	++	-	-	++	+	-	++	++
1021	++	++	+	+	++	-	-	-	-	-	-	-	-	-	-	-	-	-	-	-	-	++
1055	+++	+	+	++	++	-	-	-	-	-	-	-	-	-	++	++	-	++	+	-	++	++
1071	++	++	-	-	+	-	-	-	-	-	-	-	-	-	+	-	-	+	-	-	+	+
1098	++	+	-	++	-	-	-	-	-	-	-	-	-	-	+	-	-	++	-	-	++	+
1118	-	-	-	-	-	-	-	-	-	-	-	-	-	-	-	-	-	-	-	-	-	-
1187	++	++	-	++	++	-	-	-	-	-	-	-	-	-	++	-	-	-	-	-	++	++
1299	++	++	++	++	++	-	-	-	-	-	-	-	-	-	+	-	-	+	+	-	++	++
1356	++	-	++	++	++	++	-	-	-	-	-	-	-	-	+	-	-	++	+	-	++	++
1357	++	++	-	++	+	++	++	++	-	-	-	++	++	-	++	-	-	+	-	++	+	-
1533	++	++	-	++	++	-	-	-	-	-	-	-	-	-	++	-	-	+	+	-	++	++
1534	++	++	-	+	-	-	-	-	-	-	-	-	-	-	-	-	-	+	-	-	-	-
1540	++	+	-	+	++	++	+	-	-	-	-	-	-	-	-	-	-	-	-	-	+	++
1541	++	-	-	++	-	-	-	-	-	-	-	-	-	-	++	-	-	-	-	-	-	-
1543	++	-	-	++	-	++	++	+	+	-	-	-	-	+	-	-	-	+	-	-	-	+
1544	++	++	-	++	++	++	++	++	++	++	++	++	++	++	+	-	++	++	-	++	++	++
1545	+++	++	++	++	++	-	-	-	-	-	-	-	-	-	++	+	-	++	+	-	++	++
1547	++	++	-	++	-	++	++	++	++	-	++	-	++	++	-	-	-	-	-	++	++	-
1552	++	-	-	-	+	++	++	-	+	-	-	-	-	-	-	-	-	-	-	-	-	+

Table 2. *Cont.*

A. flavus Strains (SRRC)	CPA-Type Alkaloids																					
	α-CPA	β-CPA	α-CPA Imine	cAATrp	2-oxoCPA	Speradine A	3-OH-Speradine A	Speradine B	Speradine C	Speradine D	Speradine F	Speradine H	Cyclopiamide A	Cyclopiamide B	Cyclopiamide C	Cyclopiamide D	Cyclopiamide E	Cyclopiamide F	Cyclopiamide G	Cyclopiamide J	Compound 335	Compound 369
1553	++	+	+	+	++	-	-	-	-	-	++	-	-	-	++	-	-	++	+	-	++	++
1554	+++	+	+	++	++	+	-	-	-	-	-	-	-	-	++	-	-	-	-	-	++	++
1557	++	++	-	++	+	++	++	++	++	-	++	++	++	++	-	-	++	-	-	++	++	-
1558	++	++	-	++	++	-	-	+	-	-	-	-	-	-	++	+	-	+	+	-	++	++
1559	+++	++	+	++	++	-	-	-	-	-	-	-	-	-	++	-	-	++	+	-	++	++
1565	++	++	-	+	+	-	-	-	-	-	-	-	-	-	+	+	-	-	-	-	++	++
1566	+++	++	-	+	++	-	-	-	-	-	-	-	-	-	-	-	-	-	-	-	++	++
1568	-	-	-	-	-	-	-	-	-	-	-	-	-	-	-	-	-	-	-	-	-	-
1571	++	++	++	++	+	++	+++	++	++	+	+	-	++	++	+	-	+	+	-	++	+	++
1573	+++	++	-	+	+	-	-	-	-	-	-	-	-	-	++	+	-	+	+	-	++	++
1574	+++	++	++	+	++	-	-	-	-	-	-	-	-	-	+	+	-	++	+	-	++	++
1575	++	++	-	++	-	++	++	++	++	-	++	++	++	++	-	-	++	-	-	++	++	-
1576	++	++	+	++	++	++	++	-	+	-	-	-	-	-	+	+	-	+	+	-	++	++
1578	+++	++	++	++	++	-	-	-	-	-	-	-	+	-	++	+	-	+	-	-	++	++
1591	++	++	-	++	++	++	+++	++	++	-	++	++	++	++	+	+	+	+	-	++	++	++
1626	++	++	+	++	++	++	++	-	+	-	-	-	-	-	++	+	-	+	+	-	++	++
1637	++	++	+	++	++	++	++	++	++	+	++	-	++	++	++	-	++	+	+	+	+	++
2000	++	-	-	++	++	-	-	-	-	-	-	-	-	-	++	-	-	-	-	-	+	++
2001	++	++	+	+	+	-	-	-	-	-	-	-	-	-	++	-	-	++	-	-	+	++
2033	+++	++	++	+	++	-	-	-	-	-	-	-	-	-	++	+	-	+	+	-	++	++
2035	+++	++	-	-	++	-	-	-	-	-	-	-	-	++	-	+	-	+	+	-	++	++
2114	-	-	-	-	-	-	-	-	-	-	-	-	-	-	-	-	-	-	-	-	-	-
2115	+++	++	++	++	++	-	-	-	-	-	-	-	-	-	-	-	-	+	+	-	++	-
2118	++	-	++	+	++	+	-	-	-	-	-	-	-	-	+	-	-	++	+	-	++	++
2524	+++	++	-	+	++	-	-	-	+	-	-	-	-	-	++	+	-	+	+	-	++	++
2711	+++	-	-	++	++	-	-	-	-	-	-	-	-	-	+	-	-	+	-	-	++	++

"+", Peak area $\leq 10^4$; "++", $10^4 <$ Peak area $< 10^6$; "+++", Peak area $\geq 10^6$; "-" not detected; SRRC-Southern Regional Research Center; CPA-cyclopiazonic acid; The presence of compounds was verified with MS/MS spectra as described under Sections 2.1–2.3.

3. Discussion

Chemically, α-CPA is a hybrid prenylated indole alkaloid, structurally characterized by a rigid pentacyclic (6/5/6/5/5) skeleton bearing a unique heterocyclic pyrrolidine-2,4-dione (tetramic acid) motif as the main part of its pharmacophore. This mycotoxin does not contain any *O*-methyl or *N*-methyl groups in its structural scaffold and is characterized by the ability to form an intramolecular hydrogen bond due to its keto-enol tautomerism. Since α-CPA is a very well-known mycotoxin, its identification in *A. flavus* fungal extracts was straightforward, and was also supported by data from previous work [1,3,45,46]. The same mode of fragmentation was observed in the MS/MS spectrum of α-CPA and those of other indole derivatives (β-CPA, α-CPA imine and cAATrp) (Figure 3). The identification of these compounds was also straightforward, as they share the same scaffold with α-CPA. On the other hand, the identification of the oxindole CPA-type alkaloids was more complex due to their high chemical diversity in terms of different structural scaffolds. Moreover, this subclass of alkaloids is less studied compared to α-CPA itself and other indole derivatives. For this reason, and also to simplify the identification, we subdivided the oxindole subclass of CPA-type alkaloids in four different chemical groups: (i) 2-oxindoles with saturated ring C (2-oxo-CPA and cyclopiamide G); (ii) 2-oxindoles with unsaturated ring C (cyclopiamide C, cyclopiamide D and cyclopiamide F); (iii) *N*-methyl-2-oxindoles with saturated ring C (speradine A, 3-hydroxy-speradine A, speradines B-D, speradine F, cyclopiamide H and cyclopiamide J); and (iv) *N*-methyl-2-oxindoles with unsaturated ring C (speradine E, speradine H, cyclopiamide A, cyclopiamide B, and cyclopiamide E). The MS/MS spectra demonstrated that all the metabolites belonging to the same chemical group share two or three typical fragments that reflect the core structure of the respective chemical group (Table S2), and which can therefore be used as diagnostic ions. This chemical classification and the diagnostic ion approach facilitated the dereplication of oxindoles included in our study, and it is also a useful methodology for the identification of yet unknown CPA-type oxindoles. Besides accurate mass measurements and MS/MS data, other analytical parameters like elution order and isotope-model fitting prove to be complementary identification aids. For instance, in our study, speradine A eluted between α-CPA and 2-oxo-CPA, which actually represents the same elution pattern as reported by other researchers [33]. Similar successful dereplication approaches based on TOF technologies have been reported previously [44,47,48].

Production of polyketide-amino acid hybrid metabolites, as is the case for α-CPA, is catalyzed by PKS-NRPS enzymes. PKS-NRPSs are complex and multi-domain enzymes characterized as having the ability to synthesize highly diverse chemical groups of secondary metabolites [49]. The genome of *A. flavus* is predicted to harbor two PKS-NRPS hybrid gene clusters. One (cluster #55) is the gene cluster of CPA [9], while the other one (cluster #23) has been shown to be responsible for the production of a series of 2-pyridones [50,51]. As indicated above, the CPA gene cluster in *A. flavus* NRRL 3357 contains only three functional genes and a Zn_2Cys_6 transcription factor-encoding gene (*ctfR1*), which seems to be inactive [10]. On the other hand, our study demonstrated a huge chemical diversity within the CPA family of mycotoxins. Therefore, different *A. flavus* strains have the ability to produce many more metabolites apart from α-CPA and its precursors (Table 2). Although it is hard to predict if all the identified metabolites are linked with the CPA gene cluster, the chemical scaffold resemblance strongly suggests a convergent biochemical origin. It is likely that the genetic material reported for the CPA gene cluster in NRRL 3357 is not sufficient to explain the chemical diversity found in other strains of *A. flavus*. Hence, most likely other *A. flavus* strains like S295, S283, S1637, and S1544 possess additional genes within their CPA clusters that may catalyze extra biochemical steps in the biosynthetic pathway of CPA. This concept was already demonstrated in *A. oryzae* NBRC 4177 in the case of 2-oxoCPA. A cytochrome P450 oxidase gene (*cpaH*) located within a CPA cluster in NBRC 4177 was shown to mediate the conversion of α-CPA to 2-oxoCPA [32]. In the same fashion, it was reported that a strain of *A. tamarii* (NBRC 4099) harbors the gene *cpaM*, which encodes for a *N*-methyltransferase involved in the synthesis of 1-*N*-methyl-2-oxoCPA (speradine A) from 2-oxoCPA [33]. The inability of speradine A production in *A. oryzae* is caused by mutations or partial deletions in *cpaM*. In analogy with aforementioned findings, we can speculate that the homolog genes of *cpaH* and *cpaM* may be also present in the CPA gene cluster of 2-oxo-CPA and speradine A-producing strains of *A. flavus*. In this regard, the CPA gene cluster seems more genetically diverse than expected, especially when we consider the identification of other complex CPA-type alkaloids like speradine F, cyclopiamide E, cyclopiamide J, the aspergillines, etc. Additional genetic data are needed to support the occurrence of these metabolites in *A. flavus* fungal extracts. Nevertheless, we cannot exclude the possibility of involvement in the CPA-biosynthetic network of enzymes that are encoded by genes located outside the CPA gene cluster. Moreover, there is also the possibility of non-enzymatic generation of metabolites under different culture conditions or during various sample treatment procedures.

The tetramic acid structural moiety is a very important nitrogen-containing heterocycle that is often the pharmacophore to interact with various biological targets, thus it exhibits a wide range of biological and toxicological activities [52–54]. Apart from α-CPA, 2-oxoCPA and speradine A, for which the SERCA-blocking activity was clearly demonstrated, the toxicological profile for the rest of the CPA-type metabolites is unknown [4,12,34]. In this context, it is important both to evaluate the actual toxicological potential of these CPA-related alkaloids and to get an idea about their contributions in the overall CPA toxicity. Moreover, there are no data about their occurrences in food and feed commodities, which further complicates our understanding about human and animal exposure to CPA contamination. Hence, conducting survey studies of CPA-type mycotoxins in different food and feed matrices might be necessary to decipher the real impact of CPA contamination in human and animal health, especially considering the fact that CPA is always overshadowed by concern about aflatoxin contamination.

One of the main approaches for pre-harvest control of aflatoxin contamination in crops is the introduction of non-aflatoxigenic *A. flavus* isolates into agricultural fields to displace aflatoxin-producing strains. The necessary features for a potential *A. flavus* strain to be used as a biocontrol agent include ability to grow rapidly, to be adapted for specific plant colonization, and to produce sclerotia for long-term survival in the fields [55,56]. From a toxicological point of view, the only seemingly important feature for a biocontrol strain is the inability to produce aflatoxins. However, the inability to produce other relevant mycotoxins, among which are α-CPA and our newly identified CPA analogues, is highly desirable due to the unknown long-term and cumulative toxicological effects of these metabolites. Thus, a careful screening of these metabolites is obviously very important in order to avoid any possible inadvertent mycotoxicoses from the introduction of biocontrol agents into agricultural fields. Also, increased efforts to assess the genetic stability of the *A. flavus* strains used in biocontrol products should be undertaken. This is highly relevant since there is now evidence for a sexual stage in several aflatoxigenic *Aspergillus* species [57,58]. Hence, the possibility of re-gaining aflatoxin- and/or CPA-producing properties through sexual recombination is a fact that should not be neglected. The same concern of exposure to CPA-type mycotoxins can be made regarding *A. oryzae* strains which are being used extensively in the food fermentation industry [59].

Data from the present work show a conspicuous presence of CPA and its derivatives in the different *A. flavus* strains investigated, and highlight the previously mentioned need for a thorough assessment of the actual impact of contamination of crops with CPA-type mycotoxins on human and animal health. Two commercially available biocontrol strains (AF36 and NRRL 21882) were included in this study. AF36 was originally isolated from a cotton field in Arizona and was approved for application on cotton in Arizona. Loss of aflatoxigenicity in AF36 is the result of a nonsense mutation in *pksA* (*aflC*), a pathway gene in aflatoxin biosynthesis [60]. AF36 has an otherwise full aflatoxin cluster. AF36 does have a fully functional CPA cluster, and although it is effective at excluding toxigenic strains and reducing AF levels, this strain is reported to significantly increase CPA accumulation in food and feed commodities [61]. This is consistent with our findings which demonstrated the ability of AF36 to produce α-CPA and a series of other CPA-like alkaloids, including its precursors (cAATrp and β-CPA), 2-oxoCPA, cyclopiamide C, F, G and the two other previously unreported derivatives (see 2.3). Remarkably, our data uncovered the presence of the CPA family of mycotoxins in fungal extracts of NRRL 21882, which is reported to lack both the aflatoxin and CPA gene clusters [61]. These findings could support the phenomenon of heterokaryosis and vertical transmission of cryptic alleles in *A. flavus* in NRRL 21882. Olarte et al. [62] reported that it is possible for some *A. flavus* to possess cryptic alleles, particularly when typically masked genes are amplified because the dominant genome lacks the gene(s) of interest. Moore et al. [63] were able to amplify and sequence a portion of the *aflW/aflX* region in NRRL 21882. This indicates NRRL 21882 may be heterokaryotic with a low-copy genome possessing cryptic alleles for a functional CPA gene cluster, thereby making this particular strain a CPA producer even though it is missing the entire subtelomeric region of chromosome 3. Existence of similar non-parental cryptic alleles has been reported also in other fungal species [64,65].

It is worth noting that besides their toxicological relevance, there is currently a revival of interest for pharmacological application of CPA-type pharmacophores. As stated above, α-CPA is a specific nanomolar inhibitor of mammalian SERCA-1. This SERCA-inhibiting activity of α-CPA is being exploited extensively for various pharmacological purposes. Thus, Kotsubei et al. [66], by using α-CPA as an experimental inhibitor, demonstrated sufficient differences in active binding sites between mammalian and bacterial SERCA. These differences in the CPA pocket between mammalian and bacterial Ca^{2+}-ATPase suggest that a bacterial-specific CPA derivative could be developed, hence making the bacterial SERCA a potential drug target. Moreover, α-CPA has been shown to possess clear antiviral activity as well as causing perturbations in cardiac ventricular myocytes [67,68]. Based on these concepts, CPA-type alkaloids identified in this study, and/or other derivatives which may be discovered or synthesized in the future, could be useful bioactive scaffolds to be employed in future pharmacological applications.

4. Conclusions

Accurate mass high resolution mass spectrometry (HRMS), combined with a careful investigation of fragmentation patterns, proved to be a suitable dereplication strategy for the identification of cyclopiazonic acid (CPA)-type alkaloids. This approach resulted in our finding 22 CPA-type alkaloids in *A. flavus* cultures. Two of these metabolites are new discoveries as they have never been reported. Though the other 20 compounds have been previously identified in different fungi, 13 of these metabolites were identified for the first time in *A. flavus*. These results provide a better insight into the diversity of CPA-type alkaloids in *A. flavus* and raise concerns about the extent of the overall *A. flavus* mycotoxin problem. The described identification strategy can be applied in programs aiming to assess the occurrence of this type of mycotoxin in food and feed commodities. Our results also demonstrate the great potential of *A. flavus* in producing a myriad of secondary metabolites, highlighting the need for a more thorough investigation into potential mycotoxicity of non-aflatoxigenic strains that are currently being used in biocontrol strategies.

5. Materials and Methods

5.1. Chemicals and Materials

Methanol (MeOH) and acetonitrile (ACN), LC-MS grade, were obtained from Biosolve (Valkenswaard, the Netherlands), whereas HPLC-grade MeOH was from VWR International (Zaventem, Belgium). Ethyl acetate (EtOAc), dichloromethane (DCM) and acetone (dimethyl ketone (DMK)) were purchased from Acros Organics (Geel, Belgium). Sigma-Aldrich (Bornem, Belgium) supplied ammonium formate ($HCOONH_4$). Formic acid (HCOOH, Merck, Darmstadt, Germany) was used. Ultrapure H_2O was produced by a Milli-Q Gradient System (Millipore, Brussels, Belgium). Ultrafree®-MC centrifugal filter units (0.22 μm) from Millipore (Bedford, MA, USA) were used. Sigma-Aldrich supplied agar, corn steep solids, dextrose, peptone, sucrose, yeast extract, dipotassium hydrogen phosphate trihydrate ($K_2HPO_4 \cdot 3H_2O$), magnesium sulfate heptahydrate ($MgSO_4 \cdot 7H_2O$), and iron(II) sulfate heptahydrate ($FeSO_4 \cdot 7H_2O$). Triton X-100, potassium chloride (KCl), and sodium nitrate ($NaNO_3$) were from Merck.

5.2. Strains and Growth Conditions

A. flavus strains used in this study are listed in Table S1 as supplementary data. Conidia of each respective strain were inoculated on solid Wickersham media (≈25 mL of medium per plate, D = 10 cm) which contains 2.0 g yeast extract, 3.0 g peptone, 5.0 g corn steep solids, 2.0 g dextrose, 30.0 g sucrose, 2.0 g $NaNO_3$, 1.0 g $K_2HPO_4 \cdot 3H_2O$, 0.5 g $MgSO_4 \cdot 7H_2O$, 0.2 g KCl, 0.1 g $FeSO_4 \cdot 7H_2O$, 15.0 g agar per litre (pH 5.5). All cultures were incubated at 28 °C in the dark for 7 days.

5.3. Sample Preparation

The fungal colonies and agar were cut into small pieces with a scalpel and these were subsequently transferred to a 500 mL screw-cap Ehrlenmeyer flask. Metabolites were extracted with 30 mL MeOH:DCM:EtOAc 10:20:30, (*v*/*v*/*v*). The samples were agitated for 60 min on an Agitelec overhead shaker (J. Toulemonde and Cie, Paris, France). A total of 4 mL of extract was transferred to a glass tube and evaporated under a stream of nitrogen. The residue was reconstituted with 200 µL MeOH:ACN:H_2O 30:30:40, (*v*/*v*/*v*), and centrifuged in an Ultrafree®-MC centrifugal device for 5 min at 14,000× *g*.

5.4. UHPLC-qTOF-MS Analysis

The experiments were carried out using a hybrid Q-TOF MS instrument, the AB SCIEXTripleTOF®4600 (AB Sciex, Concord, ON, Canada), equipped with a DuoSpray™ and coupled to an Eksigent ekspert™ ultraLC 100-XL system. The DuoSpray™ ion source (consisting of both electrospray ionization (ESI) and atmospheric pressure chemical ionization (APCI) probes) was operated in the positive ESI mode (ESI^+). The APCI probe was used for automated mass calibration using the Calibrant Delivery System (CDS). The CDS injects a calibration solution matching the polarity of ionization, and calibrates the mass axis of the TripleTOF® system in all scan functions used (MS and/or MS/MS). The Q-TOF HRMS method consisted of a full scan TOF survey (dwell time 100 ms, 100–1600 Da) and a maximum number of eight IDA MS/MS scans (dwell time 50 ms). The MS parameters were as follows: curtain gas (CUR) 25 psi, nebulizer gas (GS 1) 50 psi, heated gas (GS 2) 60 psi, ion spray voltage (ISVF) 5.5 kV, interface heater temperature (TEM) 500 °C, Collision Energy (CE) 10 V and declustering potential (DP) 70 V. For the IDA MS/MS experiments, a CE of 35 V was applied with a collision energy spread (CES) of 15 V. An Eksigentekspert™ ultraLC 100-XL system was used for separation. The column was a ZORBAX RRHD Eclipse Plus C18 (1.8 µm, 2.1 × 100 mm) from Agilent Technologies (Diegem, Belgium). The mobile phase consisted of H_2O:MeOH (95:5, *v*/*v*) containing 0.1% HCOOH and 10 mM $HCOONH_4$ (solvent A) and MeOH:H_2O (95:5, *v*/*v*) containing 0.1% HCOOH and 10 mM $HCOONH_4$ (solvent B). The gradient elution program for LC-qTOF HRMS analyses was applied as follows: 0–0.5 min: 0% B, 0.5–7 min: 0%–99% B, 7–9 min: 99% B, 9–10 min: 99%–0% B, 10–14 min: 0% B. The flow rate was 0.4 mL/min. The column temperature was set at 40 °C and temperature of the autosampler was 4 °C. 5 µL of sample were injected. The instrument was controlled by Analyst® TF 1.6 software, while data processing was carried out using PeakView® software version 2.0 and MasterView™ software version 1.0 (all from AB Sciex).

Supplementary Materials: The following are available online at www.mdpi.com/2072-6651/9/1/35/s1, Figure S1: Cyclization mechanism of the ions at *m/z* 156, 155 and 154 in the fragmentation pathway of β-CPA, Figure S2: MS/MS spectra and structural fragments of: A. 2-oxo-CPA; B. speradine A, and C. 3-hydroxy-speradine A, Figure S3: MS/MS spectra and structural fragments of: A. speradine B; B. speradine C; C. speradine D; D. speradine F, and E. speradine H, Figure S4: MS/MS spectra and structural fragments of: A. cyclopiamide A; B. cyclopiamide B; C. cyclopiamide C, and D. cyclopiamide D, Figure S5: MS/MS spectra and structural fragments of: A. cyclopiamide E; B. cyclopiamide F; C. cyclopiamide G, and D. cyclopiamide J, Figure S6: MS/MS spectra and structural fragments of: A. 11,12-Dehydro α-CPA (Compound 335); B: 3-Hydroxy-2-oxoCPA (compound 369), Table S1: List of *A. flavus* strains used in this study, Table S2: Characteristic fragments of oxindole CPA-type alkaloids.

Acknowledgments: Valdet Uka was financially supported by Project Basileus V (Erasmus Mundus Action 2) funding from the European Commission.

Author Contributions: V.U., G.G.M., S.D.S. and J.D.D.M. conceived and designed the experiments; V.U., N.A.-M. and J.D.D.M. performed the experiments; V.U., D.N. and J.D.D.M. analyzed the data; V.U., G.G.M. and J.D.D.M. wrote the paper.

Conflicts of Interest: The authors declare no conflict of interest.

References

1. Holzapfel, C.W. The isolation and structure of cyclopiazonic acid, a toxic metabolite of *Penicillium cyclopium* Westling. *Tetrahedron* **1968**, *24*, 2101–2119. [CrossRef]
2. Ohmomo, S.; Sugita, M.; Abe, M. Isolation of cyclopiazonic acid, cyclopiazonic acid imine and bissecodehydrocyclopiazonic acid from the cultures of *Aspergillus versicolor* (Vuill.) Tiraboshi. *J. Agric. Chem. Soc.* **1973**, *47*, 57–63.
3. Luk, K.C.; Kobbe, B.; Townsend, J.M. Production of cyclopiazonic acid by *Aspergillus flavus* Link. *Appl. Environ. Microbiol.* **1977**, *33*, 211–212. [PubMed]
4. Burdock, G.A.; Flamm, W.G. Review Article: Safety assessment of the mycotoxin cyclopiazonic acid. *Int. J. Toxicol.* **2000**, *19*, 195–218. [CrossRef]
5. Frisvad, J.C. The connection between the *Penicillia* and *Aspergillus* and mycotoxins with special emphasis on misidentified isolates. *Arch. Environ. Contam. Toxicol.* **1989**, *18*, 452–467. [CrossRef] [PubMed]
6. El-Banna, A.A.; Pitt, J.I.; Leistner, L. Production of mycotoxins by *Penicillium* species. *Syst. Appl. Microbiol.* **1987**, *1*, 42–46. [CrossRef]
7. Dorner, J.W. Production of cyclopiazonic acid by *Aspergillus tamarii* Kita. *Appl. Environ. Microbiol.* **1983**, *46*, 1435–1437. [PubMed]
8. Liu, X.; Walsh, C.T. Cyclopiazonic acid biosynthesis in *Aspergillus* sp.: Characterization of a reductase-like R* domain in cyclopiazonate synthetase that forms and releases cyclo-acetoacetyl-L-tryptophan. *Biochemistry* **2009**, *48*, 8746–8757. [CrossRef] [PubMed]
9. Seshime, Y.; Juvvadi, P.R.; Tokuoka, M.; Koyama, Y.; Kitamoto, K.; Ebizuka, Y.; Fujii, I. Functional expression of the *Aspergillus flavus* PKS–NRPS hybrid CpaA involved in the biosynthesis of cyclopiazonic acid. *Bioorg. Med. Chem. Lett.* **2009**, *19*, 3288–3292. [CrossRef] [PubMed]
10. Chang, P.K.; Ehrlich, K.C.; Fujii, I. Cyclopiazonic acid biosynthesis of *Aspergillus flavus* and *Aspergillus oryzae*. *Toxins* **2009**, *1*, 74–99. [CrossRef] [PubMed]
11. Chang, P.K.; Ehrlich, K.C. Cyclopiazonic acid biosynthesis by *Aspergillus flavus*. *Toxin Rev.* **2011**, *30*, 79–89. [CrossRef]
12. Seidler, N.W.; Jonaz, I.; Vegh, M.; Martonosi, A. Cyclopiazonic Acid is a specific inhibitor of the Ca^{2+}-ATPase of sarcoplasmic reticulum. *J. Biol. Chem.* **1989**, *264*, 17816–17823. [PubMed]
13. Lytton, J.; Westlin, M.; Hanleyll, M.R. Thapsigargin inhibits the sarcoplasmic or endoplasmic reticulum Ca-ATPase family of calcium pumps. *J. Biol. Chem.* **1991**, *266*, 17067–17071. [PubMed]
14. Purchase, I.F. The acute toxicity of the mycotoxin cyclopiazonic acid to rats. *Toxicol. Appl. Pharmacol.* **1971**, *18*, 114–123. [CrossRef]
15. Norred, W.P.; Morrissey, R.E.; Rilley, R.T.; Cole, R.J.; Dorner, J.W. Distribution, excretion, and skeletal muscle effects of the mycotoxin [^{14}C]cyclopiazonic acid in rats. *Food Chem. Toxicol.* **1985**, *23*, 1069–1076. [CrossRef]
16. Nishie, K.; Cole, R.J.; Dorner, J.W. Toxicity and neuropharmacology of cyclopiazonic acid. *Food Chem. Toxicol.* **1985**, *23*, 831–839. [CrossRef]
17. Nishie, K.; Cole, R.J.; Dorner, J.W. Toxic effects of cyclopiazonic acid in the early phase of pregnancy in mice. *Res. Commun. Chem. Pathol. Pharmacol.* **1987**, *55*, 303–315. [PubMed]
18. Antony, M.; Shukla, Y.; Janardhanan, K.K. Potential risk of acute hepatotoxicity of kodo poisoning due to exposure to cyclopiazonic acid. *J. Ethnopharmacol.* **2003**, *87*, 211–214. [CrossRef]
19. Njobeh, P.B.; Dutton, M.F.; Koch, S.H.; Chuturgoon, A.; Stoev, S.; Seifert, K. Contamination with storage fungi of human food from Cameroon. *Int. J. Food Microbiol.* **2009**, *135*, 193–198. [CrossRef] [PubMed]
20. Finoli, C.; Vecchio, A.; Galli, A.; Franzetti, L. Production of cyclopiazonic acid by molds isolated from Taleggio cheese. *J. Food Prot.* **1999**, *62*, 1198–1202. [CrossRef] [PubMed]
21. Sosa, M.J.; Cordoba, J.J.; Diaz, C.; Rodriguez, M.; Bermudez, E.; Asensio, M.A.; Nunez, F. Production of cyclopiazonic acid by *Penicillium commune* isolated from dry-cured ham on ameatextract-based substrate. *J. Food. Prot.* **2002**, *65*, 988–992.

22. Vaamonde, G.; Patriarca, A.; Fernandez Pinto, V.; Comerio, R.; Degrossi, C. Variability of aflatoxin and cyclopiazonic acid production by *Aspergillus* section *Flavi* from different substrates in Argentina. *Int. J. Food Microbiol.* **2003**, *88*, 79–84. [CrossRef]
23. Dorner, J.W.; Cole, R.J.; Erlington, D.J.; Suksupath, S.; McDowell, G.H.; Bryden, W.L. Cyclopiazonic acid residues in milk and eggs. *J. Agric. Food Chem.* **1994**, *42*, 1516–1518. [CrossRef]
24. Lee, Y.J.; Hagler, W.M.J. Aflatoxin and cyclopiazonic acid production by *Aspergillus flavus* isolated from contaminated maize. *J. Food Sci.* **1991**, *56*, 871–872. [CrossRef]
25. Urano, T.; Trucksess, M.W.; Beaver, R.W.; Wilson, D.M.; Dorner, J.W.; Dowell, F.E. Co-occurrence of cyclopiazonic acid and aflatoxins in corn and peanuts. *J. Off. Anal. Chem. Int.* **1992**, *75*, 838–841.
26. Heperkan, D.; Somuncuoglu, S.; Karbancioglu-Güler, F.; Mecik, N. Natural contamination of cyclopiazonic acid in dried figs and co-occurrence of aflatoxin. *Food Control* **2012**, *23*, 82–86. [CrossRef]
27. Zorzete, P.; Baquiao, A.C.; Atayde, D.D.; Reis, T.A.; Goncalez, E.; Correa, B. Mycobiota, aflatoxins and cyclopiazonic acid in stored peanut cultivars. *Food Res. Int.* **2013**, *52*, 380–386. [CrossRef]
28. Ezekiel, C.N.; Sulyok, M.; Somorin, Y.; Odutayo, F.I.; Nwabekee, S.U.; Balogun, A.T.; Krska, R. Mould and mycotoxin exposure assessment of melon and bush mango seeds, two common soup thickeners consumed in Nigeria. *Int. J. Food Microbiol.* **2016**, *237*, 83–91. [CrossRef] [PubMed]
29. Holzapfel, C.W.; Hutchinson, R.D.; Wilkins, D.C. The isolation and structure of two new indole derivatives from *Penicillium cyclopium* Westling. *Tetrahedron* **1970**, *26*, 5239–5246. [CrossRef]
30. Lin, A.Q.; Lin, D.; Fang, Y.C.; Wang, F.Z.; Zhu, T.J.; Gu, Q.Q.; Zhu, W.M. *iso*-α-Cyclopiazonic acid, a new natural product isolated from the marine-derived fungus *Aspergillus flavus* C-F-3. *Chem. Nat. Compd.* **2009**, *45*, 677–680. [CrossRef]
31. Lan, W.J.; Wang, K.T.; Xu, M.Y.; Zhang, J.J.; Lam, C.K.; Zhong, G.H.; Xu, J.; Yang, D.P.; Li, H.J.; Wang, L.Y. Secondary metabolites with chemical diversity from the marine-derived fungus *Pseudallescheria boydii* F19-1 and their cytotoxic activity. *RSC Adv.* **2016**, *6*, 76206–76213. [CrossRef]
32. Kato, N.; Tokuoka, M.; Shinohara, Y.; Kawatani, M.; Uramoto, M.; Seshime, Y.; Fujii, I.; Kitamoto, K.; Takahashi, T.; Takahashi, S.; et al. Genetic safeguard against mycotoxin cyclopiazonic acid production in *Aspergillus oryzae*. *ChemBioChem* **2011**, *12*, 1376–1382. [CrossRef] [PubMed]
33. Tokuoka, M.; Kikuchi, T.; Shinohara, Y.; Koyama, A.; Iio, S.; Kubota, T.; Kobayashi, J.; Koyama, Y.; Totsuka, A.; Shindo, H.; et al. Cyclopiazonic acid biosynthesis gene cluster gene *cpaM* is required for speradine A biosynthesis. *Biosci. Biotechnol. Biochem.* **2015**, *79*, 2081–2085. [CrossRef] [PubMed]
34. Tsuda, M.; Mugishima, T.; Komatsu, K.; Sone, T.; Tanaka, M.; Mikami, Y.; Shiro, M.; Hirai, M.; Ohizumie, Y.; Kobayashi, J. Speradine A, a new pentacyclic oxindole alkaloid from a marine-derived fungus *Aspergillus tamarii*. *Tetrahedron* **2003**, *59*, 3227–3230. [CrossRef]
35. Wang, N.; Hu, J.C.; Liu, W.; Wang, S.J. Cyclopiazonic Acid Compound, and Preparation and Application Thereof. CN Patent 103,183,666 A, 3 July 2013.
36. Hu, X.; Xia, Q.W.; Zhao, Y.Y.; Zheng, Q.H.; Liu, Q.Y.; Chen, L.; Zhang, Q.Q. Speradines B-E, four novel tetracyclic oxindole alkaloids from the marine-derivied fungus *Aspergillus oryzae*. *Heterocycles* **2014**, *89*, 1662–1669.
37. Hu, X.; Xia, Q.W.; Zhao, Y.Y.; Zheng, Q.H.; Liu, Q.Y.; Chen, L.; Zhang, Q.Q. Speradines F-H, three new oxindole alkaloids from the marine-derived fungus *Aspergillus oryzae*. *Chem. Pharm. Bull.* **2014**, *62*, 942–946. [CrossRef] [PubMed]
38. Zhu, H.; Chena, C.; Wang, J.; Li, X.N.; Wei, G.; Guo, Y.; Yao, G.; Luo, Z.; Zhang, J.; Xue, Y.; et al. Penicamedine A, a highly oxygenated hexacyclic indole alkaloid from *Penicillium camemberti*. *Chem. Biodivers.* **2015**, *12*, 1547–1553. [CrossRef] [PubMed]
39. Ma, X.; Peng, J.; Wu, G.; Zhu, T.; Li, G.; Gu, Q.; Li, D. Speradines B-H, oxygenated cyclopiazonic acid alkaloids from the sponge-derived fungus *Aspergillus flavus* MXH-X104. *Tetrahedron* **2015**, *71*, 3522–3527. [CrossRef]
40. Zhou, M.; Miao, M.M.; Du, G.; Li, X.N.; Shang, S.Z.; Zhao, W.; Lu, Z.H.; Yang, G.Y.; Che, C.T.; Hu, Q.F.; et al. Aspergillines A-E, highly oxygenated hexacyclic indole-tetrahydrofuran-tetramic acid derivatives from *Aspergillus versicolor*. *Org. Lett.* **2014**, *16*, 5016–5019. [CrossRef] [PubMed]
41. Xu, X.; Zhang, X.; Nong, X.; Wei, X.; Qi, S. Oxindole alkaloids from the fungus *Penicillium commune* DFFSCS026 isolated from deep-sea-derived sediments. *Tetrahedron* **2015**, *71*, 610–615. [CrossRef]

42. Holzapfel, C.W.; Bredenkamp, M.W.; Snyman, R.M.; Boeyens, J.C.A.; Allen, C.C. Cyclopiamide, an isoindolo[4,6-cd]indole from *Penicillium cyclopium*. *Phytochemistry* **1990**, *29*, 639–642. [CrossRef]
43. Abbas, H.K.; Zablotowicz, R.M.; Horn, B.W.; Phillips, N.A.; Johnson, B.J.; Jin, X.; Abel, C.A. Comparison of major biocontrol strains of non-aflatoxigenic *Aspergillus flavus* for the reduction of aflatoxins and cyclopiazonic acid in maize. *Food Addit. Contam.* **2011**, *28*, 198–208. [CrossRef] [PubMed]
44. Nielsen, K.F.; Mansson, M.; Rank, C.; Frisvad, J.C.; Larsen, T.O. Dereplication of microbial natural products by LC-DAD-TOFMS. *J. Nat. Prod.* **2011**, *74*, 2338–2348. [CrossRef] [PubMed]
45. Losito, I.; Monaci, L.; Aresta, A.; Zambonin, C.G. LC-ion trap electrospray MS-MS for the determination of cyclopiazonic acid in milk samples. *Analyst* **2002**, *127*, 499–502. [CrossRef] [PubMed]
46. Moldes-Anaya, A.S.; Asp, T.N.; Eriksen, G.S.; Skaar, I.; Rundberget, T. Determination of cyclopiazonic acid in food and feeds by liquid chromatography-tandem mass spectrometry. *J. Chromatogr. A* **2009**, *1216*, 3812–3818. [CrossRef] [PubMed]
47. Klitgaard, A.; Iversen, A.; Andersen, M.R.; Larsen, T.O.; Frisvad, J.C.; Nielsen, K.F. Aggressive dereplication using UHPLC-DAD-QTOF: Screening extracts for up to 3000 fungal secondary metabolites. *Anal. Bioanal. Chem.* **2014**, *406*, 1933–1943. [CrossRef] [PubMed]
48. Arroyo-Manzanares, N.; Diana di Mavungu, J.; Uka, V.; Gámiz-Gracia, L.; García-Campaña, A.M.; de Saeger, S. An integrated targeted and untargeted approach for the analysis of ergot alkaloids in cereals using UHPLC-hybrid quadrupole time-of-flight mass spectrometry. *World Mycotoxin J.* **2015**, *8*, 653–666. [CrossRef]
49. Boettger, D.; Hertweck, C. Molecular diversity sculpted by fungal PKS-NRPS hybrids. *ChemBioChem* **2013**, *14*, 28–42. [CrossRef] [PubMed]
50. Cary, J.W.; Uka, V.; Han, Z.; Buyst, D.; Harris-Coward, P.Y.; Ehrlich, K.C.; Wei, Q.; Bhatnagar, D.; Dowd, P.F.; Martens, S.L.; et al. An *Aspergillus flavus* secondary metabolic gene cluster containing a hybrid PKS-NRPS is necessary for synthesis of the 2-pyridones, leporins. *Fungal Genet. Biol.* **2015**, *81*, 88–97. [CrossRef] [PubMed]
51. Arroyo-Manzanares, N.; Diana di Mavungu, J.; Uka, V.; Malysheva, S.V.; Cary, J.W.; Ehrlich, K.C.; Vanhaecke, L.; Bhatnagar, D.; de Saeger, S. Use of UHPLC high-resolution Orbitrap mass spectrometry to investigate the genes involved in the production of secondary metabolites in *Aspergillus flavus*. *Food Addit. Contam. Part A* **2015**, *32*, 1656–1673. [CrossRef] [PubMed]
52. Royles, B.J.L. Naturally Occurring Tetramic Acids: Structure, Isolation, and Synthesis. *Chem. Rev.* **1995**, *95*, 1981–2001. [CrossRef]
53. Spatz, J.H.; Welsch, S.J.; Duhaut, D.; Jäger, N.; Boursier, T.; Fredrich, M.; Allmendinger, L.; Ross, G.; Kolb, J.; Burdack, C.; et al. Tetramic acid derivatives via Ugi–Dieckmann reaction. *Tetrahedron Lett.* **2009**, *50*, 1705–1707. [CrossRef]
54. Yang, Y.L.; Lu, C.P.; Chen, M.Y.; Chen, K.Y.; Wu, Y.C.; Wu, S.H. Cytotoxic polyketides containing tetramic acid moieties isolated from the fungus *Myceliophthora thermophila*: Elucidation of the relationship between cytotoxicity and stereoconfiguration. *Chemistry* **2007**, *13*, 6985–6991. [CrossRef] [PubMed]
55. Abbas, H.K.; Wilkinson, J.R.; Zablotowicz, R.M.; Accinelli, C.; Abel, C.A.; Bruns, H.A.; Weaver, M.A. Ecology of *Aspergillus flavus*, regulation of aflatoxin production, and management strategies to reduce aflatoxin contamination of corn. *Toxin Rev.* **2009**, *28*, 142–153. [CrossRef]
56. Ehrlich, K.C. Non-aflatoxigenic *Aspergillus flavus* to prevent aflatoxin contamination in crops: Advantages and limitations. *Front. Microbiol.* **2014**, *5*, 50. [CrossRef] [PubMed]
57. Moore, G.G. Sex and recombination in aflatoxigenic *Aspergilli*: Global implications. *Front. Microbiol.* **2014**, *5*, 32. [CrossRef] [PubMed]
58. Olarte, R.A.; Horn, B.W.; Dorner, J.W.; Monacell, J.T.; Singh, R.; Stone, E.A.; Carbone, I. Effect of sexual recombination on population diversity in aflatoxin production by *Aspergillus flavus*. *Phytopathology* **2012**, *103*, 8.
59. Couto, S.R.; Sanroma, M.A. Application of solid-state fermentation to food industry-A review. *J. Food Eng.* **2006**, *76*, 291–302. [CrossRef]
60. Ehrlich, K.C.; Cotty, P.J. An isolate of *Aspergillus flavus* used to reduce aflatoxin contamination in cottonseed has a defective polyketide synthase gene. *Appl. Microbiol. Biotech.* **2004**, *65*, 473–478. [CrossRef] [PubMed]
61. Chang, P.K.; Horn, B.W.; Dorner, J.W. Sequence breakpoints in the aflatoxin biosynthesis gene cluster and flanking regions in nonaflatoxigenic *Aspergillus flavus* isolates. *Fungal Genet. Biol.* **2005**, *42*, 914–923. [CrossRef] [PubMed]

62. Olarte, R.A.; Horn, B.W.; Dorner, J.W.; Monacell, J.T.; Singh, R.; Stone, E.A.; Carbone, I. Effect of sexual recombination on population diversity in aflatoxin production by *Aspergillus flavus* and evidence for cryptic heterokaryosis. *Mol. Ecol.* **2012**, *21*, 1453–1476. [CrossRef] [PubMed]
63. Moore, G.G.; Singh, R.; Horn, B.W.; Carbone, I. Recombination and lineage-specific gene loss in the aflatoxin gene cluster of *Aspergillus flavus*. *Mol. Ecol.* **2009**, *18*, 4870–4887. [CrossRef] [PubMed]
64. Nevzglyadova, O.V.; Gaivoronskii, A.A.; Artemov, A.V.; Smirnova, T.I.; Soidla, T.R. Detection of concealed "illegitimate" nuclei in tetrad analysis of the diploid progeny of heterokaryons in *Saccharomyces cerevisiae*. *Russian J. Genet.* **2001**, *37*, 617–623. [CrossRef]
65. Vercauteren, A.; Boutet, X.; D'hondt, L.; van Bockstaele, E.; Maes, M.; Leus, L.; Chandelier, A.; Heungens, K. Aberrant genome size and instability of *Phytophthora ramorum* oospore progenies. *Fungal Genet. Biol.* **2011**, *48*, 537–543. [CrossRef] [PubMed]
66. Kotsubei, A.; Gorgel, M.; Morth, J.P.; Nissen, P.; Andersen, J.L. Probing determinants of cyclopiazonic acid sensitivity of bacterial Ca^{2+}-ATPases. *FEBS J.* **2013**, *280*, 5441–5449. [CrossRef] [PubMed]
67. Cui, R.; Wang, Y.; Wang, L.; Li, G.; Lan, K.; Altmeyer, R.; Zou, G. Cyclopiazonic acid, an inhibitor of calcium-dependent ATPases withantiviral activity against human respiratory syncytial virus. *Antivir. Res.* **2015**, *132*, 38–45. [CrossRef] [PubMed]
68. Kistamas, K.; Szentandrassy, N.; Hegyi, B.; Vaczi, K.; Ruzsnavszky, F.; Horvath, B.; Banyasz, T.; Nanasi, P.P.; Magyar, J. Changes in intracellular calcium concentration influence beat-to-beat variability of action potential duration in canine ventricular myocytes. *J. Physiol. Pharmacol.* **2015**, *66*, 73–81. [PubMed]

© 2017 by the authors. Licensee MDPI, Basel, Switzerland. This article is an open access article distributed under the terms and conditions of the Creative Commons Attribution (CC BY) license (http://creativecommons.org/licenses/by/4.0/).

Article

Mycotoxigenic Potentials of *Fusarium* Species in Various Culture Matrices Revealed by Mycotoxin Profiling

Wen Shi [1], Yanglan Tan [2], Shuangxia Wang [2], Donald M. Gardiner [3], Sarah De Saeger [4], Yucai Liao [5], Cheng Wang [6], Yingying Fan [6], Zhouping Wang [1,*] and Aibo Wu [2,*]

1 State Key Laboratory of Food Science and Technology, School of Food Science and Technology, Jiangnan University, Wuxi 214122, China; sw1596321@sina.com

2 SIBS-UGENT-SJTU Joint Laboratory of Mycotoxin Research, Key Laboratory of Food Safety Research, Institute for Nutritional Sciences, Shanghai Institutes for Biological Sciences, Chinese Academy of Sciences, University of Chinese Academy of Sciences, 294 Taiyuan Road, Shanghai 200031, China; yltan@sibs.ac.cn (Y.T.); shuangxiawang@163.com (S.W.)

3 Commonwealth Scientific and Industrial Research Organisation (CSIRO), 306 Carmody Road, St Lucia QLD 4067, Australia; Donald.Gardiner@csiro.au

4 Laboratory of Food Analysis, Faculty of Pharmaceutical Sciences, Ghent University, Ottergemsesteenweg 460, Gent 9000, Belgium; Sarah.DeSaeger@UGent.be

5 College of Plant Science and Technology, Huazhong Agricultural University, Wuhan 430000, China; yucailiao@mail.hzau.edu.cn

6 Institute of Quality Standards & Testing Technology for Agro-Products, Laboratory of Quality and Safety Risk Assessment for Agro-Products (Urumqi), Ministry of Agriculture, Xinjiang Academy of Agricultural Sciences, 403 Nanchang Road, Urumqi 830091, China; wangcheng312@sina.com (C.W.); fyyxaas@sina.com (Y.F.)

* Correspondence: wangzp@jiangnan.edu.cn (Z.W.); abwu@sibs.ac.cn (A.W.); Tel.: +86-510-8532-6195 (Z.W.); +86-21-5492-0716 (A.W.)

Academic Editor: Antonio Moretti

Received: 31 August 2016; Accepted: 21 December 2016; Published: 26 December 2016

Abstract: In this study, twenty of the most common *Fusarium* species were molecularly characterized and inoculated on potato dextrose agar (PDA), rice and maize medium, where thirty three targeted mycotoxins, which might be the secondary metabolites of the identified fungal species, were detected by liquid chromatography–tandem mass spectrometry (LC-MS/MS). Statistical analysis was performed with principal component analysis (PCA) to characterize the mycotoxin profiles for the twenty fungi, suggesting that these fungi species could be discriminated and divided into three groups as follows. Group I, the fusaric acid producers, were defined into two subgroups, namely subgroup I as producers of fusaric acid and fumonisins, comprising of *F. proliferatum*, *F. verticillioides*, *F. fujikuroi* and *F. solani*, and subgroup II considered to only produce fusaric acid, including *F. temperatum*, *F. subglutinans*, *F. musae*, *F. tricinctum*, *F. oxysporum*, *F. equiseti*, *F. sacchari*, *F. concentricum*, *F. andiyazi*. Group II, as type A trichothecenes producers, included *F. langsethiae*, *F. sporotrichioides*, *F. polyphialidicum*, while Group III were found to mainly produce type B trichothecenes, comprising of *F. culmorum*, *F. poae*, *F. meridionale* and *F. graminearum*. A comprehensive picture, which presents the mycotoxin-producing patterns by the selected fungal species in various matrices, is obtained for the first time, and thus from an application point of view, provides key information to explore mycotoxigenic potentials of *Fusarium* species and forecast the *Fusarium* infestation/mycotoxins contamination.

Keywords: *Fusarium* fungi; mycotoxin profiles; principal component analysis; culture substrates

1. Introduction

Fusarium spp. are a large complex genus, known as worldwide plant pathogens which infect and colonize various cereal crops such as maize, rice, wheat and oats in temperate and semi-tropical areas, including China, North America, South Africa and all European cereal-growing areas [1–5]. *Fusarium* spp. have been found to cause significant reduction in quality and yield in many food and feed crops, estimated at between 10% and 30%. The worst affected crops are wheat, maize and rice, where *Fusarium* spp. are known to cause *Fusarium* head blight (FHB) of wheat, sheath rot disease of maize and bakanae disease of rice [6–9].

The widespread presence of fungi and mycotoxins in pre-harvest infected plants or in-store grains are of great concern for human and animal health. The most occurring *Fusarium* mycotoxins are deoxynivalenol (DON), 3-acetyl deoxynivalenol (3-ADON), 15-acetyl deoxynivalenol (15-ADON), nivalenol (NIV) and fusarenon X (Fus-X); T-2 toxin, HT-2 toxin, neosolaniol (NEO) and diacetoxyscirpenol (DAS); zearalenone (ZEN), fumonisin B1 (FB1), fumonisin B2 (FB2) and fusaric acid [10–15]. Acute and chronic exposure to these mycotoxins exhibits various toxic effects to plants and animals, and poses a potential health risk for humans [16,17]. Due to the high toxicity and worldwide occurrence of the mycotoxins, maximum levels concerning some major mycotoxins have been set in the European countries [18] and also in China [19].

The phase of maize fusariosis with the highest toxicological concern is the ear rot, but large amounts of mycotoxins can also be formed in infected leaves (NIV), rotted stalks (notably ZEN and DON) and whole plants (ZEN) [20]. The variability in the fungal strains is an important issue for food safety, as multiple mycotoxins with different toxicities could be produced. So far, the risks of combined toxicity have been poorly understood, but generally it can be concluded that co-exposure to several different mycotoxins often results in synergistic effects [21]. In addition, the matrix significantly influences the toxin-producing abilities of the mycotoxigenic fungi, leading to complex mycotoxin contamination situations. Therefore, it is a critical issue to investigate the mycotoxin profiles and reveal mycotoxigenic potentials of various *Fusarium* spp. in different substrates.

Several studies have been performed to investigate the relationship between *Fusarium* spp. and mycotoxin production. In Germany, as well as in many other central European countries, *F. graminearum* is the predominant *Fusarium* fungi in wheat followed by *F. culmorum*, both of which have been associated with occurrence of ZEN and DON in wheat and other crops [22,23]. In China, 3-ADON, 15-ADON and NIV are the main mycotoxins produced by *F. graminearum* isolated from wheat ears with clear FHB symptoms [24]. Several other surveys also suggested that *F. solani*, *F. graminearum* and *F. sambucinum* could produce one or more mycotoxins, such as DON in north-central United States [25] and ZEN, NIV, 15-ADON in Argentina [12,26]. However, most of the studies only focused on the main important *Fusarium* fungi isolated from cereal grains, with very little attention paid to other fungal species, such as *Fusarium musae*, *Fusarium fujikuroi*, *Fusarium concentricum*, *Fusarium lateritium*, *Fusarium incarnatum-equiseti*, *Fusarium meridionale* and *Fusarium polyphialidicum*. This study therefore took in account the less studied species for the following reasons: (1) they play an important role in spoilage of grain cereals during storage and marketing; (2) these *Fusarium* species can potentially produce mycotoxins in maize and rice matrices even though they were isolated from other substrates such as banana, green pepper and barley. No previous attempts have been made to study the distributions of all frequently occurring mycotoxins (such as ZEN and its derivatives, type B trichothecenes, type A trichothecenes, FB1, FB2 and fusaric acid), along with some other less studied *Fusarium* metabolites produced by various *Fusarium* spp.

The major focus of this study is to thoroughly investigate the mycotoxin-producing capabilities of twenty *Fusarium* species in different culture substrates. A definitive understanding of the prevalence of *Fusarium* spp. and their associated mycotoxigenic potential is not only critical for the development of strategies for monitoring and managing mycotoxin contamination, but also to obtain a precise picture of the toxicological risks related to maize and rice consumption by humans and animals.

2. Results and Discussion

2.1. Molecular Characterization of Fusarium Species

The electrophoresis chromatograms of the *EF-1α* gene from twenty strains (Table 1) collected from different areas are shown in Figure 1A. The single band observed for all the selected strains demonstrated the purity of the fungi and the species to be *Fusarium* strains.

The phylogenetic tree constructed based on *EF-1α* gene is shown in Figure 1B. After comparison of the targeted gene sequences with the standard sequences in GenBank, the identity of all the *Fusarium* strains was clearly confirmed since the similarities of the sequence between the targeted fungi and the standard one were equal to or above 96%.

Table 1. The information of the *Fusarium* strains used in this study.

Strain No.	Code	*Fusarium* Species	Origin	Host
1	MUCL [1] 52463	*Fusarium temperatum*	Belgium	Maize
2	MUCL 43485	*Fusarium subglutinans*	United States	Maize
3	MUCL 42823	*Fusarium culmorum*	Belgium	Wheat
4	MUCL 51036	*Fusarium fujikuroi*	Philippines	Rice
5	MUCL 34988	*Fusarium langsethiae*	-	Wheat
6	MUCL 52574	*Fusarium musae*	Honduras	Banana
7	MUCL 53395	*Fusarium poae*	Belgium	Maize
8	MUCL 43483	*Fusarium proliferatum*	-	-
9	F-1	*Fusarium graminearum*	China	Wheat
10	MUCL 53602	*Fusarium sporotrichioides*	Belgium	Maize
11	MUCL 42821	*Fusarium tricinctum*	Belgium	Wheat
12	MUCL 43478	*Fusarium verticillioides*	United States	Maize
13	B40 = F50/1-i1-B	*Fusarium oxysporum*	China	Barley
14	MC1_30	*Fusarium meridionale*	China	Maize
15	M-12-0203-A1	*Fusarium equiseti*	China	Maize
16	M-12-0501-J1	*Fusarium sacchari*	China	Maize
17	M-12-0601-D12	*Fusarium solani*	China	Maize
18	Q29	*Fusarium concentricum*	China	Green pepper
19	W21	*Fusarium andiyazi*	China	Maize
20	XB4-1	*Fusarium polyphialidicum*	China	Barley

[1] MUCL Mycothèque de l'Université catholique de Louvain (Louvain-la-Neuve, Belgium).

2.2. Applicability of LC-MS/MS Method

The utilized LC-MS/MS method was established for simultaneous determination of multiple mycotoxins in *Lentinula edodes* in the previous study, and its applicability on PDA, rice and maize was validated. In the present study, the recoveries at concentration levels of 50 $\mu g \cdot kg^{-1}$ for all mycotoxins spiked into each sample were tested. The experiment was done in quintuplicate. The results showed that satisfactory recoveries with mean values in the range of 72.5%–119.8% in PDA, 72.5%–119.5% in rice and 72.3%–119.6% in maize were obtained for all 33 mycotoxins (Table S1), verifying the suitability of the method employed for determination of the targeted mycotoxins in the above matrices. MRM chromatograms of mycotoxins detected in the media by the selected *Fusarium* species are presented in Figures S1–S4, showing that these mycotoxins can be identified by their retention times and two selective monitoring transitions.

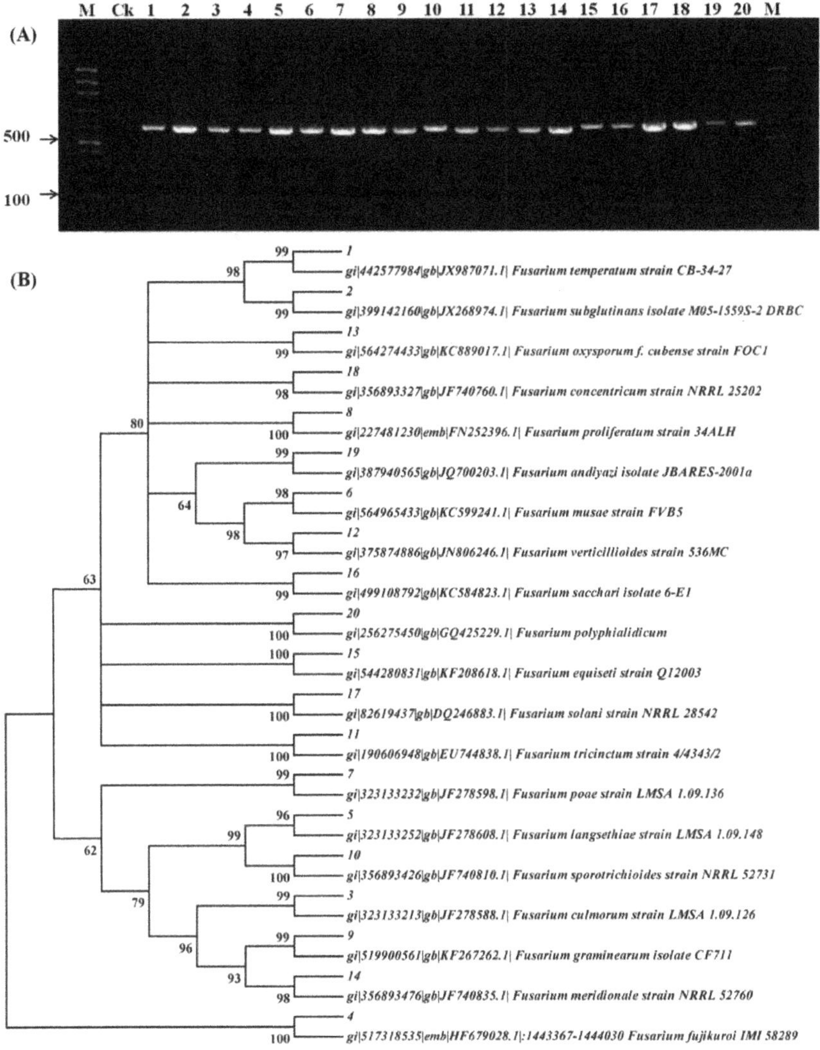

Figure 1. Electrophoresis chromatographs of the *EF-1α* gene from different purified *Fusarium* strains (**A**) and subsequently constructed phylogenetic tree; (**B**). M indicates the 100-bp molecular marker; CK indicates negative control; 1–20 indicates the *Fusarium* strains as described in Table 1.

2.3. Principal Component Analysis

Mycotoxin production of *Fusarium* species are influenced greatly by culture conditions [27–29]. For instant, too high or low temperature showed inhibition of toxin biosynthesis [28] and thus the appropriate temperature such as 25 °C was adopted in some of related in vitro experiments when investigating multiple *Fusarium* species [30,31]. The temperature effects on mycotoxin production (PDA, 21 days) were initially evaluated between 5 °C and 40 °C in our study, and the results indicated that the sensitivity to temperature varied for different *Fusarium* species. Given that the optimum temperature for majority of the investigated twenty species ranged from 20 °C to 30 °C (results shown in Table S2), 25 °C was chosen and kept constant for the following experiments. The culture medium is also the predominant effect on mycotoxin production by fungi, such as carbohydrate and nitrogen sources [30,32]. Previous reports showed that the carbohydrate-rich media were apparently more favorable for toxin producing [30,33], which was consistent with our results that significantly more mycotoxins were observed in rice and maize media than those in PDA medium (Tables S3–S5), except fusaric acid produced with highest abundance by some of *Fusarium* species on PDA medium. In order to further investigate the mycotoxigenic abilities of various *Fusarium* species on different culture media, PCA was carried out to classify the *Fusarium* strains based on their mycotoxin production. The score plots including PC1 (direction of largest variance) and PC2 (perpendicular to PC1 and against the largest variance) were extracted from the first two principal components, which presented the maximum variability in the data and made it easier to visually discriminate the differences [34]. The value of the loading plot reflects the contribution of each variable to the sample classification in the PCA. The farther from the origin a variable is placed, the higher contribution of that variable made to the PCA model [35].

In this study, thirteen mycotoxins were detected in the growth media by selected *Fusarium* species and then set as the variables for evaluation. As shown in Figure 2, the first principal component (PC1) and the second principal component (PC2) accounted for 67.22% and 25.82% of the variation for PDA (Figure 2A1), 68.46% and 23.20% for rice medium (Figure 2B1), 72.29% and 23.94% for maize medium (Figure 2C1), respectively. It could be obviously seen that for all the above three culture media the cumulative variance contribution of PC1 and PC2 was more than 90%, proving the significant variability of mycotoxin profiles of different *Fusarium* strains studied.

For PDA medium (Figure 2A2), fusaric acid correlated positively with PC1 while the fumonisins (FB1 and FB2) correlated negatively, indicating the critical role of these three mycotoxins in the differentiation of the *Fusarium* strains. Type A trichothecenes (T-2 toxin, HT-2 toxin, NEO and DAS) contribute negatively to PC1, but act as significantly positive contributors for PC2, verifying the important role of these four mycotoxins in further discrimination of the *Fusarium* strains. Similarly, the contributors for PC1 for rice medium (Figure 2B2) were fusaric acid, three type A trichothecenes (T-2 toxin, HT-2 toxin and NEO) and fumonisins (FB1 and FB2) while NIV, Fus-X and DAS contributed for PC2. In regard to maize medium, fusaric acid, type A trichothecenes (T-2 toxin, HT-2 toxin, DAS and NEO) and fumonisins (FB1 and FB2) contributed significantly to PC1, and NIV and Fus-X were the contributors for PC2 (Figure 2C2). Based on the mycotoxin profiles in PCA factor loading plots, the detected mycotoxins were divided into three major groups, including the Group I for fusaric acid, which could be subsequently divided into subgroup I for co-occurrence of fusaric acid and fumonisins (FB1 and FB2) and subgroup II for fusaric acid only, Group II for type A trichothecenes (T-2, HT-2, NEO, DAS) and Group III for type B trichothecenes (DON, 15-ADON, 3-ADON, NIV, Fus-X). Consequently, the targeted twenty toxigenic *Fusarium* fungal strains were grouped as shown in Table 2. The mycotoxin profiles of representative strains (*F. proliferatum* (A) for Group I, *F. langsethiae* (B) for Group II, and *F. graminearum* (C) and *F. meridionale* (D) for Group III) in each group cultivated on PDA, maize and rice medium are shown in Figure 3.

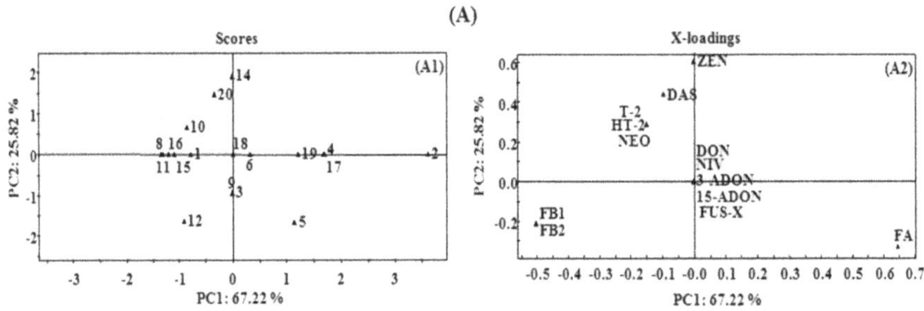

Figure 2. *Cont.*

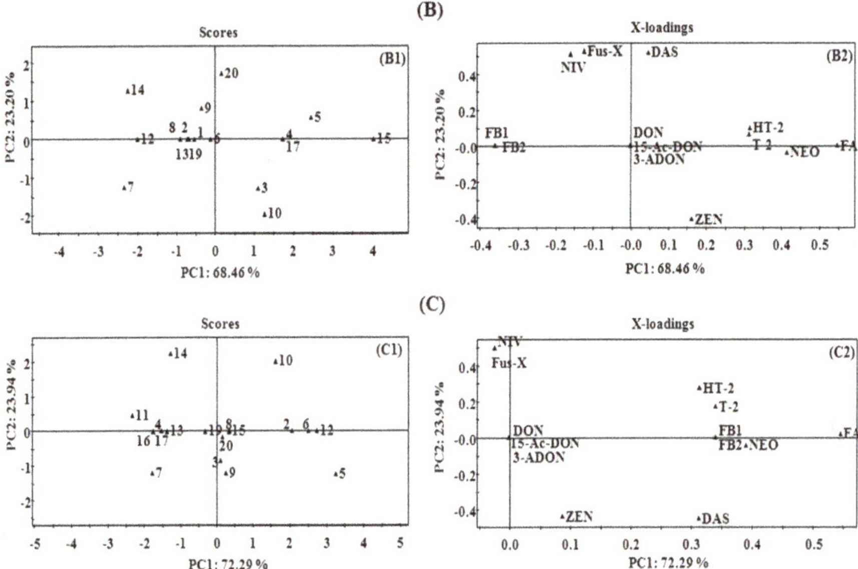

Figure 2. Statistical results of principal component analysis (PCA) of the detected mycotoxins by twenty *Fusarium* species in PDA (**A**), rice (**B**) and maize (**C**) medium. (**A1**), (**B1**) and (**C1**) on the left refer to the score plots showing the locations of the *Fusarium* species; (**A2**), (**B2**) and (**C2**) on the right were designed to the loading plot interpreting the relationships between mycotoxins produced by *Fusarium* species.

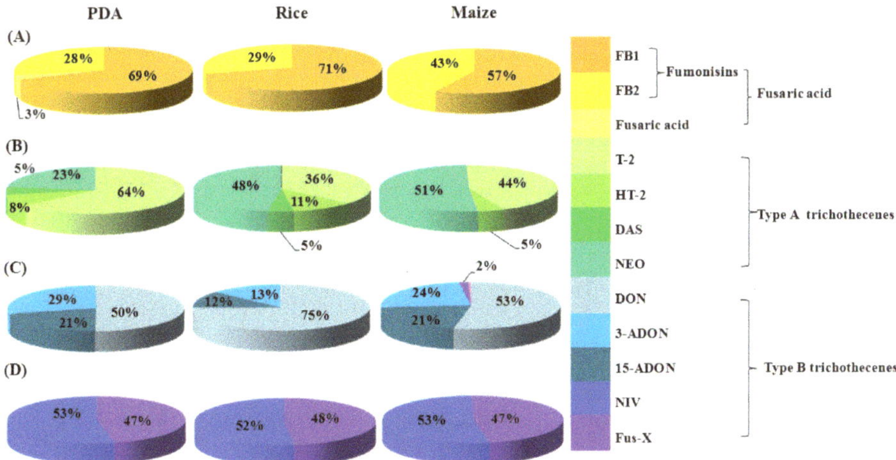

Figure 3. Pie charts of mycotoxin profiling produced by 4 representative *Fusarium* species including *F. proliferatum* (**A**), *F. langsethiae* (**B**), *F. graminearum* (**C**) and *F. meridionale* (**D**) in PDA, rice and maize medium.

Table 2. Grouping of mycotoxigenic *Fusarium* species using mycotoxin profiles.

Group		Strain No.	*Fusarium* Species	Major Mycotoxins Produced	Other Mycotoxins Produced
Group I (Fusaric acid)	Subgroup I (Fumonisins and fusaric acid)	8	*F. proliferatum*	FB1, FB2, Fusaric acid	-
		12	*F. verticillioides*	FB1, FB2, Fusaric acid	-
		4	*F. fujikuroi*	FB1, FB2, Fusaric acid	-
		17	*F. solani*	FB1, FB2, Fusaric acid	-
	Subgroup II (Fusaric acid only)	1	*F. temperatum*	Fusaric acid	-
		2	*F. subglutinans*	Fusaric acid	-
		6	*F. musae*	Fusaric acid	-
		11	*F. tricinctum*	Fusaric acid	-
		13	*F. oxysporum*	Fusaric acid	-
		15	*F. equiseti*	Fusaric acid	-
		16	*F. sacchari*	Fusaric acid	-
		18	*F. concentricum*	Fusaric acid	-
		19	*F. andiyazi*	Fusaric acid	-
Group II (Type A trichothecenes)		5	*F. langsethiae*	T-2, HT-2, NEO, DAS	-
		10	*F. sporotrichioides*	T-2, HT-2, NEO, DAS	-
		20	*F. polyphialidicum*	DAS	-
Group III (Type B trichothecenes)		3	*F. culmorum*	NIV, Fus-X	T-2, HT-2, NEO, ZEN
		7	*F. poae*	NIV, Fus-X	T-2, HT-2, NEO, DAS
		14	*F. meridionale*	NIV, Fus-X	NEO, ZEN
		9	*F. graminearum*	DON, 15-ADON, 3-ADON	ZEN

2.4. *Mycotoxin-Producing Capacities of Fusarium Species in Different Growth Media*

2.4.1. Group I/Fusaric Acid Producers

Thirteen *Fusarium* strains, including *F. proliferatum*, *F. verticillioides*, *F. fujikuroi*, *F. solani*, *F. temperatum*, *F. subglutinans*, *F. musae*, *F. tricinctum*, *F. oxysporum*, *F. equiseti*, *F. sacchari*, *F. concentricum* and *F. andiyazi*, belonged to Group I due to their fusaric acid producing abilities. Among them, *F. proliferatum*, *F. verticillioides*, *F. fujikuroi* and *F. solani* belonged to subgroup I as producers of both fumonisins and fusaric acid, and the other nine *Fusarium* species were considered to merely produce fusaric acid (Table 2).

F. proliferatum and *F. verticillioides* are the major fumonisin producers with average concentrations for FB1 being 10,085 and 15,168 $\mu g \cdot kg^{-1}$ in PDA, 146,726 and 273,894 $\mu g \cdot kg^{-1}$ in rice and 104,810 and 237,208 $\mu g \cdot kg^{-1}$ in maize medium, while the average concentrations for FB2 were 354 and 594 $\mu g \cdot kg^{-1}$ on PDA, 60,378 and 98,523 $\mu g \cdot kg^{-1}$ on rice, 77,939 and 180,778 $\mu g \cdot kg^{-1}$ on maize, respectively (Tables S3–S5 and Figure 4). The high mycotoxin-producing abilities of these two *Fusarium* species found in this study are in good agreement with the previous studies [36,37]. *F. fujikuroi* produced fumonisins as well, but showed much lower concentration levels with values less than 200 $\mu g \cdot kg^{-1}$ on all the three media, the mycotoxigenic potential of which have been reported to be greatly dependent on the isolated hosts and inoculation conditions [36,38]. Meanwhile, this is the first report about fumonisin production by *F. solani*.

With regard to the individual mycotoxin, relatively higher contents of FB1 were generated compared to FB2 by the same fungi on the three media (Figures 3A and 4), which have been reported previously [33,39]. The ratios between the two fumonisins (FB1/FB2) for *F. fujikuroi* and *F. solani* were in the range of 1.4–3.1 in all media, but particularly 25.5 and 28.5 in PDA for *F. verticillioides* and *F. proliferatum*, respectively. As expected, the amounts of fumonisins produced in maize and rice were relatively higher than that in PDA, proving the influential role of the composition in different media in fumonisin-producing capabilities of *Fusarium* strains [40,41]. Fusaric acid were also detected with the above four *Fusarium* fungi but the amount was lower than that of fumonisins.

Nine *Fusarium* strains were found to only produce fusaric acid in our study at average concentration levels ranging from 15 to 12,435 $\mu g \cdot kg^{-1}$ (Tables S3–S5), and the production levels in PDA were higher than that in rice and maize media especially for *F. subglutinans*, *F. musae*, *F. concentricum* and *F. andiyazi*. In previous studies, *F. sacchari* and *F. andiyazi* were detected to produce low amounts of fumonisin [42,43]. Note that *F. equiseti* was considered as trichothecene producer (DON, 15-ADON, NIV, FUS-X, HT-2 and DAS) [31,44], but it showed a considerable intraspecies variation in profiles of trichothecene production, and even trichothecenes were not observed with some isolates of *F. equiseti* [45].

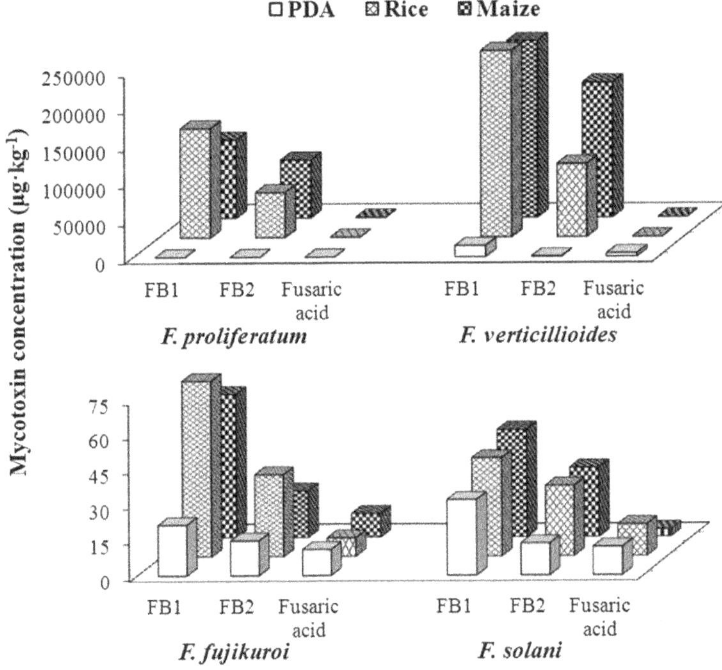

Figure 4. Investigation of the fumonisin B1 (FB1), fumonisin B2 (FB2) and fusaric acid producing abilities of *F. proliferatum*, *F. verticillioides*, *F. fujikuroi* and *F. solani* after incubation for 21 days at 25 °C on PDA, rice and maize media.

2.4.2. Group II/Type A Trichothecene Producers

Group II was defined as type A trichothecene producer including *F. langsethiae*, *F. sporotrichioides* and *F. polyphialidicum*, which mainly produced one or several of type A trichothecenes, such as T-2, HT-2, NEO and DAS (Table 2).

Among the Group II *Fusarium* strains, *F. langsethiae* and *F. sporotrichioides* were found to be prolific producers of T-2, which was also demonstrated in Kokkonen et al.'s studies [27,30]. In general, high concentrations of T-2, NEO, and low production of HT-2 and DAS were observed for all the studied substrates (Tables S3–S5, Figures 3B and 5). These results were consistent with the results from Yli-Mattila et al., reporting that *F. langsethiae* and *F. sporotrichioides* produced high levels of T-2 with mean concentrations about 21,700–38,600 $\mu g \cdot kg^{-1}$, and low mean concentrations of DAS with 90–2800 $\mu g \cdot kg^{-1}$ [46]. With respect to *F. polyphialidicum*, it appeared to be a rare *Fusarium* species isolated from plant debris collected in South Africa [47], and the mycotoxin producing abilities have been only limitedly investigated up to date, reporting it as FB1 producer [48]. In this study, DAS was found to be produced by this fungus for the first time with mean concentration levels of 23, 1333 and 3386 μg·kg^{-1} in PDA, rice and maize medium, respectively (Tables S3–S5). Additionally, it could be obviously seen that for Group II fungi (Type A trichothecene producers), the highest concentrations of various mycotoxins were produced in maize, followed by rice, and lowest values were observed in PDA.

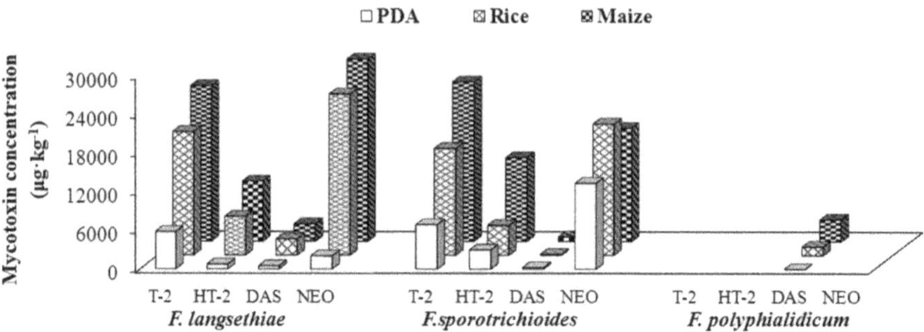

Figure 5. Investigation of the type A trichothecene mycotoxins (T-2, HT-2, NEO and DAS) producing abilities of *F. langsethiae*, *F. sporotrichioides* and *F. polyphialidicum* after incubation for 21 days at 25 °C on PDA, rice and maize media.

2.4.3. Group III/Type B Trichothecene Producers

The concentrations of NIV and Fus-X were 269 and 13 μg·kg^{-1} in PDA, 3151 and 1022 μg·kg^{-1} in rice, 2039 and 1260 μg·kg^{-1} in maize produced by *F. culmorum*; 566 and 60 μg·kg^{-1} in PDA, 21,231 and 1838 μg·kg^{-1} in rice, 979 and 137 μg·kg^{-1} in maize produced by *F. poae*; 123 and 107 μg·kg^{-1} in PDA, 120,342 and 112,167 μg·kg^{-1} in rice, 45,453 and 37,175 μg·kg^{-1} in maize produced by *F. meridionale*, respectively (Tables S3–S5, Figure 6). *F. poae* and *F. culmorum* have previously been considered as good producers for NIV and Fus-X [49–51]. In the present study, *F. graminearum* produced large amounts of DON, 3-ADON and 15-ADON, with concentration levels in PDA, rice and maize media in the range of 13,532–286,258 μg·kg^{-1}, 7700–50,344 μg·kg^{-1} and 5716–44,943 μg·kg^{-1}, respectively (Tables S3–S5, Figure 6). In previous studies, two type B trichothecence producing chemotypes were identified, i.e., the NIV and DON chemotypes [52–54]. Based on the results of this study, *F. culmorum*, *F. poae* and *F. meridionale* can be grouped into the NIV chemotype, while *F. graminearum* could be classified into the DON chemotype (Table 2).

Another feature with Group III is the co-occurrence of multiple types of mycotoxins in rice and maize media. As consistent with previous in vitro results, ZEN was produced by *F. culmorum* [27,31], *F. meridionale* [1] and *F. graminearum* [27,31] with highest levels herein in rice media. Previous studies indicated that *F. culmorum* showed intraspecies differences in the production of trichothecenes [45]. In our experiments, type A trichothecenes were detected in rice and maize media with *F. culmorum*, as well as *F. poae* and *F. meridionale*, especially significant amount of DAS produced by *F. poae*.

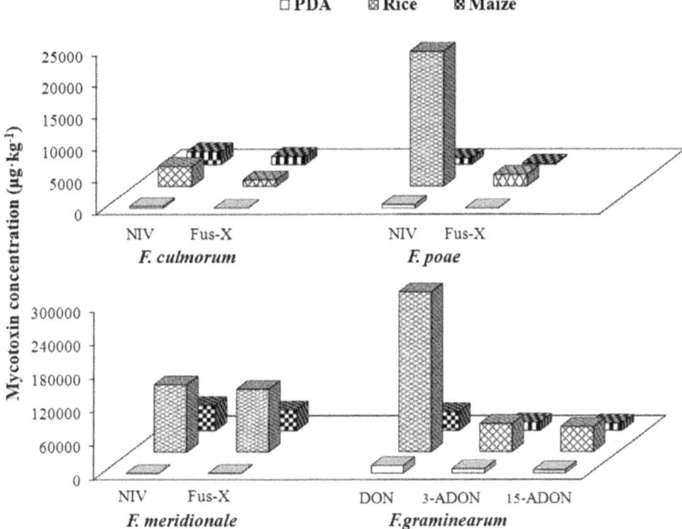

Figure 6. Investigation of the type B trichothecene mycotoxins (NIV, Fus-X, DON, 3-ADON and 15-ADON) producing abilities of *F. culmorum*, *F. poae*, *F. meridionale* and *F. graminearum* after incubation for 21 days at 25 °C on PDA, rice and maize media.

3. Conclusions

Twenty *Fusarium* species isolated from different regions were identified by molecular approaches and then inoculated on three growth media, PDA, rice and maize. The produced mycotoxins were determined quantitatively by LC-MS/MS and results were statistically analyzed using PCA. *Fusarium* species were accordingly divided into three groups, and mycotoxin profiles were thoroughly investigated to provide the direct evidences for clarification of the correlation between different mycotoxigenic fungi, mycotoxins and growth media. The targeted mycotoxin profiling in this study revealed mycotoxigenic potentials of *Fusarium* species in various culture substrates, which would contribute to further research concerning mycotoxin analysis and fungal investigations, as well as provide supporting information for controlling occurrence of fungi and their metabolic mycotoxins from farm to fork to ensure public health safety.

4. Materials and Methods

4.1. Fungal Strains, Materials and Chemicals

Twenty strains of *Fusarium* fungi were used in this study. *F. temperatum*, *F. subglutinans*, *F. culmorum*, *F. fujikuroi*, *F. langsethiae*, *F. musae*, *F. poae*, *F. proliferatum*, *F. sporotrichioides*, *F. tricinctum* and *F. verticillioides* were provided by Mycothèque de l'Université catholique de Louvain (MUCL, Louvain-la-Neuve, Belgium). The other nine *Fusarium* strains, including *F. graminearum*, *F. oxysporum*, *F. meridionale*, *F. equiseti*, *F. sacchari*, *F. solani*, *F. concentricum*, *F. andiyazi*, and *F. polyphialidicum*, were obtained by single spore isolation in our laboratory. Information about geographical location and plant hosts of all the investigated fungal species are presented in Table 1. Cereal matrices used for preparation of rice and maize medium were purchased from local suppliers, which were all mycotoxins-free as confirmed by liquid chromatography-tandem mass spectrometry (LC-MS/MS).

The mycotoxin standards of aflatoxin B1 (AFB1), aflatoxin B2 (AFB2), aflatoxin G1 (AFG1), aflatoxin G2 (AFG2), aflatoxin M1 (AFM1), aflatoxin M2 (AFM2), HT-2 toxin, T-2 toxin and ochratoxin A (OTA) were supplied by Alexisa (San Diego, CA, USA). 15-ADON and 3-ADON were purchased from

Biopure (Tulln, Austria). Fusaric acid, ZEN, zearalanone (ZAN), α-zearalenol (α-ZEL), α-zearalanol (α-ZAL), β-zearalenol (β-ZEL), β-zearalanol (β-ZAL), DON, NIV, deepoxy-DON, sterigmatocystin (SMC), Fus-X, citrinine (CIT), NEO, DAS, mycophenolic acid (MPA), cyclopiazonic acid (CPA), verruculogen (VER), FB1, FB2, patulin (PAT) and gliotoxin were purchased from Sigma-Aldrich (St. Louis, MO, USA). HPLC grade of acetonitrile and methanol were purchased from Merck (Darmstadt, Germany). Other solvents and chemicals were of HPLC or analytical grade from local suppliers. Deionized water purified by Milli-Q water (Millipore, Billerica, MA, USA) was used throughout the experiments.

4.2. Molecular Characterization of the *Fusarium* Strains

Fusarium strains were molecularly characterized by examining the sequence of the translation elongation factor 1-alpha (*EF-1α*) gene, known as one of the most pertinent genes for identification of the *Fusarium* species [55].

Mycelia plugs from 7-day old potato dextrose agar (PDA) (composition seen in Section 4.3) cultures were transferred to potato dextrose broth (PDB) medium (200 g potato and 20 g glucose per litre) and incubated while shaking (100 rpm) at 28 °C in the dark for 5 days. After incubation, the mycelia were harvested by filtration through filtering cloth, freeze-dried and ground to fine powders using a TissueLyser II system (Qiagen Tissuelyser II, Retsch, Haan, Germany).

Genomic DNA of strains was extracted based on the Cetyltriethyl Ammnonium Bromide (CTAB) protocol described by Wang et al. [56]. Portions of the *EF-1α* gene were amplified with primers EF1T (3′-ATGGGTAAGGAGGACAAGAC-5′) and EF2T (3′-GGAAGTACCAGTGATCATGTT-5′) in a thermal cycler (T100 Thermal Cycler, Bio-Rad, Foster City, CA, USA). Polymerase chain reaction (PCR) amplification was performed using a modified procedure described [57]. PCR reaction mixtures (total volume of 20 μL) contained 80 ng of fungal genomic DNA template, 1 × PCR buffer (20 mM Tris-HCl pH 8.3, 20 mM KCl, 10 mM $(NH_4)_2SO_4$, 2 mM $MgSO_4$) (TransGen Biotech, Beijing, China), 0.25 mM deoxynucleoside triphosphate (dNTPs) (Dongsheng Biotech, Guangzhou, China), 2.5 U of Easy Taq DNA polymerase (TransGen Biotech, Beijing, China) and 0.2 μM of each primer. The conditions for thermal cycler consisted of an initial denaturation step at 94 °C for 4 min, followed by 30 cycles of denaturation at 94 °C for 30 s, annealing at 56 °C for 40 s and extension at 72 °C for 30 s, then a final extension of 72 °C for 5 min. An aliquot of 8 μL of amplified products was separated by electrophoresis onto a 1% agarose gel, stained with ethidium bromide and photographed under UV light in a Bio-Imaging system (Bio-Rad, Hercules, CA, USA). The incised fragment gels were sent to Invitrogen™ (Shanghai, China) for sequencing. Then the *EF-1α* amplicon sequences (shown in Table S6) generated in this study were compared with sequences available by using the BLAST program [58]. The phylogenetic trees were made using MEGA5.0 for Neighbor-joining (N-J) analysis and the reliability was confirmed by bootstrapping using 1000 random replicates.

4.3. Preparation of Different Types of Growth Media

Three typical media including PDA, maize medium and rice medium, were prepared for inoculation and incubation of the *Fusarium* strains. PDA medium (200 g potato, 20 g glucose, and 15–20 g agar per litre) was prepared by autoclaving at 121 °C for 15 min and then 15 mL of molten media was poured into 9 cm diameter sterile Petri dishes. Maize/rice media were prepared by adding 25 mL deionized water into 50 g of mycotoxin-free maize/rice samples, vigorously shaken to prevent clumping, maintained overnight and sterilized in an autoclave for 15 min at 121 °C.

4.4. Inoculation of the Targeted Fungal Strains

Prior to the inoculation experiments, each fungal strain was cultured separately on PDA for 7 days at 25 °C for activation of the strain. The inoculation method was conducted as previously described [59,60] with minor modifications. A piece of 6 mm diameter agar disc taken from the margin of a 7-day old colony of each strain grown on PDA was placed in the centre of each test medium and incubated at 25 °C for 21 days. Control samples were prepared following the same procedure without fungal inoculation and each treatment was performed in triplicate. After 21 days of incubation, the media were harvested and dried at 40 °C–50 °C until constant weight was achieved, and then finely ground into homogenous powders and stored in the freezer for mycotoxin analysis.

4.5. *Analysis of Multiple Mycotoxins*

The mycotoxins produced by various *Fusarium* strains were extracted and simultaneously determined by LC-MS/MS covering a total of thirty three frequently occurring mycotoxins, which has been established in the previous studies [61].

4.6. *Statistical Analysis*

A pie chart and a three-dimensional histogram model were plotted using Microsoft Office Excel 2003 (Microsoft Corp., Redmond, WA, USA). Statistical analysis was performed using SPSS statistical package 17.0 (SPSS Inc., Chicago, IL, USA). One-way analysis of variance (ANOVA) was performed to determine the significance of the main factors and their interactions. $p < 0.05$ was considered statistically significant. Multivariate analysis was used to perform principal component analysis (PCA) by SIMCA-P software 11.0 (Umetrics, Umea, Sweden).

Supplementary Materials: The following are available online at www.mdpi.com/2072-6651/9/1/6/s1, Figure S1: LC-MS/MS chromatogram of FB1, FB2 and fusaric acid ((**a**): standards; (**b**): samples: rice medium, *F. proliferatum*), Figure S2: LC-MS/MS chromatogram of four type A trichothecenes ((**a**): standards; (**b**): samples: rice medium, *F. langsethiae*), Figure S3: LC-MS/MS chromatogram of five type B trichothecenes ((**a**): standards; (**b**): samples: NIV, Fus-X: maize medium, *F. poae*; DON, ADONs: maize medium, *F. graminearum*), Figure S4: LC-MS/MS chromatogram of ZEN ((**a**): standards; (**b**): samples: rice medium, *F. graminearum*), Table S1: Relative recoveries of the thirty three targeted mycotoxins at concentration of 50 $\mu g \cdot kg^{-1}$ spiked in PDA, rice and maize samples using ^{13}C-AFB1, ^{13}C-OTA, ^{13}C-T-2, ^{13}C-DON and ^{13}C-ZEN as the internal standards ($n = 5$), Table S2: Optimum temperatures for mycotoxin production of *Fusarium* species (PDA, 21 days), Table S3: Mycotoxins profiles of 20 species of toxicogenic *Fusarium* spp. in PDA medium under culture condition of 25 °C for 21 days ($\mu g \cdot kg^{-1}$), Table S4: Mycotoxins profiles of 20 species of toxicogenic *Fusarium* spp. in rice medium under culture condition of 25 °C for 21 days ($\mu g \cdot kg^{-1}$), Table S5: Mycotoxins profiles of 20 species of toxicogenic *Fusarium* spp. in maize medium under culture condition of 25 °C for 21 days ($\mu g \cdot kg^{-1}$), Table S6: *EF-1α* Sequences of 20 *Fusarium* species.

Acknowledgments: This work was supported by the National Basic Research Program of China (Grant 2013CB127801), National Natural Science Foundation of China (31471661) and Shanghai Technical Standards Project (15DZ0503800). MUCL (Mycothèque de l'Université catholique de Louvain, Louvain-la-Neuve, Belgium), and in particular François Van Hove and Françoise Munaut are acknowleged for providing some *Fusarium* strains. We thank Dr. Yuanhong Shan in the Core Facility Centre of the Institute of Plant Physiology and Ecology, Chinese Academy of Sciences for mass spectrometry assistance.

Author Contributions: W.S., Y.T. and S.W. designed and performed the experiments and analyzed the data; W.S., Y.T., Z.W. and A.W. wrote the paper; D.M.G., S.D.S., Y.L., C.W. and Y.F. contributed materials and/or revised the manuscript; A.W. and Z.W. supervised the whole experiments.

Conflicts of Interest: The authors declare no conflict of interest.

References

1. Duan, C.X.; Qin, Z.H.; Yang, Z.H.; Li, W.X.; Sun, S.L.; Zhu, Z.D.; Wang, X.M. Identification of pathogenic fusarium spp. causing maize ear rot and potential mycotoxin production in China. *Toxins* **2016**, *8*. [CrossRef] [PubMed]
2. McMullen, M.P.; Enz, J.; Lukach, J.; Stover, R. Environmental conditions associated with *Fusarium* head blight epidemics of wheat and barley in the northern great plains, north america. *Cereal Res. Commun.* **1997**, *25*, 777–778.
3. Kemp, G.H.J.; Pretorius, Z.A.; Wingfield, M.J. *Fusarium* glume spot of wheat: A newly recorded mite-associated disease in south africa. *Plant Dis.* **1996**, *80*, 48–51. [CrossRef]
4. Gortz, A.; Oerke, E.C.; Steiner, U.; Waalwijk, C.; de Vries, I.; Dehne, H.W. Biodiversity of *Fusarium* species causing ear rot of maize in Germany. *Cereal Res. Commun.* **2008**, *36*, 617–622. [CrossRef]

5. Covarelli, L.; Stifano, S.; Beccari, G.; Raggi, L.; Lattanzio, V.M.T.; Albertini, E. Characterization of *Fusarium verticillioides* strains isolated from maize in Italy: Fumonisin production, pathogenicity and genetic variability. *Food Microbiol.* **2012**, *31*, 17–24. [CrossRef] [PubMed]
6. McMullen, M.; Bergstrom, G.; De Wolf, E.; Dill-Macky, R.; Hershman, D.; Shaner, G.; Van Sanford, D. A unified effort to fight an enemy of wheat and barley: *Fusarium* head blight. *Plant Dis.* **2012**, *96*, 1712–1728. [CrossRef]
7. Desjardins, A.E.; Plattner, R.D. Fumonisin b(1)-nonproducing strains of *Fusarium verticillioides* cause maize (zea mays) ear infection and ear rot. *J. Agric. Food Chem.* **2000**, *48*, 5773–5780. [CrossRef] [PubMed]
8. Abbas, H.K.; Cartwright, R.D.; Xie, W.; Mirocha, C.J.; Richard, J.L.; Dvorak, T.J.; Sciumbato, G.L.; Shier, W.T. Mycotoxin production by *Fusarium proliferatum* isolates from rice with *Fusarium* sheath rot disease. *Mycopathologia* **1999**, *147*, 97–104. [CrossRef] [PubMed]
9. Kelly, A.C.; Clear, R.M.; O'Donnell, K.; McCormick, S.; Turkington, T.K.; Tekauz, A.; Gilbert, J.; Kistler, H.C.; Busman, M.; Ward, T.J. Diversity of *Fusarium* head blight populations and trichothecene toxin types reveals regional differences in pathogen composition and temporal dynamics. *Fungal Genet. Biol.* **2015**, *82*, 22–31. [CrossRef] [PubMed]
10. Bottalico, A.; Perrone, G. Toxigenic *Fusarium* species and mycotoxins associated with head blight in small-grain cereals in Europe. *Eur. J. Plant Pathol.* **2002**, *108*, 611–624. [CrossRef]
11. Tian, Y.; Tan, Y.L.; Liu, N.; Liao, Y.C.; Sun, C.P.; Wang, S.X.; Wu, A.B. Functional agents to biologically control deoxynivalenol contamination in cereal grains. *Front. Microbiol.* **2016**, *7*. [CrossRef] [PubMed]
12. Molto, G.A.; Gonzalez, H.H.L.; Resnik, S.L.; Gonzalez, A.P. Production of trichothecenes and zearalenone by isolates of *Fusarium* spp. from argentinian maize. *Food Addit. Contam.* **1997**, *14*, 263–268. [CrossRef] [PubMed]
13. Quarta, A.; Mita, G.; Haidukowski, M.; Santino, A.; Mule, G.; Visconti, A. Assessment of trichothecene chemotypes of *Fusarium culmorum* occurring in Europe. *Food Addit. Contam.* **2005**, *22*, 309–315. [CrossRef] [PubMed]
14. Schollenberger, M.; Muller, H.M.; Rufle, M.; Suchy, S.; Plank, S.; Drochner, W. Natural occurrence of 16 *Fusarium* toxins in grains and feedstuffs of plant origin from Germany. *Mycopathologia* **2006**, *161*, 43–52. [CrossRef] [PubMed]
15. Tian, Y.; Tan, Y.L.; Liu, N.; Yan, Z.; Liao, Y.C.; Chen, J.; De Saeger, S.; Hua, Y.; Zhang, Q.; Wu, A.B. Detoxification of deoxynivalenol via glycosylation represents novel insights on antagonistic activities of trichoderma when confronted with *Fusarium graminearum*. *Toxins* **2016**, *8*. [CrossRef] [PubMed]
16. Sudakin, D.L. Trichothecenes in the environment: Relevance to human health. *Toxicol. Lett.* **2003**, *143*, 97–107. [CrossRef]
17. Stoev, S.D. Food safety and increasing hazard of mycotoxin occurrence in foods and feeds. *Crit. Rev. Food Sci.* **2013**, *53*, 887–901. [CrossRef] [PubMed]
18. Commission, E. Commission regulation (EC) 1881/2006 of december 19th 2006 replacing regulation (EC) 466/2001 setting maximum levels for certain contaminants in foodstuffs. *Off. J. Eur. Commun.* **2006**, *L364*, 5–24.
19. The National Food Safety Standard of Maximum Levels of Mycotoxin in Foods (GB 2761–2011). Available online: http://gain.fas.usda.gov/Recent%20GAIN%20Publications/Maximum%20Levels%20of%20Mycotoxins%20in%20Foods_Beijing_China%20-%20Peoples%20Republic%20of_12-29-2014.pdf (accessed on 24 December 2016).
20. Logrieco, A.; Mule, G.; Moretti, A.; Bottalico, A. Toxigenic *Fusarium* species and mycotoxins associated with maize ear rot in Europe. *Eur. J. Plant. Pathol.* **2002**, *108*, 597–609. [CrossRef]
21. van Egmond, H.P.; Schothorst, R.C.; Jonker, M.A. Regulations relating to mycotoxins in food: Perspectives in a global and European context. *Anal. Bioanal. Chem.* **2007**, *389*, 147–157. [CrossRef] [PubMed]
22. Bosch, U.; Mirocha, C.J. Toxin production by *Fusarium* species from sugar-beets and natural occurrence of zearalenone in beets and beet fibers. *Appl. Environ. Microb.* **1992**, *58*, 3233–3239.
23. Christ, D.S.; Marlander, B.; Varrelmann, M. Characterization and mycotoxigenic potential of *Fusarium* species in freshly harvested and stored sugar beet in Europe. *Phytopathology* **2011**, *101*, 1330–1337. [CrossRef] [PubMed]

24. Zhang, J.B.; Li, H.P.; Dang, F.J.; Qu, B.; Xu, Y.B.; Zhao, C.S.; Liao, Y.C. Determination of the trichothecene mycotoxin chemotypes and associated geographical distribution and phylogenetic species of the *Fusarium graminearum* clade from China. *Mycol. Res.* **2007**, *111*, 967–975. [CrossRef] [PubMed]
25. Delgado, J.A.; Schwarz, P.B.; Gillespie, J.; Rivera-Varas, V.V.; Secor, G.A. Trichothecene mycotoxins associated with potato dry rot caused by *Fusarium graminearum*. *Phytopathology* **2010**, *100*, 290–296. [CrossRef] [PubMed]
26. Castillo, M.; Samar, M.; Molto, G.; Resnik, S.; Pacin, A. Trichothecenes and zearalenone production by *Fusarium* species isolated from argentinean black beans. *Mycotoxin Res.* **2002**, *18*, 31–36. [CrossRef] [PubMed]
27. Kokkonen, M.; Ojala, L.; Parikka, P.; Jestoi, M. Mycotoxin production of selected *Fusarium* species at different culture conditions. *Int. J. Food Microbiol.* **2010**, *143*, 17–25. [CrossRef] [PubMed]
28. Llorens, A.; Mateo, R.; Hinojo, M.J.; Valle-Algarra, F.M.; Jimenez, M. Influence of environmental factors on the biosynthesis of type B trichothecenes by isolates of *Fusarium* spp. from Spanish crops. *Int. J. Food Microbiol.* **2004**, *94*, 43–54. [CrossRef] [PubMed]
29. Martins, M.L.; Martins, H.M. Influence of water activity, temperature and incubation time on the simultaneous production of deoxynivalenol and zearalenone in corn (zea mays) by *Fusarium graminearum*. *Food Chem.* **2002**, *79*, 315–318. [CrossRef]
30. Kokkonen, M.; Jestoi, M.; Laitila, A. Mycotoxin production of *Fusarium langsethiae* and *Fusarium sporotrichioides* on cereal-based substrates. *Mycotoxin Res.* **2012**, *28*, 25–35. [CrossRef] [PubMed]
31. Richard, E.; Heutte, N.; Sage, L.; Pottier, D.; Bouchart, V.; Lebailly, P.; Garon, D. Toxigenic fungi and mycotoxins in mature corn silage. *Food Chem. Toxicol.* **2007**, *45*, 2420–2425. [CrossRef] [PubMed]
32. Brzonkalik, K.; Herrling, T.; Syldatk, C.; Neumann, A. The influence of different nitrogen and carbon sources on mycotoxin production in *Alternaria alternata*. *Int. J. Food Microbiol.* **2011**, *147*, 120–126. [CrossRef] [PubMed]
33. Mateo, J.J.; Jimenez, M. Trichothecenes and fumonisins produced in autoclaved tiger nuts by strains of *Fusarium sporotrichioides* and *Fusarium moniliforme*. *Food Microbiol.* **2000**, *17*, 167–176. [CrossRef]
34. Azira, T.N.; Man, Y.B.C.; Hafidz, R.N.R.M.; Aina, M.A.; Amin, I. Use of principal component analysis for differentiation of gelatine sources based on polypeptide molecular weights. *Food Chem.* **2014**, *151*, 286–292. [CrossRef] [PubMed]
35. Marina, A.M.; Man, Y.B.C.; Amin, I. Use of the saw sensor electronic nose for detecting the adulteration of virgin coconut oil with rbd palm kernel olein. *J. Am. Oil Chem. Soc.* **2010**, *87*, 263–270. [CrossRef]
36. Wulff, E.G.; Sorensen, J.L.; Lubeck, M.; Nielsen, K.F.; Thrane, U.; Torp, J. *Fusarium* spp. Associated with rice bakanae: Ecology, genetic diversity, pathogenicity and toxigenicity. *Environ. Microbiol.* **2010**, *12*, 649–657. [CrossRef] [PubMed]
37. Stepien, L.; Koczyk, G.; Waskiewicz, A. Fum cluster divergence in fumonisins-producing *Fusarium* species. *Fungal Biol.* **2011**, *115*, 112–123. [CrossRef] [PubMed]
38. Suga, H.; Kitajima, M.; Nagum, R.; Tsukiboshi, T.; Uegaki, R.; Nakajima, T.; Kushiro, M.; Nakagawa, H.; Shimizu, M.; Kageyama, K.; et al. A single nucleotide polymorphism in the translation elongation factor 1 alpha gene correlates with the ability to produce fumonisin in Japanese *Fusarium fujikuroi*. *Fungal Biol.* **2014**, *118*, 402–412. [CrossRef] [PubMed]
39. Hinojo, M.J.; Medina, A.; Valle-Algarra, F.M.; Gimeno-Adelantado, J.V.; Jimenez, M.; Mateo, R. Fumonisin production in rice cultures of *Fusarium verticillioides* under different incubation conditions using an optimized analytical method. *Food Microbiol.* **2006**, *23*, 119–127. [CrossRef] [PubMed]
40. Ung-Soo, L.; Myong-Yur, L.; Kwang-Sop, S.; Yun-Sik, M.; Chae-Min, C.; Ueno, Y. Production of fumonisin B_1 and B_2 by *Fusarium moniliforme* isolated from Korean corn kerneis for feed. *Mycotoxin Res.* **1994**, *10*, 67–72. [CrossRef] [PubMed]
41. Fadl-Allah, E.; Stack, M.; Goth, R.; Bean, G. Production of fumonisins B_1, B_2 and B_3 by *Fusarium proliferatum* isolated from rye grains. *Mycotoxin Res.* **1997**, *13*, 43–48. [CrossRef] [PubMed]
42. Leslie, J.F.; Plattner, R.D.; Desjardins, A.E.; Klittich, C.J.R. Fumonisin B_1 production by strains from different mating populations of *Gibberella fujikuroi* (*Fusarium* section *liseola*). *Phytopathology* **1992**, *82*, 341–345. [CrossRef]
43. Marasas, W.F.O.; Rheeder, J.P.; Lamprecht, S.C.; Zeller, K.A.; Leslie, J.F. *Fusarium andiyazi* sp nov., a new species from sorghum. *Mycologia* **2001**, *93*, 1203–1210. [CrossRef]

44. Abramson, D.; Clear, R.M.; Smith, D.M. Trichothecene production by *Fusarium* spp isolated from manitoba grain. *Can. J. Plant Pathol.* **1993**, *15*, 147–152. [CrossRef]
45. Hestbjerg, H.; Nielsen, K.F.; Thrane, U.; Elmholt, S. Production of trichothecenes and other secondary metabolites by *Fusarium culmorum* and *Fusarium equiseti* on common laboratory media and a soil organic matter agar: An ecological interpretation. *J. Agric. Food Chem.* **2002**, *50*, 7593–7599. [CrossRef] [PubMed]
46. Yli-Mattila, T.; Ward, T.J.; O'Donnell, M.; Proctor, R.H.; Burkin, A.A.; Kononenko, G.P.; Gavrilova, O.P.; Aoki, T.; McCormick, S.P.; Gagkaeva, T.Y. *Fusarium sibiricum* sp. nov, a novel type a trichothecene-producing *Fusarium* from Northern Asia closely related to *F-sporotrichioides* and *F-langsethiae*. *Int. J. Food Microbiol.* **2011**, *147*, 58–68. [CrossRef] [PubMed]
47. Guarro, J.; Rubio, C.; Gene, J.; Cano, J.; Gil, J.; Benito, R.; Moranderia, M.J.; Miguez, E. Case of keratitis caused by an uncommon *Fusarium* species. *J. Clin. Microbiol.* **2003**, *41*, 5823–5826. [CrossRef] [PubMed]
48. Abbas, H.K.; Ocamb, C.M. First report of production of fumonisin B-1 by *Fusarium polyphialidicum* collected from seeds of *Pinus strobus*. *Plant Dis.* **1995**, *79*, 642. [CrossRef]
49. Thrane, U.; Adler, A.; Clasen, P.E.; Galvano, F.; Langseth, W.; Logrieco, A.; Nielsen, K.F.; Ritieni, A. Diversity in metabolite production by *Fusarium langsethiae*, *Fusarium poae*, and *Fusarium sporotrichioides*. *Int. J. Food Microbiol.* **2004**, *95*, 257–266. [CrossRef] [PubMed]
50. Wagacha, J.M.; Muthomi, J.W. *Fusarium culmorum*: Infection process, mechanisms of mycotoxin production and their role in pathogenesis in wheat. *Crop Prot.* **2007**, *26*, 877–885. [CrossRef]
51. Scoz, L.B.; Astolfi, P.; Reartes, D.S.; Schmale, D.G.; Moraes, M.G.; Del Ponte, E.M. Trichothecene mycotoxin genotypes of *Fusarium graminearum* sensu stricto and *Fusarium meridionale* in wheat from southern Brazil. *Plant Pathol.* **2009**, *58*, 344–351. [CrossRef]
52. Sydenham, E.W.; Marasas, W.F.O.; Thiel, P.G.; Shephard, G.S.; Nieuwenhuis, J.J. Production of mycotoxins by selected *Fusarium-graminearum* and *F-crookwellense* isolates. *Food Addit. Contam.* **1991**, *8*, 31–41. [CrossRef] [PubMed]
53. Bakan, B.; Pinson, L.; Cahagnier, B.; Melcion, D.; Semon, E.; Richard-Molard, D. Toxigenic potential of *Fusarium culmorum* strains isolated from French wheat. *Food Addit. Contam.* **2001**, *18*, 998–1003. [CrossRef] [PubMed]
54. Burlakoti, R.R.; Ali, S.; Secor, G.A.; Neate, S.M.; McMullen, M.P.; Adhikari, T.B. Comparative mycotoxin profiles of gibberella zeae populations from barley, wheat, potatoes, and sugar beets. *Appl. Environ. Microbiol.* **2008**, *74*, 6513–6520. [CrossRef] [PubMed]
55. O'Donnell, K.; Kistler, H.C.; Cigelnik, E.; Ploetz, R.C. Multiple evolutionary origins of the fungus causing panama disease of banana: Concordant evidence from nuclear and mitochondrial gene genealogies. *Proc. Natl. Acad. Sci. USA* **1998**, *95*, 2044–2049. [CrossRef] [PubMed]
56. Wang, J.H.; Li, H.P.; Qu, B.; Zhang, J.B.; Huang, T.; Chen, F.F.; Liao, Y.C. Development of a generic pcr detection of 3-acetyldeoxynivalenol-, 15-acetyldeoxynivalenol- and nivalenol-chemotypes of *Fusarium graminearum* clade. *Int. J. Mol. Sci.* **2008**, *9*, 2495–2504. [CrossRef] [PubMed]
57. BLAST program. Avaliable online: https://blast.ncbi.nlm.nih.gov/Blast.cgi (accessed on 5 August 2016).
58. Van Poucke, K.; Monbaliu, S.; Munaut, F.; Heungens, K.; De Saeger, S.; Van Hove, F. Genetic diversity and mycotoxin production of *Fusarium* lactis species complex isolates from sweet pepper. *Int. J. Food Microbiol.* **2012**, *153*, 28–37. [CrossRef] [PubMed]
59. Busko, M.; Chelkowski, J.; Popiel, D.; Perkowski, J. Solid substrate bioassay to evaluate impact of trichoderma on trichothecene mycotoxin production by *Fusarium* species. *J. Sci. Food Agric.* **2008**, *88*, 536–541. [CrossRef]
60. Medina, A.; Magan, N. Temperature and water activity effects on production of T-2 and HT-2 by *Fusarium langsethiae* strains from north European countries. *Food Microbiol.* **2011**, *28*, 392–398. [CrossRef] [PubMed]
61. Han, Z.; Feng, Z.H.; Shi, W.; Zhao, Z.H.; Wu, Y.J.; Wu, A.B. A quick, easy, cheap, effective, rugged, and safe sample pretreatment and liquid chromatography with tandem mass spectrometry method for the simultaneous quantification of 33 mycotoxins in *lentinula edodes*. *J. Sep. Sci.* **2014**, *37*, 1957–1966. [CrossRef] [PubMed]

© 2016 by the authors. Licensee MDPI, Basel, Switzerland. This article is an open access article distributed under the terms and conditions of the Creative Commons Attribution (CC BY) license (http://creativecommons.org/licenses/by/4.0/).

Article

Occurrence of *Fusarium* Mycotoxins in Cereal Crops and Processed Products (*Ogi*) from Nigeria

Cynthia Adaku Chilaka [1,2,*], Marthe De Boevre [1], Olusegun Oladimeji Atanda [3] and Sarah De Saeger [1]

1 Laboratory of Food Analysis, Department of Bioanalysis, Faculty of Pharmaceutical Sciences, Ghent University, Ottergemsesteenweg 460, 9000 Ghent, Belgium; marthe.deboevre@ugent.be (M.D.B.); sarah.desaeger@ugent.be (S.D.S.)
2 Department of Food Science and Technology, College of Applied Food Science and Tourism, Michael Okpara University of Agriculture, Umuahia-Ikot Ekpene Road, Umudike, PMB 7267 Umuahia, Abia State, Nigeria
3 Department of Biological Sciences, McPherson University, KM 96 Lagos-Ibadan Expressway, 110117 Seriki Sotayo, Ogun State, Nigeria; olusegunatanda@yahoo.co.uk
* Correspondence: cynthia.chilaka@ugent.be; Tel.: +32-9264-8133

Academic Editor: HJ (Ine) van der Fels-Klerx
Received: 30 September 2016; Accepted: 13 November 2016; Published: 18 November 2016

Abstract: In Nigeria, maize, sorghum, and millet are very important cash crops. They are consumed on a daily basis in different processed forms in diverse cultural backgrounds. These crops are prone to fungi infestation, and subsequently may be contaminated with mycotoxins. A total of 363 samples comprising of maize (136), sorghum (110), millet (87), and *ogi* (30) were collected from randomly selected markets in four agro-ecological zones in Nigeria. Samples were assessed for *Fusarium* mycotoxins contamination using a multi-mycotoxin liquid chromatography-tandem mass spectrometry (LC-MS/MS) method. Subsequently, some selected samples were analysed for the occurrence of hidden fumonisins. Overall, 64% of the samples were contaminated with at least one toxin, at the rate of 77%, 44%, 59%, and 97% for maize, sorghum, millet, and *ogi*, respectively. Fumonisins were the most dominant, especially in maize and *ogi*, occurring at the rate of 65% and 93% with mean values of 935 and 1128 μg/kg, respectively. The prevalence of diacetoxyscirpenol was observed in maize (13%), sorghum (18%), and millet (29%), irrespective of the agro-ecological zone. Other mycotoxins detected were deoxynivalenol, zearalenone, and their metabolites, nivalenol, fusarenon-X, HT-2 toxin, and hidden fumonisins. About 43% of the samples were contaminated with more than one toxin. This study suggests that consumption of cereals and cereal-based products, *ogi* particularly by infants may be a source of exposure to *Fusarium* mycotoxins.

Keywords: *Fusarium* mycotoxins; occurrence; cereal; *ogi*; LC-MS/MS; Nigeria

---

1. Introduction

Mycotoxins are secondary metabolites produced by a wide diversity of toxigenic fungi, which often contaminate crops worldwide [1]. These fungi are ubiquitous in nature, and may contaminate crops in the field or during storage, thus producing mycotoxins under favourable environmental conditions. *Fusarium* fungi are of high significance because of their ability to cause several devastating plant diseases, and being responsible for economic losses and trade barriers, while having potential in producing a wide range of mycotoxins. *Fusarium* mycotoxins have been linked to several health related problems in animals and humans ranging from acute (such as anorexia and diarrhoea) to chronic disease conditions (such as cancer and immunosuppression) [2,3]. For instance, fumonisins when ingested are carcinogenic, neurotoxic, and hepatotoxic and may possibly lead to death [1,2,4]. Efforts to understand the production and behaviour of mycotoxins, and to protect consumers from mycotoxicoses have led to extensive investigation of these toxins across the globe as well as establishment of regulatory maximum limits by the developed countries. Although it is estimated that several *Fusarium* mycotoxins do exist in nature, those mostly studied are the fumonisins (FB), trichothecenes (TH), and zearalenone (ZEN). This is due to the high toxic effects they exert on humans and animals, and their frequent occurrence in agricultural products especially cereals and cereal-based food products.

Cereals such as maize (*Zea mays*), sorghum (*Sorghum bicolor*), and millet (*Pennisetum glaucum*) serve as major staple crops consumed especially by the middle and low income earners in Nigeria. These crops are often processed into different food forms including processing of traditional weaning meal in the region. *Ogi* (also known as akamu) is a fermented cereal-based product used as a major traditional weaning food for infants, food for the convalescent and the elderly as well as consumed by different age groups especially as breakfast meal in Nigeria. It is produced by submerge fermentation of cereal grains (maize, sorghum, or millet) for two to three days followed by wet milling and sieving through a mesh. The fermentation process of *ogi* is usually initiated by chance inoculation under uncontrolled environmental conditions thereby resulting in variable quality of the final product. Studies have reported the prevalence of *Fusarium* mycotoxins, particularly TH, ZEN, and FB, in cereal crops and cereal-based products globally [5–8]. In most cases, these mycotoxins may co-exist in food and food products which often results to a synergistic, additive or antagonistic toxic effect on the host [9]. The mixed effects of mycotoxins have been revealed by the study of Kouadio et al. [10] on the effects of combinations of ZEN, fumonisin B_1 (FB_1), and deoxynivalenol (DON) on the human intestinal cell line (Caco-2). FB_1 in combination with ZEN showed lesser effect on the reduction of cell viability when compared to the combined effect of FB_1 with other mycotoxins because of the antagonistic effect of FB_1 on ZEN [10]. Similarly, Speijer and Speijer [9] observed the antagonistic effect of DON on T-2 in the inhibition of human lymphocytes proliferation. It is noteworthy to mention that ternary combination of type B TH (fusarenon-X (FUS-X), nivalenol (NIV), and DON) exhibited an antagonistic interaction on the intestinal epithelial cells which is possibly linked to a lower toxicity of FUS-X in the mixture [11]. There exist a potential relationship in the reduction of FUS-X toxicity and the competition between DON and NIV at the substrate binding sites of the de-acetylase thus leading to a reduced deacetylation of FUS-X [11]. Cases of synergistic interaction exhibited by combination of mycotoxins such as ZEN, DON, and FB_1 have also been reported [10,12]. Harvey et al. [13] and Kubena et al. [14] demonstrated the synergistic and additive effects resulting to growth depression in pigs and broiler chicks, respectively because of co-occurrence of mixed mycotoxins (DON and FB_1). A synergistic interaction between several combinations of type B TH on epithelial cell toxicity has also been recorded [11,15]. Recently, issues of possible co-existence of these *Fusarium* mycotoxins and their modified forms have become of great concern. Modification of mycotoxins may be triggered by food processing, or matrix related, or through conjugation by either plant, fungi or animal [16,17]. These modified mycotoxins often escape routine analysis leading to underestimation of actual mycotoxin levels in products and may possibly hydrolysed into the parent toxins during digestion [18]. Several studies on the occurrence of *Fusarium* mycotoxins in cereals and cereal-based products have reported the natural occurrence and co-occurrence of modified mycotoxins such as DON-3-glucoside (DON-3G), ZEN-4-glucoside (ZEN-14G), and α- and β-zearalenol-4-glucoside (α- and β-ZEL-4G) [5,19]. The possible underestimation of FB concentration in cereals and cereal-based products as a result of presences of hidden FB has been demonstrated [20–22]. Hidden FB cannot be directly analysed as they have to be released from the matrix into extractable form (hydrolysed FB) by sample treatment often by alkaline hydrolysis [23,24].

The increasing rate of climate change, which is characterised by significant increase or decrease in temperature and/or alteration of rainfall during planting season in sub-Saharan Africa (especially Nigeria), may have predisposed this region to *Fusarium* mycotoxins contamination. Evidence of possible occurrence of *Fusarium* mycotoxins in Nigeria is revealed by the frequent incidence of major mycotoxin producing *Fusarium* species such as *F. verticillioides*, *F. graminearum*, *F. poae*, *F. proliferatum*, and *F. sporotrichioides* in Nigerian food commodities [25,26]. In spite of this obvious evidence, limited study has been undertaken to ascertain the possible occurrence of *Fusarium* mycotoxins and their modified forms in Nigeria food products. This has resulted to the lack of regulatory maximum levels governing the control of *Fusarium* mycotoxins in Nigeria. Sub-Saharan African countries including Nigeria solely depend on maximum levels set by the European Union and the Codex Alimentarius Commission on control of *Fusarium* mycotoxins without considering the feeding habits and other socio-economic dynamics faced by this region. The main objective of the present paper is to investigate the occurrence of *Fusarium* mycotoxins and their modified forms including fumonisin B_1, B_2, and B_3; hidden FB; DON; 3-acetyl-DON (3ADON); 15-acetyl-DON (15ADON); DON-3G; ZEN; α-zearalenol (α-ZEL); β-zearalenol (β-ZEL); ZEN-14G; NIV; FUS-X; T-2 toxin (T-2); HT-2 toxin (HT-2); diacetoxyscirpenol (DAS); and neosolaniol (NEO) in Nigerian cereals—maize, sorghum, millet, and the processed products (*ogi*).

2. Results and Discussion

2.1. *Fusarium* Mycotoxins Contamination in Cereals (Maize, Sorghum, and Millet) and Processed Products (*Ogi*) from Nigeria

A total of 363 samples comprising maize (n = 136), sorghum (n = 110), millet (n = 87), and *ogi* (n = 30) were evaluated for the occurrence of *Fusarium* mycotoxins and modified forms including FB_1, FB_2, FB_3, hidden FB, DON, 3ADON, 15ADON, DON-3G, ZEN, α-ZEL and β-ZEL, ZEN-14G, NIV, FUS-X, T-2, HT-2, DAS, and NEO. These samples were collected from randomly selected markets from four agro-ecological zones in Nigeria between September 2015 and October 2015. Out of the 18 *Fusarium* mycotoxins analysed in the samples, 15 toxins were present in at least one of the samples. Data on the incidence and occurrence level of individual *Fusarium* mycotoxins in the cereals (maize, sorghum, and millet) and processed products (*ogi*) are illustrated in Tables 1 and 2. Over 40% prevalence rate of the mycotoxins was recorded in all sample types, with the individual rate of 77%, 44%, 59%, and 97% for maize, sorghum, millet, and *ogi*, respectively.

Maize, sorghum, millet, and *ogi* contained 13, 13, 10, and 14 *Fusarium* secondary metabolites, respectively, of which only four (FB_1, FB_2, DON, and ZEN) are regulated by the European Union (EU). Fumonisins were the most dominant mycotoxins occurring at high level and incidence rate in all the food types especially in maize and *ogi* samples. The sum of fumonisins (FB_1 + FB_2 + FB_3 (FB)) were in the ranges of 32–8508 μg/kg (65%), 45–180 μg/kg (8%), 74–22,064 μg/kg (14%), and 125–3557 μg/kg (93%) in maize, sorghum, millet, and *ogi*, respectively. Except for sorghum, most of the maize and millet samples in this study exceeded the maximum regulatory limit set for the sum of FB_1 and FB_2 (1000 μg/kg) by the European Union (EU) [27] suggesting the high exposure of the population to this toxin. A similar high FB incidence rate has been reported in several studies from sub-Saharan Africa [28–34] at concentrations ranging up to 53,863 μg/kg [30]. High incidence of FB, especially in maize may be explained by the susceptibility of the maize crop to FB producing fungi (*F. verticillioides* and *F. proliferatum*) [35]. Sorghum and millet had a much lower incidence rate, however an extreme concentration of FB was recorded in one of the millet samples (22,064 μg/kg). Lower concentrations and incidence rate in sorghum and millet from Ethiopia have previously been reported [36]. However, the reversed trend was observed by Ayalew et al. [37], who recorded higher levels of FB (range: 1370–2117 μg/kg) in sorghum samples. Of the FB, FB_1 occurred at a more frequent rate than FB_2 and FB_3. Although, we observed that some of the millet (n = 4) and maize (n = 16) samples were contaminated with only FB_2. Such trend has previously been reported in cereals suggesting the possible contamination of *Aspergillus niger,* which is a principal producer of FB_2 [38]. The study of Ezekiel et al. [39] on sorghum grain confirms the possible occurrence of only FB_2 in cereals from Nigeria. The incidence and levels of FB as observed in *ogi* is of concern. This present study reveals for the first time the occurrence of *Fusarium* mycotoxins in *ogi* from Nigerian market. The maximum concentration and percentage incidence of FB_1, FB_2, and FB_3 detected in *ogi* samples were 1903 μg/kg (93%), 1,283 μg/kg (87%), and 371 μg/kg (77%), respectively (Table 1). About 83% of the *ogi* samples exceeded the EU maximum limit of 200 μg/kg for processed maize-based foods for infants and young children [27]. Interestingly, out of the 30 *ogi* samples analysed, the only two FB negative samples were of sorghum-base. This confirms the previous study that sorghum is less prone to fungal infestation than maize [39]. Although there are no available data on the occurrence of *Fusarium* mycotoxins in *ogi*, studies from the same country reported the occurrence of FB in two fermented traditional cereal-based beverages (kunu-zaki and pito) [39].

The next group of dominating mycotoxins were the TH. They have been associated with the temperate regions, however studies emerging from sub-Saharan Africa have revealed the possible occurrence of these toxins in the tropics. Type B TH DON, 15ADON, DON-3G, and NIV were detected in our samples. DON was present in 16%, 3%, 13%, and 13% of maize, sorghum, millet, and *ogi* samples at a maximum level of 225 μg/kg, 119 μg/kg, 583 μg/kg, and 74 μg/kg, respectively. Interestingly no sample, irrespective of the food type, exceeded the EU maximum limit (1750 μg/kg, maize; 1250 μg/kg, other cereals; and 200 μg/kg, cereal-based infant foods) for DON [27]. Incidence of DON, as observed in this study, was similar to that reported in previous studies on Nigerian cereals [26,40], but much less than that reported in maize by Adetunji et al. [32] and Ediage et al. [34]. The same trend was reported in millet, sorghum, and cereal-based food samples from Burkina Faso [29]. Studies have shown the occurrence of acetylated DON forms and DON-3G in maize and its products [31,41]. Maize samples in our study were negative for 3ADON, 15ADON, and DON-3G. This is in agreement with a previous study on maize from Burkina Faso [29]. Sorghum, millet, and *ogi* were contaminated with 15ADON, and were negative of 3ADON. The production of acetylated derivatives (15ADON and 3ADON) by *F. graminearum* have been reported and the potential of the isolates to produce 15ADON or 3ADON as the major isomer is dependent on the geographic origin [42,43]. Although the information on the regional relationship between *F. graminearum* and the production of 15ADON or 3ADON is still lacking in Africa, Li et al. [44] and Mirocha et al. [42] reported the predominant of 3ADON in New Zealand, Austrialia, and China while 15ADON chemotype is predominant in North America. The glucoside of DON (DON-3G) was observed to contaminate sorghum and *ogi* samples in the present study. A comparable result on DON-3G in sorghum and millet from Ethiopia have also been reported [36].

Samples of maize (n = 3) and *ogi* (n = 2) were contaminated with NIV at concentration ranges of 163–271 μg/kg and 136–160 μg/kg, respectively. Occurrence of NIV in Nigerian maize has previously been reported, although at a higher incidence rate (54%) [32]. While similar result in cereal-based products as shown in this study was reported by Castillo et al. [45]. Contrary to the result reported on the occurrence of NIV in sorghum and millet by Chala et al. [36], sorghum and millet were negative for NIV in the present study. The trend observed in this study with NIV was also seen with FUS-X contamination. NEO was not detected in any of the samples analysed.

With regards to type A TH, DAS and HT-2 were present in all the sample types except for the cereal-based products (*ogi*) which was negative for DAS, while T-2 was negative in all the sample types. DAS was detected in maize, sorghum, and millet at a rate of 13%, 18%, and 29%, respectively (Table 1). The concentrations of DAS in the cereals ranged between 2 μg/kg and 25 μg/kg. The occurrence of DAS in the samples is probably associated with the occurrence of major DAS-producing fungi in this region [25,26]. DAS and HT-2 are synthesised by a wide range of *Fusarium* species, and they are alleged to be among the most toxic TH occurring in different food products. Several studies have reported the occurrence of DAS in cereals and cereal-based products [46,47]. Despite its association with the temperate weather, previous studies revealed the occurrence of DAS in the tropical regions. Adejumo et al. [26] and Adetunji et al. [32] recorded the occurrence of DAS in Nigerian maize at maximum concentrations of 51 μg/kg (9%) and 30 μg/kg (19%), respectively . Besides maize, DAS has been found to contaminate sorghum and millet from Ethiopia with maximum concentrations of 64.2 μg/kg (mean value, 11.9 μg/kg), and 1.43 μg/kg (mean value, 1.43 μg/kg), respectively [36]. A total of 1%, 8%, 5%, and 3% of maize, sorghum, millet, and *ogi*, respectively, were positive of HT-2 (Table 1). Beside the low incidence rate, none of the cereals or *ogi* samples exceeded the EU recommendation level of 100 μg/kg and 15 μg/kg for cereal and infant foods, respectively [48].

Table 1. Mean and maximum concentration (μg/kg) of *Fusarium* mycotoxins found in cereals and cereal-based products (*ogi*) from Nigeria.

Mycotoxin [1]	Maize (n = 136)			Sorghum (n = 110)			Millet (n = 87)			*Ogi* (n = 30)		
	% + ve Samples [2]	Mean [3]	Max [4]	% + ve Samples	Mean	Max	% + ve Samples	Mean	Max	% + ve Samples	Mean	Max
FB_1	65	541	8222	8	64	78	9	2333	18,172	93	590	1903
FB_2	54	376	2885	2	48	55	13	609	3892	87	472	1283
FB_3	43	117	445	2	38	46	0	na	na	77	121	371
∑FB	65	935	8508	8	83	180	14	2113	22,064	93	1128	3557
DON	16	99	225	3	100	119	13	151	543	13	61	74
15 ADON	0	na [5]	na	2	39	44	1	11	11	3	60	60
DON-3G	0	na	na	23	24	63	0	na	na	17	30	44
ZEN	1	65	65	1	38	38	14	419	1399	3	39	39
ZEN-14G	9	21	24	3	19	22	6	23	34	3	31	31
α-ZEL	1	20	20	3	33	33	0	na	na	7	20	22
β-ZEL	2	20	21	1	21	21	1	39	39	10	19	20
HT-2	1	20	20	8	20	31	5	36	36	3	13	13
NIV	2	206	271	0	na	na	0	na	na	7	148	160
FUS-X	1	154	154	0	na	na	0	na	na	7	133	137
DAS	13	3	8	18	5	16	29	5	25	0	na	na

[1] FB_1, B_2, and B_3 = fumonisin B_1, B_2, and B_3; ∑FB = sum of FB_1, B_2, and B_3; DON = deoxynivalenol; 15ADON = 15-acetyl-deoxynivalenol; DON-3G = deoxynivalenol-3-glucoside; ZEN = zearalenone; α-ZEL = α-zearalenol; β-ZEL = β-zearalenol; ZEN-14G = zearalenone-14-glucoside; NIV = nivalenol; FUS-X = fusarenon-X; HT-2 = HT-2 toxin; DAS = diacetoxyscirpenol, [2] % + ve Samples = percentage positive samples, [3] Mean = mean concentration, [4] Max = maximum concentration, [5] na = not applicable.

Table 2. Contamination levels of fumonisins, total fumonisins, and hidden fumonisin in selected samples.

Food Type	FB (μg/kg)			Total FB (μg/kg)			Hidden FB (μg/kg)		
	Median	Mean	Maximum	Median	Mean	Maximum	Median	Mean	Maximum
Maize (n = 10)	358	835	3514	543	1636	4568	144	801	2923
Sorghum (n = 10)	41	61	180	95	182	502	50	120	323
Millet (n = 10)	118	277	840	302	776	3059	179	499	2254
Ogi (n = 10)	247	531	1496	391	672	1795	117	141	313

Hidden FB concentration = the difference between the concentration of FB and the concentration of total FB after hydrolysis.

Other mycotoxins detected in the study include ZEN, α-ZEL, β-ZEL, and ZEN-14G. Recent studies have shown the prevalence of ZEN in food products from sub-Saharan Africa [36,49], however in the present study, ZEN was rarely detected. ZEN was detected in maize, sorghum, and millet at 1%, 1%, and 14%, respectively (Table 1) with the concentrations in all the sample types being less than the EU maximum limit of ZEN, except for millet with 8 samples (9%) exceeding 100 μg/kg [27]. Further, only one sample (3%) of *ogi* was positive for ZEN with the value exceeding the maximum limit of 20 μg/kg set by EU for processed cereal-based foods for infants and young children [50]. About 1% and 2% of maize were contaminated with α-ZEL and β-ZEL, respectively. Although there exist only limited studies on the occurrence of these metabolites in food products from sub-Saharan Africa, available data show their possible occurrence in Nigerian maize [32,49]. The present study is in agreement with the result of Adetunji et al. [32]. The maximum levels for β-ZEL in sorghum, millet, and *ogi* were 21 μg/kg, 39 μg/kg, and 20 μg/kg, respectively. Millet samples were negative of α-ZEL while sorghum and *ogi* had 3% and 7% incidence rate with maximum levels of 33 μg/kg and 22 μg/kg, respectively (Table 1). Chala et al. [36] reported a higher incidence rate of α-ZEL and β-ZEL in sorghum and millet compared to the current study, however, the levels reported by these authors were lower. With regards to ZEN-14G, all food type samples analysed showed positive samples with maximum concentrations of 24 μg/kg (maize), 22 μg/kg (sorghum), 34 μg/kg (millet), and 31 μg/kg (*ogi*). Occurrence of this modified form of ZEN in cereals and *ogi* in the current study is supported by a study which detected a wide range of modified forms of ZEN in cereal-based food products [18,19]. Although, there are no recommendation nor regulation limit of ZEN-14G in cereals and cereal-based products because of the non availability of toxicological data, the occurrence of ZEN-14G as observed in the present study is presumed to add additional toxic effect to the host. De Boevre et al. [51] reported the possible hydrolysis of ZEN-14G into its parent form (ZEN) in the digestive tract of mammals suggesting an additional toxicity.

2.2. Fumonisins and Hidden Fumonisins Contamination in Cereals (Maize, Sorghum and Millet) and Processed Products (Ogi) from Nigeria

To determine the occurrence of hidden fumonisins, samples were selected from each food type based on the FB result obtained from the multi-mycotoxin analysis (Table 1). Five FB positive and five FB negative samples of each food type (maize, sorghum, and millet) were selected for analysis. Note that eight positive samples and two negative samples of *ogi* were used for the analysis because only two samples of *ogi* were negative. Each of the samples were analysed simultaneously for FB (FB_1, FB_2, and FB_3) as well as total fumonisins after hydrolysis as described in Section 4. Calculation of hidden FB concentration was based on the difference between the concentration of FB and the concentration of total FB after hydrolysis [52]. The maximum FB and total FB concentration in the selected samples of maize, sorghum, millet, and *ogi* samples were 3514 and 4568 μg/kg, 180 and 502 μg/kg, 840 and 3059 μg/kg, and 1496 and 1795 μg/kg, respectively. After hydrolysis, we observed an increment ranging from 1.3 to 5.2 times higher levels of total FB in maize samples. The same trend was observed in sorghum, millet, and *ogi* samples. Hidden FB have been alleged to occur in processed products especially nixtamalised and thermally processed foods [52,53]. However, recent studies have revealed the occurrence of these toxins in unprocessed food products especially in raw maize samples [20,54] which suggest the possible transformation of FB to bound derivatives by natural phenomena due to plant metabolism [24]. The presence of hidden FB as observed in the current study may pose an additional health risk to consumers especially to the consumers of *ogi* analysed in this study. FB has been alleged to cause a range of toxic health effect on humans and animals especially in sub-Saharan Africa where cases of very high levels of FB have been recorded. Cases of human oesophageal cancer in South Africa and other parts of the World have been linked to the consumption of food contaminated with FB. Since it is obvious that hidden FB may cause additional toxic effect on the host as observed when low FB contaminated feed was fed to animals [55], the occurrence of hidden FB in cereals and cereal-based products should no longer be neglected especially in Nigeria where these products serve as major staple food.

2.3. Distribution of *Fusarium* Mycotoxins in Major Cereals across the Different Agro-Ecological Zones of Nigeria

Mycotoxin occurrence and distribution is influenced by different factors including crop species, climatic, and environmental conditions of a given region. The mean and maximum concentrations of individual mycotoxins in the different food types and AEZ are shown in Table 3. Fumonisins contaminations were observed in all the cereal types irrespective of the AEZ. Sudan Savanna (SS) and Northern Guinea Savanna (NGS) zones had the highest incidence rate of FB_1 in maize with a highest FB_1 concentration of 2443 μg/kg and 8222 μg/kg, respectively, when compared to Southern Guinea Savanna (SGS) and Derived Savanna (DS). A similar trend with FB_1 contamination was also seen when the sum of FB (FB_1, FB_2, and FB_3) was considered. This is also similar with the result obtained from the sorghum samples, with the SS zone registering the highest FB_1 followed by NGS and SGS. This observation could be linked to the high mycotoxins production potentials of *Fusarium* fungi in warmer climates [56] and the significant change in climatic conditions in this region characterised by increase in rainfall and longer raining seasons [57]. Among the other mycotoxins detected, DAS was the next most common metabolite contaminating all the food types across the AEZ, although at lower concentrations. While there are no existing regulatory limits set for DAS in food products, DAS has been implicated in a wide range of toxic effects in animals as well as human, ranging from acute to chronic. It has been linked to a human fatal disease (alimentary toxic aleukia), exhibiting several symptoms such as inflammation of the skin, vomiting, and damage to hematopoietic tissues [1,3]. DON also occurred in the respective cereal types across the AEZ, except for the sorghum samples from the SGS zone (Table 3). The highest incidence rate of DON was observed in maize samples from DS zone characterized by a lower temperature and higher average annual rainfall of 25–35 °C and 1300–1500 mm, respectively when compared to the other zones. In millet samples, the NGS zone registered the highest incidence rate of DON. Comparing the incidence of ZEN across the different food types and AEZ, samples from SS zones were negative of ZEN regardless of the cereal type. This result could be related to the prevailing local weather conditions of this region which is between 30 and 40 °C, which is above the optimum temperature of 25 °C for the production of ZEN [58].

In general, it is postulated that *Fusarium* fungi and subsequent mycotoxins occurrence is higher in colder regions. We observed the trend in the incidence of *Fusarium* toxins in this study with approximately colder region having multiple mycotoxins (Table 3). Further analysis to assess the significant difference in the distribution of *Fusarium* mycotoxins across the AEZ was done using the Kruskal Wallis test (Figure 1). With respect to maize, there was a significant difference in FB_1 and DAS contamination across the AEZ, whereas in millet samples, DAS contamination was significantly different across the different zones. The less difference observed across the AEZ in the current study may be attributed to the quality of the sampled food products because of the premium placed on high quality (visual observation) food products which arises through sorting and cleaning in order to add more value to the products, thus increasing the market value of the food product.

Table 3. *Fusarium* mycotoxins occurrence in cereals (maize, sorghum, and millet) across the different agro ecological zones of Nigeria.

Mycotoxin [1]	Maize (μg/kg)								Sorghum (μg/kg)						Millet (μg/kg)					
	DS (n = 30)		SGS (n = 36)		NGS (n = 40)		SS (n = 30)		SGS (n = 30)		NGS (n = 40)		SS (n = 40)		SGS (n = 30)		NGS (n = 30)		SS (n = 27)	
	Mean [2] (% + ve) [3]	Max [4]	Mean (% + ve)	Max	Mean (% + ve)	Max	Mean (% + ve)	Max	Mean (% + ve)	Max	Mean (% + ve)	Max	Mean (% + ve)	Max	Mean (% + ve)	Max	Mean (% + ve)	Max	Mean (% + ve)	Max
FB_1	117 (67)	366	249 (33)	876	928 (83)	8222	505 (77)	2443	70 (7)	71	59 (10)	76	67 (8)	78	3700 (17)	18,172	54 (10)	84	na	na
FB_2	289 (57)	1011	350 (28)	677	508 (65)	2885	295 (70)	1107	na	na	41 (3)	41	55 (3)	55	417 (37)	3892	56 (13)	103	44 (7)	47
FB_3	114 (47)	353	147 (25)	445	126 (53)	441	91 (50)	213	na	na	31 (3)	31	46 (3)	46	na	na	na	na	na	na
DON	78 (27)	147	99 (17)	180	98 (10)	151	140 (13)	225	119 (3)	119	91 (5)	92	na	na	140 (7)	200	171 (20)	543	118 (11)	118
15 ADON	na [5]	na	na	na	na	na	na	na	34 (3)	34	44 (3)	44	na	na	na	na	na	na	11 (4)	11
DON-3G	na	na	na	na	na	na	na	na	12 (27)	16	30 (40)	63	22 (3)	22	na	na	na	na	na	na
ZEN	na	na	na	na	65 (3)	65	na	na	38 (3)	38	na	na	na	na	481 (33)	1399	109 (7)	198	na	na
ZEN-14G	20 (23)	24	21 (6)	22	23 (8)	23	na	na	20 (3)	20	19 (5)	22	na	na	29 (7)	34	19(7)	20	23 (4)	23
α-ZEL	na	na	20 (3)	20	na	na	na	na	na	na	33 (8)	33	na	na	na	na	na	na	na	na
β-ZEL	20 (3)	20	na	na	21 (3)	21	na	na	21 (3)	21	na	na	na	na	na	na	39	39	na	na
HT-2	20 (3)	20	na	na	na	na	na	na	19 (17)	19	24 (8)	31	11 (3)	11	35 (7)	35	36 (7)	36	na	na
NIV	228 (7)	271	na	na	na	na	163 (3)	163	na	na	na	na	na	na	na	na	na	na	na	na
FUS-X	154 (3)	154	na	na	na	na	na	na	na	na	na	na	na	na	na	na	na	na	na	na
DAS	2 (10)	2	3 (17)	6	3 (18)	4	8 (7)	8	5 (27)	13	4 (13)	5	5 (18)	16	12 (17)	25	4 (37)	6	3 (33)	4

Abbreviation: DS = Derived Savanna; SGS = Southern Guinea Savanna; NGS = Northern Guinea Savanna; SS = Sudan Savanna. [1] FB_1, B_2, and B_3 = fumonisin B_1, B_2, and B_3; DON = deoxynivalenol; 15ADON = 15-acetyl-deoxynivalenol; DON-3G = deoxynivalenol-3-glucoside; ZEN = zearalenone; α-ZEL = α-zearalenol; β-ZEL = β-zearalenol; ZEN-14G = zearalenone-14-glucoside; NIV = nivalenol; FUS-X = fusarenon-X; HT-2 = HT-2 toxin; DAS = diacetoxyscirpenol; [2] Mean = mean concentration, [3] % + ve = percentage positive samples, [4] Max = maximum concentration, [5] na = not applicable.

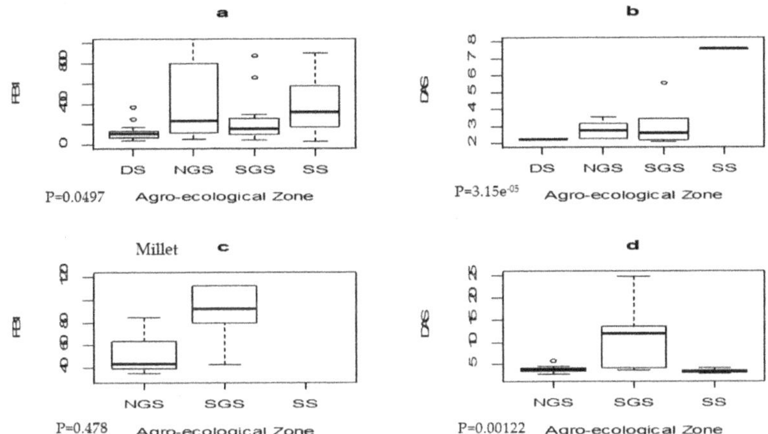

Figure 1. Differences in *Fusarium* mycotoxins in maize ((**a**) fumonisin B_1; and (**b**) diacetoxyscirpenol); and millet ((**c**) fumonisin B_1; and (**d**) diacetoxyscirpenol) across agro-ecological zones.

2.4. Co-Occurrence of *Fusarium* Mycotoxins in Cereals and Processed Products (*Ogi*) from Nigeria

Occurrence of multiple mycotoxins in food, especially cereals and cereal-based products has been an issue of great concern because of the synergistic and/or additive effects caused by the interaction of these toxins in humans and animals. In this study, we observed that 60%, 19%, 30%, and 93% of maize, sorghum, millet, and *ogi*, respectively, were contaminated with at least two mycotoxins (Figure 2). In maize samples, FB co-existed with DAS at 13%, being the highest level observed, followed by co-occurrence between FB and DON (11%). DAS and DON-3G co-occurred in 16% of sorghum samples, succeeded by DON-3G and HT2, while DAS and ZEN dominated in millet samples followed by DAS and DON. Previous studies have shown the frequent co-contamination of *Fusarium* toxins in cereal products in sub-Saharan Africa [32,34,36], which is linked to co-occurrence of several species of *Fusarium* fungi in crops or the potential ability of one *Fusarium* spp. to produce more than one mycotoxins [59]. However, the rates observed in sorghum and millet in the present study were in contrast with the data reported on co-occurrence of mycotoxins in sorghum (94%) and millet (85%) from South Korean retail markets [60], and may be attributed to the different sampling regions considered for these studies.

Although only limited studies exist on the co-occurrence of DON, ZEN, and their modified forms in sub-Saharan Africa, available data proves the potential of DON co-occurring with its modified forms in cereals from this region [34]. This was not the case in this study. The disparity may be attributed to the differences in sampling protocol. Ediage et al. [34] analysed household samples, while samples from the markets were analysed in the present study. Cereals purchased from the market often are of better quality compared to those from the household or at farmers disposal. Ironically, this is because high-graded cereal grains are often placed for sale by the farmers because of a better market bargain, while they retain the poor quality cereal grains since there are no provisional channels for destruction or diversion of these products.

Interestingly, the processed cereal-based products (*ogi*) analysed in this study had the highest co-occurrence rate compared to the cereals. This observation could be attributed to many factors (such as poor sanitation during processing). *Ogi* is fermented under uncontrolled conditions, and as such could be contaminated with any form of organism residing in the environment. Another possible avenue for contamination is the quality of the raw cereal used in the production of *ogi*. Although reduction of mycotoxin concentrations have been reported during processing of *ogi* [61], it is important to mention that reduction of mycotoxins during food processing is dependent on the initial concentrations of the raw produce and as such good quality cereal grains should be used for production of *ogi*. Up to 23% of *ogi* samples were contaminated with at least five mycotoxins, with the highest co-occurrence existing between FB and DON-3G, and subsequently between FB and DON. DON and DON-3G co-occurred in four of the samples while ZEN, ZEN-14G, α-ZEL, and β-ZEL co-occurred in one sample. This result is affirmed by the previous study on co-occurrence of these metabolites in cereal-based products [5]. However, the rates and concentration levels in their study were higher than the levels observed in this study.

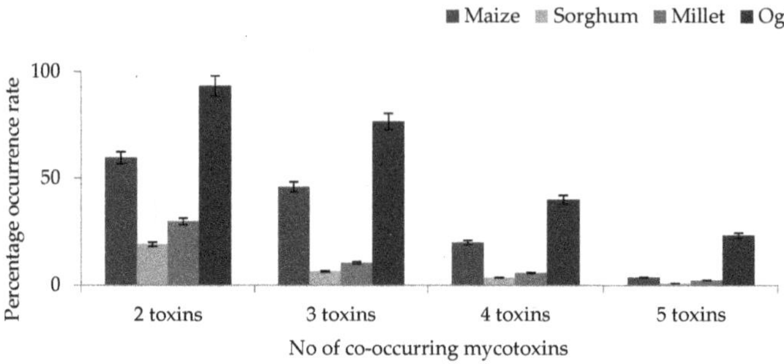

Figure 2. Percentage of co-occurrence of *Fusarium* mycotoxins in maize, sorghum, millet, and *ogi* from Nigerian markets.

3. Conclusions

The present study showed the occurrence of *Fusarium* toxins in cereals and cereal-based fermented products (*ogi*) in Nigeria. Fumonisins were the most dominating *Fusarium* mycotoxins with some of the samples exceeding the FB regulatory limits set by the EU. Although the levels of other mycotoxins detected in the samples were low, the co-occurrence of these mycotoxins presents a health risk due to the synergistic and/or additive effect, considering the fact that these food products are consumed almost on daily basis. In addition, it is worrisome that the traditional weaning food fed to a large population of infant and growing children in this region contains high levels of FB with a cocktail of other mycotoxins. This phenomenon shows that infants are affected with different variation of toxic effects. Reducing mycotoxins in *ogi* is important, and mainly determined by the quality of the raw cereal before fermentation. Therefore, there is need to educate the small-scale producers on the risk of mycotoxins and possible ways to reduce or avoid contamination. Furthermore, it is worth mentioning that though all cereals were contaminated with varying degrees of toxins, the incidence rate was higher in maize, suggesting sorghum and millet as a possible alternative. These results suggest the need for a national coordinated food safety action plan.

4. Materials and Methods

4.1. Sampling

Nigeria has seven agro-ecological zones (AEZ) (Figure 3) based on the climatic and environmental conditions, out of which four AEZ were selected for sampling which include the Derived Savanna (DS), Southern Guinea Savanna (SGS), Northern Guinea Savanna (NGS), and Sudan Savanna (SS). The sampling sites were considered on the production areas of the crops and products. One state from each AEZ was covered for each crop and product. Maize samples were collected from four AEZ while millet and sorghum were sampled from SGS, NGS, and SS; and *ogi* samples were collected from DS. Table 4 shows the detailed sampling sites.

The geographical location, temperature, and rainfall pattern of DS, SGS, NGS, and SS are documented by Udoh et al. and Atehnkeng et al. [62,63]. Briefly, the DS is characterized by an annual rainfall distribution and temperature ranging between 1300 mm and 1500 mm, and 25 °C to 35 °C, respectively. The average annual rainfall of SGS is between 1000 mm and 1300 mm with a temperature range of 26 to 38 °C. The NGS has rainfall distribution averaging between 900 mm and 1000 mm annually, and temperatures vary from 28 to 40 °C, while SS is characterised by an annual rainfall distribution between 650 mm and 1000 mm annually, and temperatures varying from 30 to 40 °C.

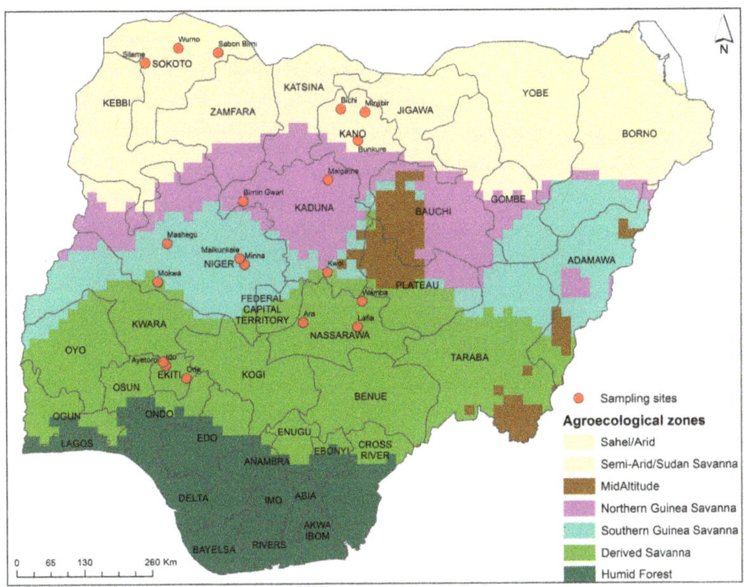

Figure 3. Map of Nigeria showing the sampling sites of maize, sorghum, millet, and *ogi*.

Table 4. Sampling sites of maize, sorghum, millet, and *ogi* from different agro-ecological zones of Nigeria.

Product Type	AEZ	State	No. of Markets	No. of Samples
Maize	SS	Kano	6	30
	NGS	Kaduna	6	36
	SGS	Niger	8	40
	DS	Nasarawa	6	30
				Total number = 136
Sorghum	SS	Kano	6	30
	NGS	Kaduna	2	40
	SGS	Niger	8	40
				Total number = 110
Millet	SS	Sokoto	6	30
	NGS	Kaduna	6	30
	SGS	Niger	8	27
				Total number = 87
Ogi	DS	Ekiti	5	30
				Total number = 30

Abbreviation: AEZ = agro-ecological zones, SS = Sudan Savanna, NGS = Northern Guinea Savanna, SGS = Southern Guinea Savanna, DS = Derived Savanna.

A total of 363 samples comprising of maize (*Zea mays*) (n = 136), sorghum (*Sorghum bicolor*) (n = 110), millet (*Pennisetum glaucum*) (n = 87), and *ogi* (n = 30) were collected from randomly selected markets between September 2015 and October 2015. All cereal samples were sorted and cleaned and had no visible mould before going to the market. Sampling was carried out as described by Adetunji et al. and European Commission Regulation with some modifications [32,64]. Briefly, the whole content of a traditional bag of 50 kg of maize, sorghum, and millet was considered as a lot. An aggregate sample size of 1 kg was composed of 5 incremental samples. Each incremental portion was about 200 g, and one was taken from different positions in the bag. In the case of *ogi*, 300 g of sample was collected from the top, middle, and bottom portions of the jute sack of 10 kg *ogi*, thoroughly mixed together, air dried, and packed in a plastic container. Each sample of the different food products (maize, sorghum, millet, and *ogi*) was thoroughly homogenised, and a representative portion of 200 g was taken, labelled, and sealed in zip lock bags, and transported to the Laboratory of Food Analysis, Ghent University, Belgium (Ottergemsesteenweg 460, Ghent, Belgium) for further analysis. Prior to analysis, maize, sorghum, millet, and *ogi* samples were milled to a sieve size of 0.5–1 mm using an IKA M20 universal mill (Sigma-Aldrich, Bornem, Belgium) and stored at −20 °C.

4.2. Chemicals and Reagents

Methanol (MeOH, LC-MS grade), glacial acetic acid (LC-MS grade), and analytical grade acetonitrile were purchased from Biosolve B.V. (Valkenswaard, The Netherlands). Analytical grade acetic acid, ammonium acetate and sodium chloride were obtained from Merck (Darmstadt, Germany). Analytical grade of n-hexane and methanol, Whatman® glass microfiber filters (GFA, 47 mm diameter) were purchased from VWR International (Zaventem, Belgium). Ultrafree®-MC centrifugal filter devices (0.22 μm) were obtained from Millipore (Bredford, MA, USA). C18 solid phase extraction (SPE) columns and MultiSep®226 AflaZon+ multifunctional columns were purchased from Alltech (Lokeren, Belgium) and Romer Labs (Gernsheim, Germany), respectively. Water was purified using a Milli-Q Gradient System (Millipore; Brussels, Belgium). All other chemicals and reagents used were of analytical grade.

The analytical mycotoxin standards including FB_1, FB_2, DON, 3ADON, 15ADON, deepoxy-deoxynivalenol (DOM), FUS-X, NIV, HT-2, NEO, ZEN, zearalanone (ZAN), α-ZEL, β-ZEL were purchased from Sigma-Aldrich (Bornem, Belgium). FB_3 was obtained from Promec Unit (Tygerberg, South Africa). DAS, DON-3G and T-2 were purchased from Biopure Referenzsubstanze (Tulln, Austria) while ZEN-14G was synthesized via an in-house validated method according to Zill et al. [65]. Stock solutions of FB_1, FB_2, FB_3, DON, 3-ADON, 15-ADON, HT-2, T-2, ZEN, ZAN, α-ZEL, β-ZEL, FUS-X, NIV, NEO, and DAS were prepared in MeOH at a concentration of 1 mg/mL. DOM (50 μg/mL), DON-3G (50.2 μg/mL) and ZEN-14G (100 μg/mL) were obtained as solution in acetonitrile. All stock solutions were stored for 1 year or until the expiration date at −18 °C. The working standard solutions were made by diluting the stock standard solutions in methanol, and were stored at −18 °C for 3 months. From the individual stock and working standard solutions, a standard mixture was prepared in methanol at the following concentrations: FB_1, FB_2, and DON (40 ng/μL), FB_3 (25 ng/μL), 3-ADON (5 ng/μL), 15-ADON (2.5 ng/μL), HT-2, T-2, ZEN, α-ZEL, β-ZEL, ZEN-14G, DON-3G, and NEO (10 ng/μL), FUS-X and NIV (20 ng/μL), and DAS (0.5 ng/μL). The mixture was stored at −18 °C, and renewed every 3 months.

4.3. Sample Extraction

Sample preparation for 17 *Fusarium* mycotoxins was carried out as described by Monbaliu et al. [66] for multi-mycotoxin analysis. Briefly, 5 g of sample was spiked with internal standards (ZAN and DOM at a concentration of 250 and 150 μg/kg, respectively). DOM was used as internal standard for DON, DON-3G, 3ADON, and 15ADON while ZAN was used for the other mycotoxins. Spiked sample was kept in the dark for 15 min and extracted with acetonitrile/water/acetic acid (79/20/1, *v*/*v*/*v*). The supernatant was passed through a preconditioned C18-SPE column, and the extract was defatted. In order to recover the 17 *Fusarium* mycotoxins, two clean-up pathways were followed. Firstly, 12.5 mL of the defatted extract was added to 27.5 mL of acetonitrile/acetic acid (99/1, *v*/*v*), and passed through a MultiSep®226 AflaZon+ multifunctional column. In the second pathway, 10 mL of defatted extract was filtered using a glass microfilter. Two milliliters of the filtered extract were combined with the Multisep226 eluate and evaporated to dryness. The residue was then redissolved in 150 µL of the mobile phase (water/methanol/acetic acid (94/5/1, *v*/*v*/*v*) + 5 mM ammonium acetate and water/methanol/acetic acid (2/97/1, *v*/*v*/*v*) + 5 mM ammonium acetate in the ratio of 3/2, *v*/*v*). The redisolved mixture was then filtered using the Ultrafree® PVDF centrifuge filters and analysed by LC-MS/MS.

4.4. Preparation of the Hydrolysed FB_1, FB_2, and FB_3 Standard

The hydrolysed FB standards were prepared as described by Dall Asta et al. [24]. Briefly, 90 µL standard solution of FB_1, FB_2, and FB_3 (50 µg/mL of each) was prepared in methanol, transferred to a 10 mL sovirel tube, and evaporated to dryness under a gentle stream of nitrogen. The residue was redissolved in 2 M NaOH (1 mL), and the reaction was incubated overnight at room temperature. After hydrolysis, the mixture was extracted by liquid-liquid partition using acetonitrile (1 mL). The extraction was repeated two more times using 1 mL fresh acetonitrile. The organic phases were pooled together, evaporated under a nitrogen stream and redissolved in 1 mL of methanol. The reaction yield was checked by LC-MS, by monitoring the conversion of FB to hydrolysed FB and the absence of side-products, and it was found to be higher than 99%. Calibration curves were prepared by proper dilution of the standard solution, assuming the total conversion of the native compounds to the hydrolysed forms.

4.5. Sample Preparation for Hydrolysed Fumonisins

Ten samples from each matrix were selected for analysis of hydrolysed fumonisin; five samples comprising of high FB contaminated samples, while the second five samples were comprised of negative samples. Sample preparation was carried out as described by Dall Asta et al. [24]. Two and half grams of each sample was weighed, and was hydrolysed at room temperature using 50 mL of 2 M NaOH, homogenised with the Ultraturrax for 3 min, and then stirred using a magnetic stirrer for 60 min. Subsequently, 50 mL of acetonitrile was added, and homogenised for 3 min, and 20 mL of the upper organic phase was transferred into a centrifuge tube, and centrifuged for 15 min at 3500 rpm. Then, 4 mL was evaporated, and the residue was redissolved in 1 mL of water/methanol (30/70, *v*/*v*), and analysed by LC-MS/MS.

4.6. Liquid Chromatography-Tandem Mass Spectrometry

A Waters Acquity UPLC system coupled to a Micromass Quattro Micro triple-quadrupole mass spectrometer (Waters, Milford, MA, USA) was used for the detection and quantification of mycotoxins in the samples. The data acquisition and processing utilities included the use of the MassLynx™ version 4.1 and QuanLynx® version 4.1 software (Micromass, Manchester, UK). The column used was a Symmetry C18 (150 mm × 2.1 mm i.d. 5 µm) column with a guard column (10 mm × 2.1 mm i.d.) of the same material (Waters, Zellik, Belgium), and was kept at room temperature. The injection volume was 20 µL. Mobile phase consisting of water/methanol/acetic acid (94/5/1, *v*/*v*/*v*) and 5 mM ammonium acetate (mobile phase A), and methanol/water/acetic acid (97/2/1, *v*/*v*/*v*) and 5 mM ammonium acetate (mobile phase B) were used at a flow rate of 0.3 mL/min with a gradient elution program. The gradient started at 95% mobile phase A with a linear decrease to 35% in 7 min. Mobile phase A decreased to 25% at 4 min, and an isocratic period of 100%, mobile phase B started at 11 min for 2 min. Initial column conditions were reached at 23 min using a linear decrease of mobile phase B, and the column was reconditioned for 5 min prior to the following injection. The mass spectrometer was operated using selected reaction monitoring (SRM) channels in positive electrospray ionization (ESI+) mode. More information and on the transition of the different mycotoxins are described by

De Boevre et al. [41] and Monbaliu et al. [67]. The capillary voltage was 3.2 kV, and nitrogen was used as the desolvation gas. Source and desolvation temperatures were set at 150 and 350 °C, respectively.

4.7. Method Validation

For each matrix, validation parameters including apparent recovery, limit of detection (LOD), limit of quantification (LOQ) and measurement uncertainty were evaluated. Five blank samples of each matrix were spiked in triplicate with the different mycotoxins (FB_1, FB_2 and FB_3, ZEN, α-ZEL, β-ZEL, ZEN-14G, DON, 3ADON, 15ADON, DON-3G, NIV, FUS-X, T-2, HT-2, DAS, NEO) on three different days at concentration levels shown on Table S1. ZAN and DOM were added as internal standards (IS). Matrix-matched calibration (MMC) plots were constructed by applying the least-squares method, and by plotting the relative peak area (peak area of toxin/peak area of IS) against the spiked concentration level of the sample. The linearity was evaluated graphically using a scatter plot, and the linear regression model was tested using a lack-of-fit test. The apparent recovery for each of the mycotoxins was evaluated by dividing the observed value from the MMC curves by the spiked level. The obtained results ranged between 75% and 110%, which is in conformity with the range set in legislation [64].

A precision study with regards to repeatability (intraday precision) and reproducibility (interday precision) within laboratory was performed using five concentration levels, and was calculated using relative standard deviation (RSD). The results are presented in Table S1. LOD and LOQ for individual mycotoxins were obtained from the signal-to-noise (S/N) ratio which have been defined and set as 3 and 10, respectively, by the International Union of Pure and Applied Chemistry (IUPAC). The LOQs of individual mycotoxins in maize, sorghum, and millet are shown in Table S2.

4.8. Statistical Analysis

Data were processed and calculated using Microsoft office Excel 2007 (Redmond, WA, USA) and the statistical package R version 3.0.3 (The R Foundation for Statistical Computing, Vienna, Austria). The Kruskal Wallis statistical test (an alternative to one-way analysis of variance (ANOVA) under nonparametric test) was used to assess the significance of the differences between the determined mycotoxin concentrations and the different AEZ.

Supplementary Materials: The following are available online at www.mdpi.com/2072-6651/8/11/342/s1, Table S1: Intraday repeatability (RSDr, %) and interday reproducibility (RSDR, %) (within laboratory) of the individual mycotoxins for maize, sorghum and millet. Table S2: Limits of detection (LOD, μg/kg) and limits of quantification (LOQ, μg/kg) of the individual mycotoxins for maize, sorghum, and millet.

Acknowledgments: The authors would like to thank Ghent University Special Research Fund (BOF 01W01014) for the financial support.

Author Contributions: C.A.C. designed the experiments, performed the sampling survey and the experiments. C.A.C. prepared the manuscript. M.D.B., O.O.A. and S.D.S. supervised the research, edited and approved the final manuscript.

Conflicts of Interest: The authors declare no conflict of interest.

References

1. Bennett, J.W.; Klich, M. Mycotoxins. *Clin. Microbiol. Rev.* **2003**, *16*, 497–516. [CrossRef] [PubMed]
2. Council for Agricultural Science and Technology (CAST). *Mycotoxins: Risks in Plant, Animal, and Human Systems*; Task force Report No. 139; Council for Agricultural Science and Technology: Ames, IA, USA, 2003.
3. Yazar, S.; Omurtag, G.Z. Fumonisins, trichothecenes and zearalenone in cereals. *Int. J. Mol. Sci.* **2008**, *9*, 2062–2090. [CrossRef] [PubMed]
4. International Agency for Research on Cancer (IARC). IARC monographs on the evaluation of carcinogenic risk to humans. *IARC* **1993**, *56*, 1–521.

5. De Boevre, M.; Di Mavungu, J.D.; Landschoot, S.; Audenaert, K.; Eeckhout, M.; Maene, P.; Haesaert, G.; De Saeger, S. Natural occurrence of mycotoxins and their masked forms in food and feed products. *World Mycotoxin J.* **2012**, *5*, 207–219. [CrossRef]
6. Rodriguez-Carrasco, Y.; Fattore, M.; Albrizio, S.; Berrada, H.; Manes, J. Occurrence of Fusarium mycotoxins and their dietary intake through beer consumption by the European population. *Food Chem.* **2015**, *178*, 149–155. [CrossRef] [PubMed]
7. De Angelis, E.; Monaci, L.; Pascale, M.; Visconti, A. Fate of deoxynivalenol, T-2 and HT-2 toxins and their glucoside conjugates from flour to bread: An investigation by high-performance liquid chromatography high-resolution mass spectrometry. *Food Addit. Contam. Part A* **2012**, *30*, 345–355. [CrossRef] [PubMed]
8. Stanciu, O.; Juan, C.; Miere, D.; Loghin, F.; Mañes, J. Occurrence and co-occurrence of *Fusarium* mycotoxins in wheat grains and wheat flour from Romania. *Food Control* **2016**, in press. [CrossRef]
9. Speijers, G.J.A.; Speijers, M.H.M. Combined toxic effects of mycotoxins. *Toxicol. Lett.* **2004**, *153*, 91–98. [CrossRef] [PubMed]
10. Kouadio, J.H.; Dano, S.D.; Moukha, S.; Mobio, T.A.; Creppy, E.E. Effects of combinations of *Fusarium* mycotoxins on the inhibition of macromolecular synthesis, malondialdehyde levels, {DNA} methylation and fragmentation, and viability in Caco-2 cells. *Toxicon* **2007**, *49*, 306–317. [CrossRef] [PubMed]
11. Alassane-Kpembi, I.; Kolf-Clauw, M.; Gauthier, T.; Abrami, R.; Abiola, F.A.; Oswald, I.P.; Puel, O. New insights into mycotoxin mixtures: The toxicity of low doses of Type B trichothecenes on intestinal epithelial cells is synergistic. *Toxicol. Appl. Pharmacol.* **2013**, *272*, 191–198. [CrossRef] [PubMed]
12. Luongo, D.; De Luna, R.; Russo, R.; Severino, L. Effects of four *Fusarium* toxins (fumonisin B_1, α-zearalenol, nivalenol and deoxynivalenol) on porcine whole-blood cellular proliferation. *Toxicon* **2008**, *52*, 156–162. [CrossRef] [PubMed]
13. Harvey, R.B.; Edrington, T.S.; Kubena, L.F.; Elissalde, M.H.; Casper, H.H.; Rottinghaus, G.E.; Turk, J.R. Effects of dietary fumonisin B_1-containing culture material, deoxynivalenol-contaminated wheat, or their combination on growing barrows. *Am. J. Vet. Res.* **1996**, *57*, 1790–1794. [PubMed]
14. Kubena, L.F.; Edrington, T.S.; Harvey, R.B.; Phillips, T.D.; Sarr, A.B.; Rottinghaus, G.E. Individual and combined effects of fumonisin B_1 present in *Fusarium moniliforme* culture material and diacetoxyscirpenol or ochratoxin A in turkey poults. *Poult. Sci.* **1997**, *76*, 256–264. [CrossRef] [PubMed]
15. Alassane-Kpembi, I.; Puel, O.; Oswald, I.P. Toxicological interactions between the mycotoxins deoxynivalenol, nivalenol and their acetylated derivatives in intestinal epithelial cells. *Arch. Toxicol.* **2015**, *89*, 1337–1346. [CrossRef] [PubMed]
16. Rychlik, M.; Humpf, H.U.; Marko, D.; Dänicke, S.; Mally, A.; Berthiller, F.; Klaffke, H.; Lorenz, N. Proposal of a comprehensive definition of modified and other forms of mycotoxins including "masked" mycotoxins. *Mycotoxin Res.* **2014**, *30*, 197–205. [CrossRef] [PubMed]
17. Berthiller, F.; Crews, C.; Dall'Asta, C.; De Saeger, S.; Haesaert, G.; Karlovsky, P.; Oswald, I.P.; Seefelder, W.; Speijers, G.; Stroka, J. Masked mycotoxins: A review. *Mol. Nutr. Food Res.* **2013**, *57*, 165–186. [CrossRef] [PubMed]
18. Broekaert, N.; Devreese, M.; De Baere, S.; De Backer, P.; Croubels, S. Modified *Fusarium* mycotoxins unmasked: From occurrence in cereals to animal and human excretion. *Food Chem. Toxicol.* **2015**, *80*, 17–31. [CrossRef] [PubMed]
19. De Boevre, M; Jacxsens, L.; Lachat, C.; Eeckhout, M.; Di Mavungu, D.M.; Audenaert, K.; Maene, P.; Haesaert, G.; Kolsteren, P.; De Meulenaer, B.; et al. Human exposure to mycotoxins and their masked forms through cereal-based foods in Belgium. *Toxicol. Lett.* **2013**, *218*, 281–292. [CrossRef] [PubMed]
20. Oliveira, M.S.; Diel, A.C.L.; Rauber, R.H.; Fontoura, F.P.; Mallmann, A.; Dilkin, P.; Mallmann, C.A. Free and hidden fumonisins in Brazilian raw maize samples. *Food Control* **2015**, *53*, 217–221. [CrossRef]
21. Dall'Asta, C.; Galaverna, G.; Mangia, M.; Sforza, S.; Dossena, A.; Marchelli, R. Free and bound fumonisins in gluten-free food products. *Mol. Nutr. Food Res.* **2009**, *53*, 492–499. [CrossRef] [PubMed]
22. Dall'Asta, C.; Mangia, M.; Berthiller, F.; Molinelli, A.; Sulyok, M.; Schuhmacher, R.; Krska, R.; Galaverna, G.; Dossena, A.; Marchelli, R. Difficulties in fumonisin determination: The issue of hidden fumonisins. *Anal. Bioanal. Chem.* **2009**, *395*, 1335–1345. [CrossRef] [PubMed]
23. Galaverna, G.; Dallsta, C.; Mangia, M.A.; Dossena, A.; Marchelli, R. Masked mycotoxins: An emerging issue for food safety. *Czech J. Food Sci.* **2009**, *27*, 89–92.

24. Dall'Asta, C.; Galaverna, G.; Aureli, G.; Dossena, A.; Marchelli, R. A LC/MS/MS method for the simultaneous quantification of free and masked fumonisins in maize and maize-based products. *World Mycotoxin J.* **2008**, *1*, 237–246. [CrossRef]
25. Ezekiel, C.N.; Odebode, A.C.; Fapohunda, S.O. Zearalenone production by naturally occurring *Fusarium* species on maize, wheat and soybeans from Nigeria. *J. Biol. Environ. Sci.* **2008**, *2*, 77–82.
26. Adejumo, T.O.; Hettwer, U.; Karlovsky, P. Occurrence of *Fusarium* species and trichothecenes in Nigerian maize. *Int. J. Food Microbiol.* **2007**, *116*, 350–357. [CrossRef] [PubMed]
27. European Commission (EC). Comission regulation (EC) No 1126/2007 of 28 September 2007 amending Regulation (EC) No 1881/2006 setting maximum levels for certain contaminants in foodstuffs as regards *Fusarium* toxins in maize and maize products. *Off. J. Eur. Union* **2007**, *L 255*, 14–17.
28. Kpodo, K.; Thrane, U.; Hald, B. Fusaria and fumonisins in maize from Ghana and their co-occurrence with aflatoxins. *Int. J. Food Microbiol.* **2000**, *61*, 147–157. [CrossRef]
29. Warth, B.; Parich, A.; Atehnkeng, J.; Bandyopadhyay, R.; Schuhmacher, R.; Sulyok, M.; Krska, R. Quantitation of mycotoxins in food and feed from burkina faso and mozambique using a modern LC-MS/MS multitoxin method. *J. Agric. Food Chem.* **2012**, *60*, 9352–9363. [CrossRef] [PubMed]
30. Phoku, J.Z.; Dutton, M.F.; Njobeh, P.B.; Mwanza, M.; Egbuta, M.A.; Chilaka, C.A. *Fusarium* infection of maize and maize-based products and exposure of a rural population to fumonisin B_1 in Limpopo Province, South Africa. *Food Addit. Contam. Part A Chem. Anal. Control Expo. Risk Assess.* **2012**, *29*, 1743–1751. [CrossRef] [PubMed]
31. Abia, W.A.; Warth, B.; Sulyok, M.; Krska, R.; Tchana, A.N.; Njobeh, P.B.; Dutton, M.F.; Moundipa, P.F. Determination of multi-mycotoxin occurrence in cereals, nuts and their products in Cameroon by liquid chromatography tandem mass spectrometry (LC-MS/MS). *Food Control* **2013**, *31*, 438–453. [CrossRef]
32. Adetunji, M.; Atanda, O.; Ezekiel, C.N.; Sulyok, M.; Warth, B.; Beltrán, E.; Krska, R.; Obadina, O.; Bakare, A.; Chilaka, C.A. Fungal and bacterial metabolites of stored maize (*Zea mays*, L.) from five agro-ecological zones of Nigeria. *Mycotoxin Res.* **2014**, *30*, 89–102. [CrossRef] [PubMed]
33. Peters, J.; Thomas, D.; Boers, E.; De Rijk, T.; Berthiller, F.; Haasnoot, W.; Nielen, M.W.F. Colour-encoded paramagnetic microbead-based direct inhibition triplex flow cytometric immunoassay for ochratoxin A, fumonisins and zearalenone in cereals and cereal-based feed rapid detection in food and feed. *Anal. Bioanal. Chem.* **2013**, *405*, 7783–7794. [CrossRef] [PubMed]
34. Ediage, E.N.; Hell, K.; De Saeger, S. A comprehensive study to explore differences in mycotoxin patterns from agro-ecological regions through maize, peanut, and cassava products: A case study, cameroon. *J. Agric. Food Chem.* **2014**, *62*, 4789–4797. [CrossRef] [PubMed]
35. Chilaka, C.A.; de Kock, S.; Phoku, J.Z.; Mwanza, M.; Egbuta, M.A.; Dutton, M.F. Fungal and mycotoxin contamination of South African commercial maize. *J. Food Agric. Environ.* **2012**, *10*, 296–303.
36. Chala, A.; Taye, W.; Ayalew, A.; Krska, R.; Sulyok, M.; Logrieco, A. Multimycotoxin analysis of sorghum (*Sorghum bicolor* L. Moench) and finger millet (*Eleusine coracana* L. Garten) from Ethiopia. *Food Control* **2014**, *45*, 29–35. [CrossRef]
37. Ayalew, A.; Fehrmann, H.; Lepschy, J.; Beck, R.; Abate, D. Natural occurrence of mycotoxins in staple cereals from Ethiopia. *Mycopathologia* **2006**, *162*, 57–63. [CrossRef] [PubMed]
38. Soares, C.; Calado, T.; Venâncio, A. Mycotoxin production by *Aspergillus niger* aggregate isolated from harvested maize in three Portuguese regions. *Rev. Iberoam. Micol.* **2012**, *30*, 9–13. [CrossRef] [PubMed]
39. Ezekiel, C.N.; Abia, W.A.; Ogara, I.M.; Sulyok, M.; Warth, B.; Krska, R. Fate of mycotoxins in two popular traditional cereal-based beverages (kunu-zaki and pito) from rural Nigeria. *LWT Food Sci. Technol.* **2015**, *60*, 137–141. [CrossRef]
40. Makun, H.A.; Dutton, M.F.; Njobeh, P.B.; Mwanza, M.; Kabiru, A.Y. Natural multi-occurrence of mycotoxins in rice from Niger State, Nigeria. *Mycotoxin Res.* **2011**, *27*, 97–104. [CrossRef] [PubMed]
41. De Boevre, M.; Di Mavungu, J.D.; Maene, P.; Audenaert, K.; Deforce, D.; Haesaert, G.; Eeckhout, M.; Callebaut, A.; Berthiller, F.; Van Peteghem, C.; et al. Development and validation of an LC-MS/MS method for the simultaneous determination of deoxynivalenol, zearalenone, T-2-toxin and some masked metabolites in different cereals and cereal-derived food. *Food Addit. Contam. Part A* **2012**, *29*, 819–835. [CrossRef] [PubMed]

42. Mirocha, C.J.; Abbas, H.K.; Windels, C.E.; Xie, W. Variation in deoxynivalenol, 15-acetyldeoxynivalenol, 3-acetyldeoxynivalenol, and zearalenone production by *Fusarium graminearum* isolates. *Appl. Environ. Microbiol.* **1989**, *55*, 1315–1316. [PubMed]
43. Abedi-tizaki, M.; Sabbagh, S. Detection of 3-Acetyldeoxynivalenol, 15-Acetyldeoxynivalenol and Nivalenol-Chemotypes of *Fusarium graminearum* from Iran using specific PCR assays a b c d. *Plant Knowl. J.* **2013**, *2*, 38–42.
44. Li, H.-P.; Wu, A.-B.; Zhao, C.-S.; Scholten, O.; Löffler, H.; Liao, Y.-C. Development of a generic PCR detection of deoxynivalenol- and nivalenol-chemotypes of *Fusarium graminearum*. *FEMS Microbiol. Lett.* **2005**, *243*, 505–511. [CrossRef] [PubMed]
45. Castillo, M.-Á.; Montes, R.; Navarro, A.; Segarra, R.; Cuesta, G.; Hernández, E. Occurrence of deoxynivalenol and nivalenol in Spanish corn-based food products. *J. Food Compos. Anal.* **2008**, *21*, 423–427. [CrossRef]
46. Gottschalk, C.; Barthel, J.; Engelhardt, G.; Bauer, J.; Meyer, K. Simultaneous determination of type A, B and D trichothecenes and their occurrence in cereals and cereal products. *Food Addit. Contam.* **2009**, *26*, 1273–1289. [CrossRef]
47. Soković, M.; Šimpraga, B. Survey of trichothecene mycotoxins in grains and animal feed in Croatia by thin layer chromatography. *Food Control* **2006**, *17*, 733–740. [CrossRef]
48. European Commission (EC). Commission recomendations on the presence of T-2 and HT-2 toxin in cereals and cereal products. *Off. J. Eur. Union* **2013**, *56*, 12–15.
49. Adejumo, T.O.; Hettwer, U.; Karlovsky, P. Survey of maize from south-western Nigeria for zearalenone, α- and β-zearalenols, fumonisin B_1 and enniatins produced by Fusarium species. *Food Addit. Contam.* **2007**, *24*, 993–1000. [CrossRef] [PubMed]
50. European Commission (EC). Commission regulation (EC) No 1881/2006 Setting maximum levels for certain contaminants in foodstuffs. *Off. J. Eur. Union* **2006**, *L 364/5*, 1–26.
51. De Boevre, M.; Graniczkowska, K.; De Saeger, S. Metabolism of modified mycotoxins studied through in vitro and in vivo models: An overview. *Toxicol. Lett.* **2015**, *233*, 24–28. [CrossRef] [PubMed]
52. Park, J.W.; Scott, P.M.; Lau, B.P.-Y.; Lewis, D.A. Analysis of heat-processed corn foods for fumonisins and bound fumonisins. *Food Addit. Contam.* **2004**, *21*, 1168–1178. [CrossRef] [PubMed]
53. Kim, E.K.; Scott, P.M.; Lau, B.P.Y. Hidden fumonisin in corn flakes. *Food Addit.Contam.* **2003**, *20*, 161–169. [CrossRef] [PubMed]
54. Dall'Asta, C.; Falavigna, C.; Galaverna, G.; Dossena, A.; Marchelli, R. In vitro digestion assay for determination of hidden fumonisins in maize. *J. Agric. Food Chem.* **2010**, *58*, 12042–12047. [CrossRef] [PubMed]
55. Shier, W.T. The fumonisin paradox: A review of research on oral bioavailability of fumonisin B_1, a mycotoxin produced by *Fusarium moniliforme*. *J. Toxicol. Toxin Rev.* **2000**, *19*, 161–187. [CrossRef]
56. Shephard, G.S.; Thiel, P.G.; Stockenstrom, S.; Sydenham, E.W. Worldwide survey of fumonisin concentration of corn and corn-based foods. *J. AOAC Int.* **1996**, *79*, 671–687. [PubMed]
57. Femine Early Warning Systems Network (FEWS NET). Visualizing Trends in 1981–2015 Rainfall in Nigeria. Nigeria Special Report. Available online: www.fews.net/sites/default/files/documents/reports/FEWSNET_NigeriaRainfallTrendsMapBook_20160601 (accessed on 28 September 2016).
58. Milani, J.M. Ecological conditions affecting mycotoxin production in cereals: A review. *Vet. Med. (Praha)* **2013**, *58*, 405–411.
59. Bottalico, A. *Fusarium* diseases of cereals: Species complex and related mycotoxin profiles in Europe. *J. Plant Pathol.* **1998**, *80*, 85–103.
60. Kim, D.-H.; Yoon, B.R.; Jeon, M.-H.; Kim, B.-C.; Cho, B.-I.; Hong, S.-Y.; Hyun, C.S. Simultaneous Determination of Multi-mycotoxins in Cereal Grains by LC-MS/MS. In *International Association for Food Protection*; IAFP: Portland, OR, USA, 2015; pp. P2–P65.
61. Okeke, C.A.; Ezekiel, C.N.; Nwangburuka, C.C.; Sulyok, M.; Ezeamagu, C.O.; Adeleke, R.A.; Dike, S.K.; Krska, R. Bacterial diversity and mycotoxin reduction during maize fermentation (steeping) for *Ogi* production. *Front. Microbiol.* **2015**, *6*, 1–12. [CrossRef] [PubMed]
62. Udoh, J.M.; Cardwell, K.F.; Ikotun, T. Storage structures and aflatoxin content of maize in five agroecological zones of Nigeria. *J. Stored Prod. Res.* **2000**, *36*, 187–201. [CrossRef]

63. Atehnkeng, J.; Ojiambo, P.S.; Donner, M.; Ikotun, T.; Sikora, R.A.; Cotty, P.J.; Bandyopadhyay, R. Distribution and toxigenicity of Aspergillus species isolated from maize kernels from three agro-ecological zones in Nigeria. *Int. J. Food Microbiol.* **2008**, *122*, 74–84. [CrossRef] [PubMed]
64. European Commission (EC). Commission regulation (EC) No 401/2006 of 23 February 2006 laying down the methods of sampling and analysis for the official control of the levels of mycotoxins in foodstuffs. *Off. J. Eur. Union* **2006**, *L70*, 12–34.
65. Zill, G.; Ziegler, W.; Engelhardt, G.; Wallnöfer, P.R. Chemically and biologically synthesized zearalenone-4-β-d-glucopyranoside: Comparison and convenient determination by gradient HPLC. *Chemosphere* **1990**, *21*, 435–442. [CrossRef]
66. Sofie, M.; Van Poucke, C.; Detavernier, C.; Dumoultn, F.; Van Velde, M.D.E.; Schoeters, E.; Van Dyck, S.; Averkieva, O.; Van Peteghem, C.; De Saeger, S. Occurrence of mycotoxins in feed as analyzed by a multi-mycotoxin LC-MS/MS method. *J. Agric. Food Chem.* **2010**, *58*, 66–71.
67. Monbaliu, S.; Van Poucke, C.; Van Peteghem, C.; Van Poucke, K.; Heungens, K.; De Saeger, S. Development of a multi-mycotoxin liquid chromatography/tandem mass spectrometry method for sweet pepper analysis. *Rapid Commun. Mass Spectrom.* **2009**, *23*, 3–11. [CrossRef] [PubMed]

© 2016 by the authors. Licensee MDPI, Basel, Switzerland. This article is an open access article distributed under the terms and conditions of the Creative Commons Attribution (CC BY) license (http://creativecommons.org/licenses/by/4.0/).

Article

Mycotoxin Contamination in Sugarcane Grass and Juice: First Report on Detection of Multiple Mycotoxins and Exposure Assessment for Aflatoxins B_1 and G_1 in Humans

Mohamed F. Abdallah [1,2], Rudolf Krska [2] and Michael Sulyok [2,*]

1 Department of Forensic Medicine and Toxicology, Faculty of Veterinary Medicine, Assiut University, Assiut 71515, Egypt; mohamed.fathi@vet.au.edu.eg

2 Center for Analytical Chemistry, Department of Agrobiotechnology (IFA-Tulln), University of Natural Resources and Life Sciences, Vienna (BOKU), Konrad Lorenz Str 20, Tulln A-3430, Austria; rudolf.krska@boku.ac.at

* Correspondence: michael.sulyok@boku.ac.at; Tel.: +43-1-47654-97312

Academic Editors: Sarah De Saeger, Siska Croubels and Kris Audenaert

Received: 10 September 2016; Accepted: 13 November 2016; Published: 18 November 2016

Abstract: This study was conducted to investigate the natural co-occurrence of multiple toxic fungal and bacterial metabolites in sugarcane grass and juice intended for human consumption in Upper Egypt. Quantification of the target analytes has been done using the "dilute and shoot" approach followed by liquid chromatography-tandem mass spectrometry (LC-MS/MS). A total number of 29 and 33 different metabolites were detected in 21 sugarcane grass and 40 juice samples, respectively, with a trend of concentrations being higher in grass than in juice. Among the regulated mycotoxins, only aflatoxin B_1 (AFB_1) and aflatoxin G_1 (AFG_1) were detected. The prevalence of AFB_1 was in 48% of grass samples and in 58% of juice with a maximum concentration of 30.6 μg/kg and 2.10 μg/kg, respectively. AFG_1 was detected in 10% of grass samples (7.76 μg/kg) and 18% of juice samples (34 μg/kg). Dietary exposure was assessed using a juice frequency questionnaire of adult inhabitants in Assiut City. The assessment revealed different levels of exposure to AFB_1 between males and females in winter and summer seasons. The estimated seasonal exposure ranged from 0.20 to 0.40 ng/kg b.w./day in winter and from 0.38 to 0.90 ng/kg b.w./day in summer.

Keywords: mycotoxins; sugarcane; sugarcane juice; aflatoxins; LC-MS/MS; exposure assessment

1. Introduction

Sugarcane, *Saccharum officinarum*, is a tropical tall perennial grass cultivated in several countries of the world. In Africa, sugarcane is the second most cultivated crop after cassava, where Egypt maintains the second position after South Africa [1]. In Egypt, around 97% of the total sugarcane production, 16 million tons in 2014, is cultivated in the upper part of the country [2,3]. From the economical point of view, sugarcane is an important cash crop beside cotton and a major contributor of income and employment for farmers. The plant has a high sucrose and low fiber content and is used mainly for raw sugar and molasses production (brownish-black viscous syrup known as black honey in Egypt), in addition to the grass left over or bagasse, which is used as an animal feed supplement or fertilizer. Furthermore, several secondary industries such as vinegar, alcohol, chipboard, paper, some chemicals, plastics, paints, fiber, insecticides and detergents are based on sugarcane and its wastes [2,4–6]. It has been estimated that 80% of the world's sugar comes from sugarcane [7,8], where Brazil and India are the largest producers worldwide [4,6]. The annual consumption of sugar in Egypt in 2010 was estimated to be 34 kg per capita [7].

During the harvesting time, chewing raw sugarcane is a common practice. In addition, sugarcane juice is considered the most popular fresh juice in Egypt, with cane juice shops spreading through all the Egyptian cities. Indian and Pakistani people share the same habit with Egyptians regarding chewing raw sugarcane and consumption of juice [9,10]. Apart from its sweet taste and being a source of energy and minerals, sugarcane juice consumption, in traditional medicine, helps in the treatment of many diseases such as jaundice, kidney stones, urogenital tract infections, and in lowering blood pressure, and healing dermal wounds; it is also reported as a natural antioxidant under various experimental conditions [10,11].

According to the Food and Agriculture Organization, 25% of the world's crops are contaminated by fungal toxic metabolites [12]. Sugarcane is a suitable host for many saprophytic fungi, especially the aflatoxigenic ones that belong to the *Aspergillus* species [13]. Products of secondary fungal metabolism, some of them being toxic and thus termed mycotoxins, can be formed either in the field and/or during storage. The most significant mycotoxins in terms of food and feed safety are aflatoxins, ochratoxin A, fumonisins, zearalenone, trichothecenes, and ergot alkaloids that have carcinogenic, mutagenic, teratogenic, cytotoxic, neurotoxic, nephrotoxic, estrogenic, dermotoxic, and immunotoxic effects [12,14–16].

Ingestion of contaminated food is the principal route for human exposure to mycotoxins [17]. One of the most important aspects in the risk analysis of food contaminants is to determine the degree of human exposure [18]. In the case of mycotoxins, this exposure is generally assessed by taking into account data on mycotoxin occurrence in foodstuffs as well as data on the dietary consumption of the concerned population [19,20], although analytical methods for determination of individual exposure by analysis of biofluids have recently been developed [21–24].

There are some methods used to assess dietary consumption, generally known as market basket, 24 h dietary recall and food record methods, food frequency methods or dietary history [25]. The degree of exposure is measured in terms of probable daily intake (PDI) per unit of body weight, and is generally expressed in ng/kg of body weight (b.w.) per day. Afterwards, the PDI value is compared with the tolerable daily intake (TDI) which is determined by certain toxicological studies for risk analysis. Several exposure and risk assessment studies for mycotoxins in different food commodities were carried out during the last few years [25–32]; nevertheless, no exposure assessment for mycotoxins from contaminated sugarcane grass and juice has been conducted yet.

The growth of toxicogenic fungi in a sugarcane crop field was documented in numerous studies [13,33]. However, these previous studies discussed only the isolation of different fungal species from the plant and the prevalence of their mycotoxins after inoculation of the isolated fungi in culture media without determining the natural (co-)occurrence of the produced mycotoxins. Ahmed et al. (2010) screened the contamination of sugarcane juice sold in Pakistan with several mycoflora including *Aspergillus flavus*, *A. fumigatus*, and *A. niger* [9], while Hariprasad et al. (2015) investigated the natural aflatoxin uptake by sugarcane from contaminated soil and its persistence in sugarcane juice and jiggery (the natural sweetener made by concentrating the sugarcane juice) using thin layer chromatography and ELISA [34]. To the best of the authors' knowledge, no reports have been published for the natural occurrence of multiple mycotoxins in cane grass and juice. Moreover, no regulations in Egypt or other countries for mycotoxins in this commodity have been established so far. Therefore, it was worthwhile to perform this survey to screen for a wide range of (toxic) fungal metabolites in sugarcane grass and juice sold in Assiut City, Egypt, using liquid chromatography-tandem mass spectrometry (LC-MS/MS) with an estimation of the seasonal human exposure to mycotoxins from sugarcane juice in order to evaluate the possible health risks.

2. Results

2.1. Occurrence of Fungal and Bacterial Metabolites in Sugarcane Grass and Juice Samples

Overall, 29 different metabolites in cane grass and 33 in cane juice were quantified and, on average, 14 and 13 different metabolites were detected per sample, respectively. In both matrixes, 20 similar or shared metabolites were detected, including, aflatoxin B_1 and G_1, averufin, 3-nitropropionic acid, kojic acid, asperglaucide, asperphenamate, and emodin, as presented in Table 1. None of the other mycotoxins addressed by regulatory limits in the European Union (EU) have been positively identified in any of the investigated samples.

Table 1. Overview of the detected analytes in sugarcane grass and juice samples.

Metabolites in Both	Metabolites only in Cane Grass	Metabolites only in Cane Juice
3-Nitropropionic acid	Alternariolmethylether	Aspinolid B
Aflatoxin B_1	Brevianamid F	Chlorocitreorosein
Aflatoxin G_1	Cyclo (L-Pro-L-Tyr)	Fusapyron
Agroclavine	Cyclo (L-Pro-L-Val)	Fusaric acid
Ascochlorin	Cytochalasin D	Gibberellic acid
Asperglaucide	Ilicicolin E	Griseofulvin
Asperphenamate	Macrosporin	Integracin A
Averufin	*N*-Benzoyl-Phenylalanine	Integracin B
Berkedrimane B	Physcion	Monocerin
Citreorosein	-	Nidurufin
Emodin	-	Versicolorin A
Ilicicolin B	-	Versicolorin C
Iso-Rhodoptilometrin	-	Xanthotoxin
Kojic acid	-	-
Norlichexanthone	-	-
Oxaline	-	-
Penicillic acid	-	-
Quinolactacin A	-	-
Skyrin	-	-
Tryptophol	-	-

Nine different metabolites were detected exclusively in grass and another 13 metabolites were only found in juice. However, the concentrations of all the shared metabolites were higher in grass than in juice; the prevalence was variable between both commodities. As an example, the maximum concentrations of aflatoxin B_1 and G_1 (30.6 and 7.76 μg/kg) in grass were higher than in juice (2.10 and 1.34 μg/kg), while the prevalence of these mycotoxins in grass (48% and 10%) was lower than in juice (58% and 18%), respectively. Data on the maximum concentration of all evaluated mycotoxins in each commodity, as well as the related median, mean and apparent recovery in the positive samples, are compiled in Tables 2 and 3.

Asperphenamate was detected in all grass and juice samples (100%). The other most prevalent metabolites in grass were emodin (100%), tryptophol (95%), citreorosein (86%), iso-rhodoptilometrin (81%), *N*-Benzoyl-Phenylalanine (81%), kojic acid (76%), and ilicicolin B (67%), while tryptophol (100%), emodin (95%), citreorosein (88%), ilicicolin B (88%), averufin (68%), and iso-rhodoptilometrin (68%) were the most frequently occurring ones in cane juice.

Kojic acid was detected in 3% of juice, which appeared to be much lower in comparison with those found in grass samples, in 76% of the analyzed samples. In addition to aflatoxins as important toxic metabolites, 3-nitropropionic acid (3-NPA) was detected in both grass and juice; however, the frequencies and concentrations were lower in juice (Tables 2 and 3). The co-occurrence of *Aspergillus flavus* metabolites (AFB_1, averufin, 3-NPA, and kojic acid) was detected in ≥28% of grass, while no juice samples were co-contaminated with all of these metabolites.

It was noticeable that the prevalence of the non-shared metabolites in grass was higher than in juice, in which four metabolites occurred in more than half of the samples, cyclo (L-Pro-L-Val) (100%), cyclo (L-Pro-L-Tyr) (67%), *N*-Benzoyl-Phenylalanine (81%), and physcion (81%)), while in juice only one metabolite occurred in more than the half the samples, versicolorin C in 73% of juice.

Table 2. Overview on the occurrence, concentrations and performance characteristics of the analytical method for the detected analytes in natural sugarcane grass samples.

			Concentration of Positive Samples (μg/kg)						
Detected Analytes	**P/N**	**Prevalence**	**Median**	**Mean**	**Minimum**	**Maximum**	**R_a**	**LOD [a]**	**LOQ [b]**
3-Nitropropionic acid	13/21	62%	5.48	27.5	<LOQ	193	81.4%	0.9	2.8
Aflatoxin B_1	10/21	48%	11.7	13.6	<LOQ	30.6	53.4%	2.8	9.2
Aflatoxin G_1	2/21	10%	5.10	5.10	<LOQ	7.76	41.6%	2.2	7.4
Agroclavine	2/21	10%	161	161	<LOQ	300	71.4%	13	44
Alternariolmethylether	9/21	43%	0.23	0.45	<LOQ	1.26	60%	0.05	0.2
Ascochlorin	6/21	29%	2.63	11.5	0.8	55.4	80.2%	0.1	0.4
Asperglaucide	9/21	43%	0.41	0.66	<LOQ	2.07	54.6%	0.1	0.4
Asperphenamate	21/21	100%	17.2	283	1.81	3998	87%	0.08	0.3
Averufin	11/21	52%	0.36	0.48	<LOQ	1.91	85.6%	0.05	0.2
Berkedrimane B	4/21	19%	4.39	18.3	<LOQ	64.2	54.2%	0.2	0.7
Brevianamid F	4/21	19%	13.8	13.5	6.30	20.0	45%	0.8	2.7
Citreorosein	18/21	86%	8.33	22.1	<LOQ	229	163%	0.8	2.7
Cyclo (L-Pro-L-Tyr)	14/21	67%	7.99	15.4	<LOQ	55.3	90.4%	1.7	5.6
Cyclo (L-Pro-L-Val)	21/21	100%	16.1	31.6	<LOQ	198	94.6%	1	3.4
Cytochalasin D	2/21	10%	71.5	71.5	5.40	137	80.4%	0.5	1.8
Emodin	21/21	100%	4.31	7.37	<LOQ	34.1	121%	0.2	0.8
Ilicicolin B	14/21	67%	28.8	43.6	5.44	141	42.2%	1.3	4.2
Ilicicolin E	1/21	5%	0.39	0.39	0.39	0.39	76.8%	0.1	0.3
Kojic acid	16/21	76%	4160	20,176	617	117,815	99.6%	75.8	250
Iso-Rhodoptilometrin	17/21	81%	0.89	6.00	0.19	72.9	110%	0.04	0.1
Macrosporin	5/21	24%	0.08	5.91	<LOQ	29.2	94.2%	0.02	0.05
N-Benzoyl-Phenylalanine	17/21	81%	8.29	66.1	<LOQ	856	93.6%	2.3	7.4
Norlichexanthone	8/21	38%	26.6	149.7	<LOQ	1025	65%	1.1	3.6
Oxaline	2/21	10%	13.0	13.0	<LOQ	25.7	70%	0.17	0.57
Penicillic acid	10/21	48%	70.5	333	13.2	2683	97%	2.5	8.1
Physcion	17/21	81%	60.4	58.2	<LOQ	73.5	98.4%	39	130
Quinolactacin A	7/21	33%	0.51	0.74	0.12	2.33	64%	0.03	0.1
Skyrin	4/21	19%	2.50	2.67	<LOQ	4.71	79.4%	0.8	2.7
Tryptophol	20/21	95%	318	887	27.6	4370	64.4%	7.08	23

Calculation of mean, median and range values was based on positive samples. P/N, number positive samples over the number of total samples; R_a, apparent recovery; [a] LOD, limit of detection; [b] LOQ, limit of quantification.

Table 3. Overview on the occurrence, concentrations and performance characteristics of the analytical method for the detected analytes in sugarcane juice samples.

			Concentration of Positive Samples (μg/kg)						
Detected Analytes	**P/N**	**Prevalence**	**Median**	**Mean**	**Minimum**	**Maximum**	**R_a**	**LOD [a]**	**LOQ [b]**
3-Nitropropionic acid	12/40	30%	2.84	3.58	<LOQ	13.6	67.6%	0.2	0.7
Aflatoxin B_1	23/40	58%	0.56	0.72	<LOQ	2.10	73.9%	0.14	0.4
Aflatoxin G_1	7/40	18%	0.10	0.30	<LOQ	1.34	62.4%	0.06	0.2
Agroclavine	8/40	20%	3.05	3.96	<LOQ	9.49	68%	0.004	0.01
Ascochlorin	18/40	45%	0.22	0.28	<LOQ	0.63	78.4%	0.1	0.3
Asperglaucide	7/40	18%	0.03	0.21	<LOQ	0.83	93.6%	0.01	0.02
Asperphenamate	40/40	100%	5.23	12.5	0.65	91.4	96.7%	0.2	0.6
Aspinolid B	1/40	3%	7.92	7.92	7.92	7.92	96.5%	0.04	0.13
Averufin	27/40	68%	0.12	0.13	<LOQ	0.32	85.2%	0.06	0.2
Berkedrimane B	26/40	65%	0.41	5.75	<LOQ	41.9	100%	0.004	0.01
Chlorocitreorosein	19/40	48%	3.26	8.65	<LOQ	51.9	102%	0.8	1.2
Citreorosein	35/40	88%	1.28	1.99	<LOQ	12.6	107%	0.1	0.4
Emodin	38/40	95%	0.45	0.82	<LOQ	3.90	92.9%	0.01	0.02
Fusapyron	7/40	18%	0.98	7.49	<LOQ	44.4	71.3%	0.1	0.4

Table 3. *Cont.*

		Concentration of Positive Samples (μg/kg)							
Detected Analytes	**P/N**	**Prevalence**	**Median**	**Mean**	**Minimum**	**Maximum**	R_a	**LOD** [a]	**LOQ** [b]
Fusaric acid	12/40	30%	67.9	258	25.4	2214	92.4%	1.33	4.39
Gibberellic acid	10/40	25%	2.30	18.7	<LOQ	134	128%	0.2	0.7
Griseofulvin	2/40	5%	0.36	0.36	0.20	0.51	97.5%	0.05	0.17
Ilicicolin B	35/40	88%	0.59	2.44	<LOQ	17.6	52.6%	0.07	0.2
Kojic acid	1/40	3%	208	208	208	208	62.7%	13.8	45.6
Integracin A	2/40	5%	1.73	1.73	1.60	1.87	67.1%	0.3	0.9
Integracin B	2/40	5%	1.12	1.19	<LOQ	1.97	64.2%	0.2	0.6
Iso-Rhodoptilometrin	27/40	68%	0.03	0.06	<LOQ	0.36	103%	0.01	0.02
Monocerin	5/40	13%	0.28	1.26	<LOQ	4.97	100%	0.02	0.08
Nidurufin [c]	14/40	35%	0.10	0.15	0.001	0.61	n.d [e]	-	-
Norlichexanthone	11/40	28%	0.84	1.67	<LOQ	6.05	82.5%	0.03	0.10
Oxaline	1/40	3%	1.61	1.61	1.61	1.61	64.6%	0.10	0.32
Penicillic acid	16/40	40%	19.9	45.7	3.02	212	77.2%	0.2	0.7
Quinolactacin A	22/40	55%	0.03	0.14	0.005	1.40	80.2%	0.001	0.004
Skyrin	7/40	18%	0.17	0.35	0.05	1.60	79.8%	0.03	0.1
Tryptophol	40/40	100%	58.4	110	<LOQ	581	54.9%	0.5	1.8
Versicolorin A	17/40	43%	0.05	0.06	<LOQ	0.23	80.7%	0.02	0.06
Versicolorin C [d]	29/40	73%	7.74	11.31	1.55	53.2	n.d [e]	-	-
Xanthotoxin	4/40	10%	2.28	1.99	<LOQ	3.32	94.6%	0.03	0.1

Calculation of mean, median and range values was based on positive samples. P/N, number positive samples over the number of total samples; R_a, apparent recovery; [a] LOD, limit of detection; [b] LOQ, limit of quantification; [c] No standard available, estimation of concentration based on response and recovery of averufin; [d] No standard available, estimation of concentration based on response and recovery of versicolorin A; [e] n.d.: not determined.

2.2. Exposure Assessment

The average daily consumption of juice for adult females (n = 90) and males (n = 91) was estimated in winter at 34.1 and 64.2 mL/day, and in summer at 66.3 and 150.4 mL/day, respectively. Due to the significant variability in juice consumption patterns between males and females in winter and summer seasons, each group for each season was considered alone for the exposure assessment.

The probable daily intakes (PDI) were obtained by integration of the results of analyzed mycotoxins with the juice consumption estimation with a body weight of 70 kg for adults. The PDI values (ng/kg) for AFB_1 and AFG_1 (Table 4) were calculated according to the following equation: PDI = (C × K)/b.w., where C is the mean content of a mycotoxin (μg/kg) and K is the average consumption of the commodity (mL/day). Mean values of all juice samples were 0.42 μg/kg for AFB_1 and 0.05 μg/kg for AFG_1.

Table 4. The probable daily intakes of AFB_1 and AFG_1 from sugarcane juice in Assiut City.

Season	Winter		Summer	
Mycotoxin/Gender	**Females**	**Males**	**Females**	**Males**
AFB_1 (ng/kg)	0.20	0.38	0.40	0.90
AFG_1 (ng/kg)	0.02	0.05	0.05	0.11

Due to the fact that aflatoxins (AFs) are potent liver carcinogens [35], there is no TDI for AFs (no safe level can be established as AFs can induce cancer even at very low doses). Therefore, most agencies, including the Joint Expert Committee on Food Additives and the US Food and Drug Administration, have not set a TDI for AFs. Of note, the obtained rate of exposure is expected be higher for children (due to the difference in body weight) and for adult persons (in case of a high rate of consumption).

As shown above in Table 4, the PDI values appear to be more than double during summer and male inhabitants are likely to be more exposed throughout the year than females.

3. Discussion

The sugarcane crop and its by-products have a great agro-industrial and economic value. Upper Egypt governorates are the main home to sugarcane cultivation and its industry in Egypt, particularly in the governorates of Sohag, Qena and Aswan [36]. The choice of Assiut City, not a major state for sugarcane cultivation in Egypt, was based on the availability of the samples and collection of data by the authors.

During harvest time, cutting of the sugarcane stalk disrupts the physiology of the plant and acts as a portal of entry for pathogenic bacteria and fungi. Some of these microorganisms, under elevated temperatures and high humidity, use the sugar as a source of energy for growth and produce different metabolic by-products that, on one hand, may cause processing problems in the mill and refinery and, on the other hand, can be toxic to animals and humans [33,37]. Red rot, smut, wilt and sett rot are the most important fungal diseases affecting sugarcane agriculture and causing remarkable economic losses [38–40]. This necessitates a detailed investigation of the probable effect of mycotoxins on the sugar industry and the other by-products such as alcohol as one of the main products derived from sugarcane.

In the light of the aforementioned results, the level of AFB_1 in sugarcane was higher than the highest maximum level regulated by the EU (12 μg/kg for unprocessed almonds, pistachios, and apricot kernels) [41] and Egyptian standards for peanuts intended for human consumption (5 μg/kg) [42]. AFs in sugarcane crops can be derived from two sources, either through the natural AF uptake from contaminated soil in the field [34] or from fungal attack to the outer fiber layer, especially after insect invasion or other parasites pre- or post-harvest [43–45]. Also, it is worth mentioning that chewing raw sugarcane is a common practice, especially during the harvesting time. However, no regulations for AFs in fruit or crop juices have been set. Recently, Hariprasad et al. (2015) detected higher concentrations of AFB_1 in sugarcane juice, 0.5–6.5 μg/kg, collected from local vendors in India [34]. In a previous study, AFB_1 was detected in mango (0.03–0.72 μg/L), guava (0.04–0.20 μg/L), apple (0.03–0.07 μg/L) and orange (0.006–0.07 μg/L) juices collected from local markets of six governorates in Egypt. The maximum concentration of AFG_1 in mango, orange, guava and apple nectar was 0.08, 0.21, 0.34, and 0.55 μg/L, respectively [46]. The detection of multiple mycotoxins in different Egyptian canned fruits juices and beverages was also investigated by Abdel-Sater et al. (2001) using thin layer chromatography. Total AFs (AFB_1 and AFG_1) in apple beverages ranged from 20–30 μg/L and AFB_1 in guava juice was at a concentration of 12 μg/L [47].

Two hypotheses for sugarcane juice contamination with the detected metabolites including AFs were developed; (1) contaminated sugarcane stems used for juice production; (2) contamination of the instruments used in juicing. A reported case in China showed that 3-NPA was the main causative agent of acute poisoning after consumption of moldy sugarcanes [48]. Moreover, 88 persons died out of the 884 exposed to moldy sugarcane during the period of 1972 until 1988 [49]. In animals the systemic administration of 3-NPA is suspected to cause Huntington's disease–like symptoms [50]. In a survey from Argentina in 2011 and 2014, the maximum concentration in native grass for grazing cattle ranged from 28.8–120 μg/kg [51] which is lower than the present report; however, higher concentrations were detected by Ezekiel et al. (2012) in poultry feed from Nigeria (up to 947 μg/kg) [52]. Therefore, the toxic effect of 3-NPA may affect humans through chewing raw sugarcane grass while it may affect animals through feeding on grass left over or on the bagasse of sugarcane.

Sugarcane juice consumption is a daily habit in Egypt, continuing throughout the year. However, the consumption rate greatly increases during the summer months. A common approach to estimate mycotoxin exposure is generally obtained through the combination of contamination results with the consumption data. So far, to the best of the authors' knowledge, there are no available data on dietary exposure to mycotoxins from sugarcane juice and/or juice consumption in Egypt or anywhere else of the world.

There are some methods used to assess dietary intake such as the food frequency questionnaire. This approach was adopted to assess the dietary exposure to different mycotoxins by several previous studies in Brazil, Spain, Japan, Malaysia [25,28,30,32,53]. In this study, the juice frequency questionnaire (online and paper based) was chosen. The participants, 91 males and 90 females, were of different ages (18–65 years old). The difference in PDI values between males and females is related to the juice intake rates. The reason may be connected with the nature of Egyptian society where males are more likely to be in contact with streets and juice shops. Also, the difference between winter and summer is due to climatic factors such as temperature.

Extensive AF exposure studies with estimation of PDI values have been reported, although little is available in the literature concerning the AFB_1 contamination in juices or similar beverages [46,47,54]. In Egypt, the PDI values of AFB_1 from contaminated corn-based snacks for adults ranged from 0.42 to 11.30 ng/kg b.w./day; however, the consumption data (50 g/day) were estimated by the authors based on the content packaging of the commercial product and PDI values were estimated with the mean value of positive samples [55]. In another study from Cairo using the same foodstuffs, the probable daily intake of aflatoxin B_1 was 3.69 ng/kg b.w./day, although the value was calculated with the mean of the positive samples as well [56]. The same author in another survey, El-Sawi in 2006, estimated the PDI of AFB_1 (0.097 ng/kg b.w./day) from corn using the mean values of contaminated samples in the calculations [57]. The mean PDI values of AFB_1 were assessed worldwide: Korea (1.19 and 5.79 ng/kg b.w./day from rice) [27], Malaysia (24.3 to 34.0 ng/kg b.w./day and for total AFs was 28.81 to 58.02 ng/kg b.w./day from 236 individual food composites consisting of 38 different types of foods) [53], Lebanon (0.63–0.66 ng/kg b.w./day from Lebanese diets) [58], Japan (0.003 to 0.004 ng/kg b.w./day, from different 24 foods categories including peanuts) [32], and Brazil (2.3 to 4.1 ng/kg b.w./day of AFs from peanuts for high consumers) [28].

Consequently, the estimated PDI levels of AFB_1 and AG1 for the Egyptian population from sugarcane juice consumption in both seasons may contribute to a public health implication due to the fact that even low levels of AFs contamination, which might fall with the permissible limits, can lead to haptic cancer in the long run, in addition to the inevitable exposure to AFs from other contaminated food commodities. Also, serious health implications can be expected in case of high rates of juice consumption.

4. Conclusions

The study aimed to detect multiple mycotoxins occurring naturally in the sugarcane crop and juice for the first time. Indeed, the present work will open the door for further studies on the occurrence of more mycotoxins in this important economic crop. Furthermore, the presence of these metabolites either in their present (parent) form or in a modified form in the secondary products, which are based on sugarcane products such as raw sugar, vinegar, alcohol, chipboard, paper, some chemicals, plastics, paints, fiber, insecticides and detergents and molasses, cannot be excluded.

The study sheds light on the non-negligible doses of aflatoxins through the consumption of juice in Upper Egypt, especially in the summer. However, the survey was limited to 181 persons in Assiut City. Also, it should be noted that chewing raw sugarcane grass is a common practice during the harvesting time which indicates an additional source of mycotoxin exposure. These first data for exposure warrant further larger-scale multi-mycotoxin-based studies aimed at providing a comprehensive assessment in other Egyptian cities as the consumption may vary among different social groups within Egypt. Due to the seasonal fluctuation of the fungal species [33,59,60] and the difference in consumption rates of juice as shown in the present survey, follow-up detection studies in summer and in other cities of the country are urgently needed to provide better insights regarding the number of contaminating mycotoxins and their quantities in both grass and juice.

It is highly recommended that a study for risk assessment in different foodstuffs be performed to estimate the contribution of sugarcane juice in comparison with other foods and beverages in the Egyptian diet. This report will assist the other countries where sugarcane grass cultivation and juice consumption are common to assess the occurrence of different mycotoxins in these commodities and the consequent possible health risks. Another important point needs to be mentioned herein regarding the contribution of grass left over from sugarcane in animal feed, as it is a very cheap supplement and a rich source of fiber. A detailed investigation must be conducted to detect the contaminating mycotoxins and the possibility of co-exposure in producing farm animals.

In order to limit the fungal attack and the potential production of toxic metabolites, appropriate pre-harvest precautions should be applied. Considering the juice, it should be produced in a hygienic media after a proper cleaning of the grass, especially the outer external cortex of the plant, as well as the wringers or the machine used for juicing, which can be a source of contamination.

5. Materials and Methods

5.1. Sample Collection

Sugarcane stem (n = 21) and fresh sugarcane juice (n = 40, each 400 mL) samples were randomly purchased from several shops as would be done by a consumer from local vendors and juice shops, respectively in January 2016 from Assiut City, Assiut Governorate, Egypt. After collection, juice samples were kept at −20 °C until LC-MS/MS analyses.

5.2. Sample Treatment, Extraction and Mycotoxins Analysis

5.2.1. Chemicals and Reagents

Acetonitrile (LC gradient grade) were purchased from VWR International (Leuven, Belgium), methanol (LC gradient grade) and glacial acetic acid (p.a.) from Merck (Darmstadt, Germany), and ammonium acetate MS grade) from Sigma-Aldrich (Vienna, Austria). Water was purified successively by reverse osmosis and an Elga Purelab ultra analytic system from Veolia Water (Bucks, UK) to 18.2 MΩ. Standards of fungal and bacterial metabolites were obtained either as gifts from various research groups or from the following commercial sources: Romer Labs® Inc. (Tulln, Austria), Sigma-Aldrich (Vienna, Austria), Iris Biotech GmbH (Marktredwitz, Germany), Axxora Europe (Lausanne, Switzerland) and LGC Promochem GmbH (Wesel, Germany), Enzo Life Sciences (Lausen, Switzerland), BioAustralis (Smithfield, Australia), AnalytiCon Discovery (Potsdam, Germany) and Toronto Research Chemicals (Toronto, ON, Canada).

5.2.2. Extraction and Estimation of Apparent Recoveries

For juice, 2 mL of each sample were extracted with 2 mL of the extraction solvent (acetonitrile/water/acetic acid, 79/20/1, $v/v/v$), centrifuged for 2 min at 4000 RPM at room temperature, and diluted 1:1 with dilution solvent (acetonitrile/water/acetic acid, 20/79/1, $v/v/v$).

For sugarcane grass, each collected stem (with no water content either the external fiber layer or collected fresh leftover or bagasse from shops and all were dried in oven at Romer lab, Tulln for 24 h at 30 °C) was ground into very small pieces (<1 mm) and 40 mL of the extraction solvent were added to 2.5 g of each sample. Afterwards, samples were shaken for 90 min at 180 RPM at room temperature, and then diluted 1:1 with dilution solvent. The volume of injection for juice and sugarcane grass was 5 μL of the diluted raw extract without further manipulation according to Sulyok et al. (2006) [61].

5.3. LC-MS/MS Parameters

The samples were analyzed using a dilute and shoot approach as described before in the literature [62]. Briefly, an Agilent 1290 Series HPLC System (Agilent, Waldbronn, Germany) coupled to a QTrap 5500 used in connection with a Turbo Ion Spray ESI source (Sciex, Foster City, CA, USA) has been used in connection with a Gemini® C18-column, 150 × 4.6 mm i.d., 5 μm particle size, protected by a C18 security guard cartridge, 4 × 3 mm i.d. (all from Phenomenex, Torrance, CA, USA). A methanol/water gradient containing 1% acetic acid and 5 mM NH_4Ac was used at 1 mL/min. Data acquisition was performed in the scheduled multiple reaction monitoring (sMRM) mode in both positive and negative polarity using two separate chromatographic runs per sample. External calibration was performed using serial dilutions of a multi-component working standard. Results were corrected for apparent recoveries that have been determined by spiking five individual samples of each matrix.

5.4. Sugarcane Juice Consumption Data

Due to lack of data for juice consumption in Egypt, juice frequency questionnaires (online and paper based) were filled in by different inhabitants in Assiut City. The participants were requested to answer the monthly consumption rate in winter and summer seasons. In order to know the exact amount per mL, two different sizes of juice cups, large and small, as sold by local shops in Egypt, were included in the questionnaires. The large cup is equal to 350 mL while the small one is 250 mL. Additional information such as age, level of education, and distance from the nearest juice shop were also documented. The total number was 181 adults inhabitants (18–65 years) of Assiut city participated in the questionnaires, females (n = 90) and males (n = 91).

Acknowledgments: The LC-MS/MS system was funded by the Federal Country Lower Austria and co-financed by the European Regional Development Fund of the European Union. The authors thank Med.Vet Menna Zakaria, research assistant at Assiut International Center of Nanomedicine (AICN), EL-Rajhy Hospital, Assuit University, for her assistance in the paper-based questionnaire. We are also grateful to all the participants in this survey.

Author Contributions: M.F.A., R.K. and M.S. co-conceived and designed the study; M.F.A. performed the collection, preparation and extraction of the samples; M.F.A. performed the questionnaires. M.F.A. and M.S. performed LC-MS/MS analysis, analyzed the data and wrote the manuscript. M.S. and R.K. supervised the whole study and approved the manuscript.

Conflicts of Interest: The authors declare no conflict of interest.

References

1. Tarimo, A.J.P.; Takamura, Y.T. Sugarcane Production, Processing and Marketing in Tanzania. *Afr. Study Monogr.* **1998**, *19*, 1–11.
2. Center for Agriculture Research, Egyptian Ministry of Agriculture and Land Reclamation. Available online: http://www.vercon.sci.eg/indexUI/uploaded/kasbalsoker946/kasbalsoker.htm (accessed on 7 September 2016). (In Arabic)
3. Food and Agriculture Organization of the United Nations (FAO). Statistics Division, FAOSTAT, 2014. Available online: http://www.faostat3.fao.org/compare/E (accessed on 8 September 2016).
4. Lee, T.S.G.; Bressan, E.A. The potential of ethanol production from sugarcane in Brazil. *Sugar Tech* **2006**, *8*, 195–198. [CrossRef]
5. El-Kholi, M.M.A. Sugar Crops Research Institute, Giza (Egypt): A Profile. *Sugar Tech* **2008**, *10*, 189–196. [CrossRef]
6. Solomon, S. Sugarcane By-Products Based Industries in India. *Sugar Tech* **2011**, *13*, 408–416. [CrossRef]
7. Center for Sugarcane Agriculture Services, Central Administration of Agricultural and Environmental Guidance. Available online: http://www.caae-eg.com/index.php/2012-12-25-10-49-19/2010-09-18-17-00-51/2011-01-10-19-57-23/604-2012-03-12-10-26-32.html (accessed on 7 September 2016). (In Arabic)
8. European Commission. Agriculture and Rural Development, Sugar. 2016. Available online: http://www.ec.europa.eu/agriculture/sugar/index_en.htm (accessed on 7 September 2016).
9. Ahmed, A.; Dawar, S.; Tariq, M. Mycoflora associated with sugar cane juice in Karachi city. *Pak. J. Bot.* **2010**, *42*, 2955–2962.
10. Abbas, S.R.; Sabir, S.M.; Ahmad, S.D.; Boligon, A.A.; Athayde, M.L. Phenolic profile, antioxidant potential and DNA damage protecting activity of sugarcane (*Saccharum officinarum*). *Food Chem.* **2014**, *147*, 10–16. [CrossRef] [PubMed]
11. Kadam, U.S.; Ghosh, S.B.; De, S.; Suprasanna, P.; Devasagayam, T.P.A.; Bapat, V.A. Antioxidant activity in sugarcane juice and its protective role against radiation induced DNA damage. *Food Chem.* **2008**, *106*, 1154–1160. [CrossRef]
12. Reddy, K.R.N.; Salleh, B.; Saad, B.; Abbas, H.K.; Abel, C.A.; Shier, W.T. An overview of mycotoxin contamination in foods and its implications for human health. *Toxin Rev.* **2010**, *29*, 3–26. [CrossRef]

13. Kumeda, Y.; Asao, T.; Takahashi, H.; Ichinoe, M. High prevalence of B and G aflatoxin-producing fungi in sugarcane field soil in Japan: Heteroduplex panel analysis identifies a new genotype within *Aspergillus Section Flavi* and *Aspergillus nomius*. *FEMS Microbiol. Ecol.* **2003**, *45*, 229–238. [CrossRef]
14. Steyn, P.S. Mycotoxins, general view, chemistry and structure. *Toxicol. Lett.* **1995**, *82–83*, 843–851. [CrossRef]
15. Zain, M.E. Impact of mycotoxins on humans and animals. *J. Saudi Chem. Soc.* **2011**, *15*, 129–144. [CrossRef]
16. Abdallah, M.F.; Girgin, G.; Baydar, T. Occurrence, prevention and limitation of mycotoxins in feeds. *Anim. Nutr. Feed Technol.* **2015**, *15*, 471–490. [CrossRef]
17. Peraica, M.; Radic, B.; Lucic, A.; Pavlovic, M. Toxic effects of mycotoxins in humans. *Bull. World Health Org.* **1999**, *77*, 754–763. [PubMed]
18. World Health Organization. *Evaluation of Certain Mycotoxins in Food*; WHO Technical Report Series, 906; World Health Organization: Geneva, Switzerland, 2002.
19. Milićević, D.R.; Škrinjar, M.; Baltić, T. Real and perceived risks for mycotoxin contamination in foods and feeds: Challenges for food safety control. *Toxins* **2010**, *2*, 572–592. [CrossRef] [PubMed]
20. Marin, S.; Ramos, A.J.; Cano-Sancho, G.; Sanchis, V. Mycotoxins: Occurrence, toxicology, and exposure assessment. *Food Chem. Toxicol.* **2013**, *60*, 218–237. [CrossRef] [PubMed]
21. Ediage, E.N.; Di Mavungu, J.D.; Song, S.; Wu, A.; Van Peteghem, C.; De Saeger, S. A direct assessment of mycotoxin biomarkers in human urine samples by liquid chromatography tandem mass spectrometry. *Anal. Chim. Acta* **2012**, *741*, 58–69. [CrossRef] [PubMed]
22. Warth, B.; Sulyok, M.; Fruhmann, P.; Mikula, H.; Berthiller, F.; Schuhmacher, R.; Hametner, C.; Abia, W.A.; Adam, G.; Fröhlich, J.; et al. Development and validation of a rapid multi-biomarker liquid chromatography/tandem mass spectrometry method to assess human exposure to mycotoxins. *Rapid Commun. Mass Spectrom.* **2012**, *26*, 1533–1540. [CrossRef] [PubMed]
23. Warth, B.; Sulyok, M.; Krska, R. LC-MS/MS based multi-biomarker approaches for the assessment of human exposure to mycotoxins. *Anal. Bioanal. Chem.* **2013**, *405*, 5687–5695. [CrossRef] [PubMed]
24. Heyndrickx, E.; Sioen, I.; Bellemans, M.; De Maeyer, M.; Callebaut, A.; De Henauw, S.; De Saeger, S. Assessment of mycotoxin exposure in the Belgian population using biomarkers: Aim, design and methods of the BIOMYCO study. *Food Addit. Contam. Part A Chem. Anal. Control Expo. Risk Assess.* **2014**, *31*, 924–931. [CrossRef] [PubMed]
25. Cano-Sancho, G.; Marín, S.; Ramos, A.J.; Peris-Vicente, J.; Sanchis, V. Occurrence and exposure assessment of aflatoxin M1 in Catalonia (Spain). *Rev. Iberoam. Micol.* **2010**, *27*, 130–135. [CrossRef] [PubMed]
26. Zimmerli, B.; Dick, R. Ochratoxin A in table wine and grape-juice: Occurrence and risk assessment. *Food Addit. Contam.* **1996**, *13*, 655–668. [CrossRef] [PubMed]
27. Park, J.W.; Kim, E.K.; Kim, Y.B. Estimation of the daily exposure of Koreans to aflatoxin B_1 through food consumption. *Food Addit. Contam.* **2004**, *21*, 70–75. [CrossRef] [PubMed]
28. Andrade, P.D.; de Mello, M.H.; França, J.A.; Caldas, E.D. Aflatoxins in food products consumed in Brazil: A preliminary dietary risk assessment. *Food Addit. Contam. Part A Chem. Anal. Control Expo. Risk Assess.* **2013**, *30*, 127–136. [CrossRef] [PubMed]
29. Cano-Sancho, G.; Sanchis, V.; Marín, S.; Ramos, A.J. Occurrence and exposure assessment of aflatoxins in Catalonia (Spain). *Food Chem. Toxicol.* **2013**, *51*, 188–193. [CrossRef] [PubMed]
30. Jager, A.V.; Tedesco, M.P.; Souto, P.C.M.C.; Oliveira, C.A.F. Assessment of aflatoxin intake in São Paulo, Brazil. *Food Control* **2013**, *33*, 87–92. [CrossRef]
31. Rodríguez-Carrasco, Y.; Ruiz, M.J.; Font, G.; Berrada, H. Exposure estimates to *Fusarium* mycotoxins through cereals intake. *Chemosphere* **2013**, *93*, 2297–2303. [CrossRef] [PubMed]
32. Sugita-Konishi, Y.; Sato, T.; Saito, S.; Nakajima, M.; Tabata, S.; Tanaka, T.; Norizuki, H.; Itoh, Y.; Kai, S.; Sugiyama, K.; et al. Exposure to aflatoxins in Japan: Risk assessment for aflatoxin B_1. *Food Addit. Contam. Part A Chem. Anal. Control Expo. Risk Assess.* **2010**, *3*, 365–372. [CrossRef] [PubMed]
33. Abd-Elaah, A.G.; Soliman, A.S. Occurrence of fungal species and mycotoxins from decayed sugarcane (*Saccharrum officinarum*) in Egypt. *Mycobiology* **2005**, *33*, 77–83. [CrossRef] [PubMed]
34. Hariprasad, P.; Vipin, A.V.; Karuna, S.; Raksha, R.K.; Venkateswaran, G. Natural aflatoxin uptake by sugarcane (*Saccharum officinaurum* L.) and its persistence in jiggery. *Environ. Sci. Pollut. Res. Int.* **2015**, *22*, 6246–6253. [CrossRef] [PubMed]
35. International Agency for Research on Cancer (IARC). *IARC Monographs on the Evaluation of Carcinogenic Risks to Humans, Volume 82*; International Agency for Research on Cancer: Lyon, France, 2002.

36. Hassan, S.F.; Nasr, M.I. Sugar industry in Egypt. *Sugar Tech* **2008**, *10*, 204–209. [CrossRef]
37. Watt, D.A.; Cramer, M.D. Post-harvest biology of sugarcane. *Sugar Tech* **2009**, *11*, 142–145. [CrossRef]
38. Sengar, A.S.; Thind, K.S.; Kumar, B.; Pallavi, M.; Gosal, S.S. In vitro selection at cellular level for red rot resistance in sugarcane (*Saccharum* sp.). *Plant Growth Regul.* **2009**, *58*, 201–209. [CrossRef]
39. Viswanathan, R.; Rao, G.P. Disease scenario and management of major sugarcane diseases in India. *Sugar Tech* **2011**, *13*, 336–353. [CrossRef]
40. Bhuiyan, S.A.; Croft, B.J.; James, R.S.; Cox, M.C. Laboratory and field evaluation of fungicides for the management of sugarcane smut caused by *Sporisorium scitamineum* in seedcane. *Australas. Plant Pathol.* **2012**, *41*, 591–599. [CrossRef]
41. European Commission. Commission Regulation (EU) No 165/2010 of February 2010 amending regulation (EC) No 1881/2006 setting maximum levels for certain contaminants in foodstuffs as regards aflatoxins. *Off. J. Eur. Union* **2010**, *50*, 8–12.
42. The Egyptian Organization for Standardization and Quality Control. Egyptian Standard Maximum Levels for Mycotoxin in Food and Feed. Alfatoxins Part 1. 1990, No. 1875-1. Available online: http://www.eos.org.eg/en/standard/6007 (accessed on 7 September 2016).
43. Kolo, I.N.; Adesiyun, A.A.; Misari, S.M.; Wayagari, W.J. Economic losses in chewing canes caused by stem borers in Nigeria. *Sugar Tech* **1999**, *1*, 148–152. [CrossRef]
44. Karunakar, G.; Easwaramoorthy, S.; David, H. Host—Parasite interaction between two species of white grubs infesting sugarcane and two species of entomopathogenic nematodes. *Sugar Tech* **2000**, *2*, 12–16. [CrossRef]
45. Suman, A.; Solomon, S.; Yadav, D.V.; Gaur, A.; Singh, M. Post-harvest loss in sugarcane quality due to endophytic microorganisms. *Sugar Tech* **2000**, *2*, 21–25. [CrossRef]
46. Osman, M.A.; EL Badry, N.; Shreif, R.M.; Youssef, M. Safety of commercial fruit juices available on the Egyptian markets with regards their content from determined by heavy metal and aflatoxins residues. *Curr. Sci. Int.* **2014**, *3*, 159–171.
47. Abdel-Sater, M.A.; Zohri, A.A.; Ismail, M.A. Natural contamination of some Egyptian fruit juices and beverages by mycoflora and mycotoxins. *J. Food Sci. Technol.* **2001**, *38*, 407–411.
48. Ming, L. Moldy sugarcane poisoning—A case report with a brief review. *J. Toxicol. Clin. Toxicol.* **1995**, *33*, 363–367. [CrossRef] [PubMed]
49. Liu, X.; Luo, X.; Hu, W. Studies on epidemiology and etiology of moldy sugarcane poisoning in China. *Biomed. Environ. Sci.* **1992**, *5*, 161–177. [PubMed]
50. Kumar, P.; Kumar, A. Protective effect of rivastigmine against 3-nitropropionic acid-induced Huntington's disease like symptoms: Possible behavioural, biochemical and cellular alterations. *Eur. J. Pharmacol.* **2009**, *615*, 91–101. [CrossRef] [PubMed]
51. Nichea, M.J.; Palacios, S.A.; Chiacchiera, S.M.; Sulyok, M.; Krska, R.; Chulze, S.N.; Torres, A.M.; Ramirez, M.L. Presence of multiple mycotoxins and other fungal metabolites in native grasses from a wetland ecosystem in Argentina intended for grazing cattle. *Toxins* **2015**, *7*, 3309–3329. [CrossRef] [PubMed]
52. Ezekiel, C.N.; Bandyopadhyay, R.; Sulyok, M.; Warth, B.; Krska, R. Fungal and bacterial metabolites in commercial poultry feed from Nigeria. *Food Addit. Contam. Part A Chem. Anal. Control Expo. Risk Assess.* **2012**, *29*, 1288–1299. [CrossRef] [PubMed]
53. Chin, C.K.; Abdullah, A.; Sugita-Konishi, Y. Dietary intake of aflatoxins in the adult Malaysian population—An assessment of risk. *Food Addit. Contam. Part B Surveill.* **2012**, *5*, 286–294. [CrossRef] [PubMed]
54. Embaby, E.M.; Awni, N.M.; Abdel-Galil, M.M.; El-Gendy, H.I. Mycoflora and mycotoxin contaminated some juices. *J. Agric. Technol.* **2015**, *11*, 693–712.
55. Amin, A.A.; Abo-Ghalia, H.H.; Hamed, A.A. Aflatoxin B_1 and B_2 in cereal–based baby foods and corn based snacks from Egypt markets: Occurrence and estimation of the daily intake of AFB1. *Afr. J. Mycol. Biotechnol.* **2010**, *15*, 1–11.
56. El-Sawi, M.A.M.; El-Sawi, S.A.M. Monitoring of fungi producing aflatoxins, and dietary intake of aflatoxins in food consumed by Egyptian infants and young children. *Acta Hortic.* **2012**, *963*, 221–230. [CrossRef]
57. El-Sawi, A.M.M. Monitoring of aflatoxins and ochratoxin A in cereals and evaluation the health risk to consumer due to their dietary intake. *J. Biol. Chem. Environ. Sci.* **2006**, *1*, 721–734.

58. Raad, F.; Nasreddine, L.; Hilan, C.; Bartosik, M.; Parent-Massin, D. Dietary exposure to aflatoxins, ochratoxin A and deoxynivalenol from a total diet study in an adult urban Lebanese population. *Food Chem. Toxicol.* **2014**, *73*, 35–43. [CrossRef] [PubMed]
59. Mohawed, S.M.; Abdel Hafez, S.I.I.; EL-Said, A.H.M.; Gherbawy, Y.A.M.H. Seasonal fluctuations of soil and root surface fungi of sugarcane (*Saccharum officinarum* L.) in Upper Egypt. *Egypt. J. Microbiol.* **2001**, *34*, 595–611.
60. Steciow, M.M. Seasonal fluctuation of the oomycetes in a polluted environment: Santigo River and affluents (Buenos Aires, Argentina). *Rev. Iberoam. Micol.* **1998**, *15*, 40–43. [PubMed]
61. Sulyok, M.; Berthiller, F.; Krska, R.; Schuhmacher, R. Development and validation of a liquid chromatography/tandem mass spectrometric method for the determination of 39 mycotoxins in wheat and maize. *Rapid Commun. Mass Spectrom.* **2006**, *20*, 2649–2659. [CrossRef] [PubMed]
62. Malachová, A.; Sulyok, M.; Beltrán, E.; Berthiller, F.; Krska, R. Optimization and validation of a quantitative liquid chromatography-tandem mass spectrometric method covering 295 bacterial and fungal metabolites including all regulated mycotoxins in four model food matrices. *J. Chromatogr. A* **2014**, *1362*, 145–156. [CrossRef] [PubMed]

© 2016 by the authors. Licensee MDPI, Basel, Switzerland. This article is an open access article distributed under the terms and conditions of the Creative Commons Attribution (CC BY) license (http://creativecommons.org/licenses/by/4.0/).

Article

Glutathione-Conjugates of Deoxynivalenol in Naturally Contaminated Grain Are Primarily Linked via the Epoxide Group

Silvio Uhlig [1,*], Ana Stanic [1,2], Ingerd S. Hofgaard [3], Bernhard Kluger [4], Rainer Schuhmacher [4] and Christopher O. Miles [1]

1. Section for Chemistry and Toxicology, Norwegian Veterinary Institute, P.O. Box 750 Sentrum, Oslo 0106, Norway; ana.stanic@vetinst.no (A.S.); chris.miles@vetinst.no (C.O.M.)
2. Department of Chemistry, University of Oslo, P.O. Box 1033, Blindern, Oslo 0315, Norway
3. Division of Biotechnology and Plant Health, NIBIO—Norwegian Institute of Bioeconomy, Høgskoleveien 7, Ås 1430, Norway; ingerd.hofgaard@nibio.no
4. Center for Analytical Chemistry, Department of Agrobiotechnology (IFA-Tulln), University of Natural Resources and Life Sciences, Vienna (BOKU), Tulln 3430, Austria; bernhard.kluger@boku.ac.at (B.K.); rainer.schuhmacher@boku.ac.at (R.S.)

* Correspondence: silvio.uhlig@vetinst.no; Tel.: +47-2321-6264

Academic Editors: Sarah De Saeger, Siska Croubels and Kris Audenaert
Received: 31 August 2016; Accepted: 7 November 2016; Published: 11 November 2016

Abstract: A glutathione (GSH) adduct of the mycotoxin 4-deoxynivalenol (DON), together with a range of related conjugates, has recently been tentatively identified by LC-MS of DON-treated wheat spikelets. In this study, we prepared samples of DON conjugated at the 10- and 13-positions with GSH, Cys, CysGly, γ-GluCys and *N*-acetylcysteine (NAC). The mixtures of conjugates were used as standards for LC-HRMS analysis of one of the DON-treated wheat spikelet samples, as well as 19 Norwegian grain samples of spring wheat and 16 grain samples of oats that were naturally-contaminated with DON at concentrations higher than 1 mg/kg. The artificially-contaminated wheat spikelets contained conjugates of GSH, CysGly and Cys coupled at the olefinic 10-position of DON, whereas the naturally-contaminated harvest-ripe grain samples contained GSH, CysGly, Cys, and NAC coupled mainly at the 13-position on the epoxy group. The identities of the conjugates were confirmed by LC-HRMS comparison with authentic standards, oxidation to the sulfoxides with hydrogen peroxide, and examination of product-ion spectra from LC-HRMS/MS analysis. No γ-GluCys adducts of DON were detected in any of the samples. The presence of 15-*O*-acetyl-DON was demonstrated for the first time in Norwegian grain. The results indicate that a small but significant proportion of DON is metabolized via the GSH-conjugation pathway in plants. To our knowledge, this is the first report of in vivo conjugation of trichothecenes via their epoxy group, which has generally been viewed as unreactive. Because conjugation at the 13-position of DON and other trichothecenes has been shown to be irreversible, this type of conjugate may prove useful as a biomarker of exposure to DON and other 12,13-epoxytrichothecenes.

Keywords: bioconjugation; HRMS; mercapturate; mycotoxin; trichothecene; thiol; LC-MS; metabolism; wheat; oats

1. Introduction

The mycotoxin 4-deoxynivalenol (DON) is produced by some species of *Fusarium* that infect grain crops [1]. DON exerts a range of toxic effects in mammals and humans, e.g., on the intestine, the immune system and the brain [2,3]. The effects on the brain have been linked to the observed emesis and feed refusal in pigs [2]. Conjugation of DON with L-glutathione (GSH) was recently shown to be a biotransformation pathway in wheat plants that were injected with DON in controlled trials [4,5]. After incubation for up to 96 h, the presence of one putative DON–GSH adduct and several possible breakdown products including the cysteine (Cys) and cysteinylglycine (CysGly) adducts were tentatively identified using untargeted or targeted LC-HRMS approaches [4,5]. However, as no characterized reference standards were available, the exact structures of these adducts could not be determined. It was recently shown that thiols, including GSH, react with DON at both the C-10 and C-13 positions (Figure 1), giving rise to a range of mono- and di-conjugated products [6–9]. Because the C-8 ketone of DON is in equilibrium with the cyclic 8,15-hemiketal form, and four 9,10-diastereoisomers can be formed by thiol addition at C-10, ten mono- and eight di-conjugated toxin derivatives can be formed for any given thiol [8]. As the structures of the major thiol addition products now have been determined, and analytical standards are available [6,7], naturally produced DON–thiol adducts can be identified. Thus, the objective of this study was to use LC-HRMS methods in order to identify naturally occurring DON–thiol adducts.

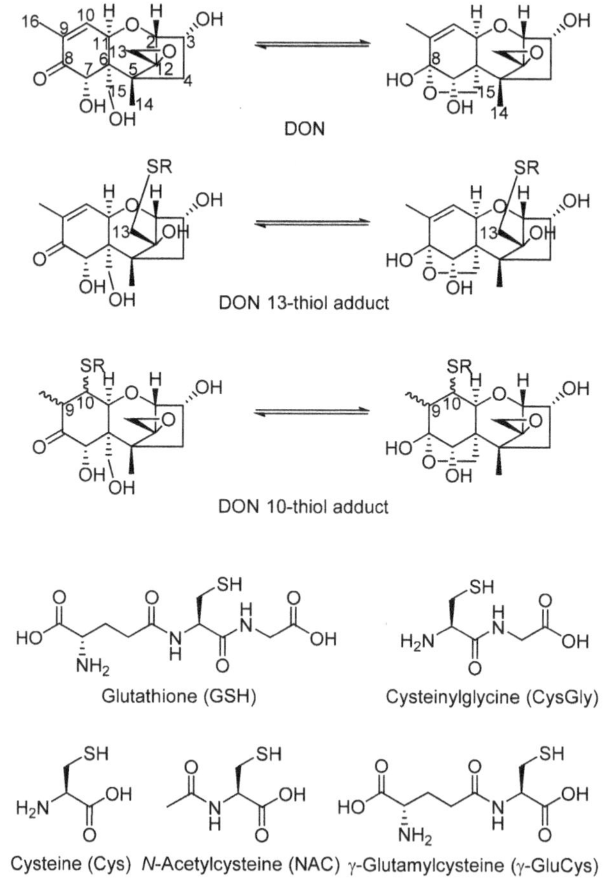

DON

DON 13-thiol adduct

DON 10-thiol adduct

Glutathione (GSH)

Cysteinylglycine (CysGly)

Cysteine (Cys) *N*-Acetylcysteine (NAC) γ-Glutamylcysteine (γ-GluCys)

Figure 1. Chemical structures of 4-deoxynivalenol (DON) and its Michael (C-10) and epoxy (C-13) adducts with thiols (RSH). In solution, these compounds exist as equilibrium mixtures of their 8-keto- (left) and 8,15-hemiketal (right) forms. The structures of thiols used in this study were glutathione (GSH), cysteinylglycine (CysGly), cysteine (Cys), γ-glutamylcysteine (γ-GluCys) and *N*-acetylcysteine (NAC). Note that GSH is the tripeptide γ-GluCysGly.

2. Results and Discussion

2.1. DON–GSH and Related Adducts in DON–Treated Wheat and Naturally Contaminated Grain

When wheat flowering ears were treated with DON and then extracted and analyzed after 12, 24, 48 and 96 h post-treatment, several putative DON conjugates, including DON–GSH and related derivatives, were detected as early as 12 h post-injection [4,5]. Although the LC-HRMS and LC-HRMS/MS data showed that these products were conjugates of DON, the approach used did not facilitate the detailed elucidation of the conjugation site, type of linkage and isomer that had been formed during the incubation. We therefore used LC-HRMS to compare an extract from a DON-treated wheat plant (*Fusarium*-head-blight-susceptible cultivar "Remus", 96 h post-exposure) with synthetic reaction mixtures containing different ratios of the DON-10- and DON-13-conjugates of GSH, CysGly, γ-GluCys and Cys (Figure 2 and Figure S1). The chromatogram from the extract of the DON-treated wheat showed one prominent peak corresponding to one of the C-10 conjugates for each of DON–GSH, DON–CysGly and DON–Cys as had already been suggested by the authors of the earlier studies [4,5]. The type of conjugate present in the extract corresponded to a kinetically favored, and thus relatively fast-forming, isomer (see chromatogram **1** for DON–GSH, –CysGly, and –Cys in Figure 2) from Michael addition at C-10 (Figure 1). It has been shown that this initial product from the reaction of DON with mercaptoethanol, Cys or GSH is replaced (under neutral or mildly basic reaction conditions) by other C-10 conjugates and a C-13 conjugate over time [6–8], and the same occurred in the present study during conjugation with GSH, CysGly, Cys (Figure 2), γ-GluCys and *N*-acetylcysteine (NAC) (Figure S1).

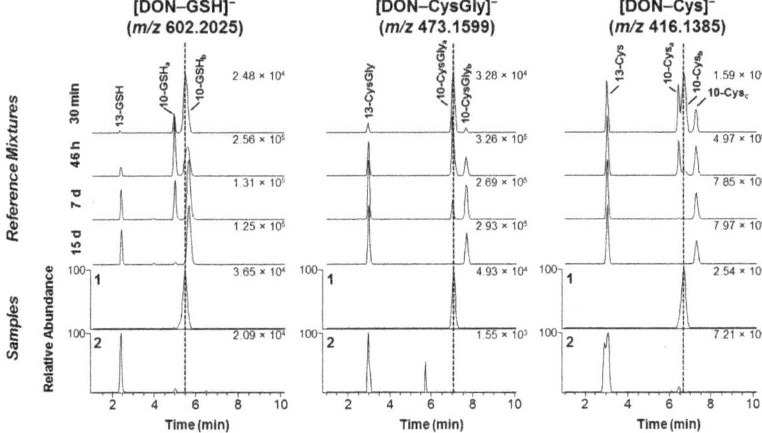

Figure 2. Extracted ion LC-HRMS chromatograms ([M − H]⁻, ±3 ppm) for: left, DON–GSH (*m/z* 602.2025); center, DON–CysGly (*m/z* 473.1599); and DON–Cys (*m/z* 416.1385). For each sample, the upper four stacked chromatograms are from the mixture of DON, GSH, CysGly and Cys (pH 10.7) at various reaction times. The two lower chromatograms are from the extract of wheat that was treated with deoxynivalenol for 96 h (**1**), and from an extract of naturally-contaminated Norwegian oats (**2**). Epoxide conjugates (addition at C-13) eluted at 2–3 min, whereas the Michael conjugates (addition at C-10) eluted at ca. 4.5–7 min. The number in the top right-hand corner of each chromatogram is the intensity of the highest peak in that chromatogram (arbitrary units). Subscripts (a–c) indicate different diastereoisomers).

In contrast to the DON-treated wheat ear sample, which had been sampled at the flowering stage, the 35 naturally-contaminated grain samples were all harvested at the end of the growing season. Thus, these plants had potentially been exposed to DON for much longer than the artificially-contaminated spring wheat (96 h). Of the 35 grain samples, 22 were found to contain at least one DON–thiol conjugate. DON–GSH conjugates were both most prevalent (found in all the 22 DON–thiol containing samples), and in general detected at the highest relative concentration based on relative peak areas, of the thiol-conjugates detected (Table S1). DON–CysGly and DON–Cys conjugates were found in two and 11 of the 35 grain samples, respectively. A few of the samples appeared to contain low levels of the same rapidly-forming diastereoisomer of the C-10 adducts of GSH, CysGly and Cys that were observed in the DON-treated wheat (Figure 2). However, the dominant isomer in the naturally-contaminated grain samples of both oats as well as wheat was the early-eluting C-13 conjugate, which was supported by comparing the LC-HRMS chromatograms with those from the synthetic reaction mixtures (Figure 2) and authentic standards of DON-13-GSH and DON-13-Cys. This finding was verified by HRMS/MS targeting the $[M-H]^-$ ions of DON–GSH (Figure 3). We previously reported that MS-fragmentation of the $[M-H]^-$ ions of DON–thiol conjugates is well suited to distinguishing between the products from Michael addition to C-10 and addition to the epoxide at C-13 [6,7]. Thus, fragmentation of the C-10 conjugates yielded primarily negatively charged amino acid or peptide fragments, while fragmentation of the C-13 conjugates gave product ions primarily related to the trichothecene moiety [6,7]. A concentrated (approximately 5:1) oat extract was used to obtain HRMS/MS spectra of the DON–GSH isomers present in the sample, and the resulting HRMS/MS spectrum of the presumed DON-13-GSH conjugate in the oats was essentially identical to the HRMS/MS spectrum of DON-13-GSH in the reference mixture (Figure 3). The signal/noise of the $[M-H]^-$ ions for the DON-10-GSH conjugate in the sample was low, and only the major product ion from targeted-HRMS/MS was observed (*m/z* 306.0774 for $[GSH-H]^-$, Δ −2.9 ppm) (Figure 3).

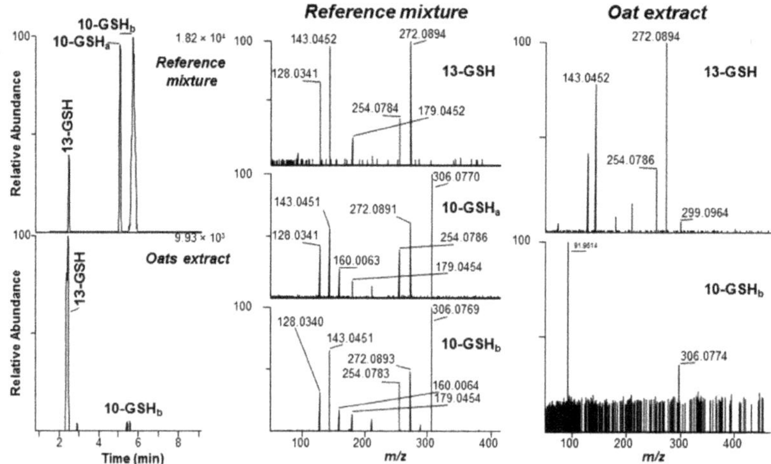

Figure 3. Left-hand panels, extracted ion LC-HRMS/MS chromatograms (*m/z* 272.0894 + 306.0769; ±5 ppm) precursor ions at m/z 602.2 ($[M-H]^-$ for DON–GSH) of: a mixture of DON, GSH, CysGly, γ-GluCys and Cys (pH 10.7) after reaction for one week (upper chromatogram); and an extract from naturally-contaminated ripe Norwegian oat grains (lower chromatogram). The early-eluting adduct is DON-13-GSH (from addition of L-glutathione to the epoxide group of DON), while the later-eluting adducts are from isomers of DON-10-GSH (from addition to the 9,10-double bond of DON). Comparison of the HRMS/MS spectrum of the early-eluting peak from the oats (right-hand upper panel) with that of DON-13-GSH in the reaction mixture (center upper panel) confirmed its identity. The *m/z* 306.0774 product-ion in the later-eluting conjugate in the oats (lower right-hand panel) suggested that the oats also contained minor amounts of DON-10-GSH, conjugate since this was also the most prominent product-ion in all the semi-synthetic isomers of DON-10-GSH (lower two panels for the reference mixture). Subscripts (a,b) indicate different diastereoisomers.

Sulfides are known to be oxidized by 30% hydrogen peroxide to sulfoxides or sulfones [10,11], and thiol-conjugates of DON are conveniently oxidized to their sulfoxides using this approach without further oxidation to sulfones [6,8]. Oxidation with hydrogen peroxide followed by LC-MS analysis may thus be applied as a simple technique to verify putative DON–thiol conjugates in complex mixtures, an approach analogous to that developed for identifying Met-containing peptides in cyanobacterial blooms [12]. Addition of hydrogen peroxide to an extract from Norwegian oats oxidized approximately 90% of the DON-13-GSH to the corresponding sulfoxides (DON-13-GS(O)H) within 90 min (Figure 4). This resulted in a decrease in the intensities of the $[M + H]^+$ and $[M - H]^-$ peaks for DON-13-GSH, and the concurrent appearance of two major peaks for the diasteroisomers of DON–GS(O)H at *m/z* 620.2137 ($[M + H]^+$, Δ2.8 ppm) and 618.2014 ($[M - H]^-$, Δ6.4 ppm) (Figure 4). Similar results were obtained when an aliquot of a mixture from reaction of DON with GSH was treated with hydrogen peroxide for 60 min (Figure S2).

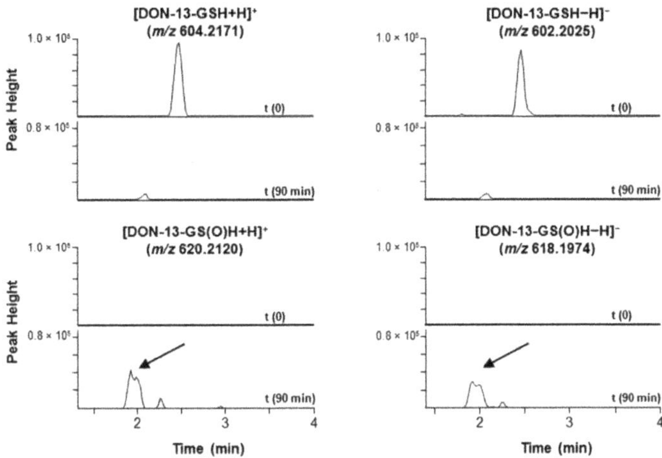

Figure 4. Extracted ion LC-HRMS chromatograms (±5 or 7.5 ppm for positive/negative ion mode, respectively) for $[M + H]^+$ and $[M - H]^-$ of DON-13-GSH and its sulfoxide, DON-13-GS(O)H, in an extract of ripe Norwegian oat grains. The upper two chromatograms show the disappearance of the DON-13-GSH peak after treatment of the extract with hydrogen peroxide for 90 min, while the lower two chromatograms show the concurrent appearance of the partially separated peaks from a pair of DON-13-GS(O)H diastereoisomers.

The composition of DON–thiol conjugates in the DON-injected wheat sample and the naturally-contaminated oat and wheat samples are in accord with observations from earlier studies on the complex and dynamic reaction of DON with various thiols [6–8]. The epoxide conjugates form irreversibly but relatively slowly, while the diastereomeric Michael addition products form reversibly with different, but generally much faster, reaction rates [6–8]. This is therefore consistent with finding the most rapidly-formed C-10 Michael adduct in the DON-injected wheat, but predominantly the corresponding C-13 epoxide adducts in the naturally-contaminated grain that potentially had been exposed to DON over a much longer time. Together with the fact that *Fusarium graminearum* is known to produce DON as a virulence factor of fungal spread already during the early stage of host infection, these findings also suggest that the DON-13-thiol conjugates detected in naturally-contaminated harvest-ripe grain might be largely the result of chemical, rather than enzymatic, reactions. If the formation of DON-13-thiol conjugates was an enzyme-catalyzed reaction it would be expected to observe such conjugates in the sample from the DON-treated wheat spikelet. In vitro, conjugation of DON with thiols generally proceeds to give primarily Michael adducts initially, but the reactions often do not go to completion due to autoxidation of the thiol [6–8]. However, in vivo produced DON would be exposed continuously to consistent endogenous physiological concentrations of GSH that would allow the slow but irreversibly-formed DON-13-GSH to accumulate.

The peak areas for DON-13-GSH and DON-13-Cys in the LC-HRMS chromatograms from the grain samples correlated well with those of DON (Figure 5). However, the peak areas of the conjugates were typically 100–1000-fold lower than that of DON in the same sample. The peak area of DON-3-β-D-glucoside likewise also correlated well with the peak area of DON (Figure 5).

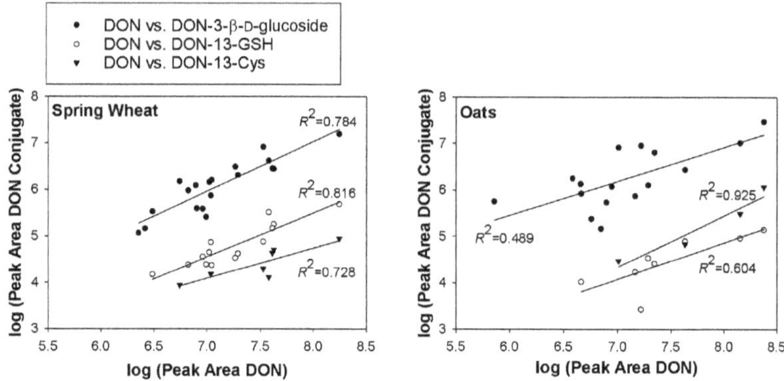

Figure 5. Plots of the log-transformed LC-HRMS peak areas of DON ([M + formate]$^-$) versus those for DON-3-β-D-glucoside ([M + formate]$^-$), DON-13-GSH ([M − H]$^-$), and DON-13-Cys ([M − H]$^-$) in extracts from naturally contaminated Norwegian spring wheat (left) and oats (right). Curves and squared correlation coefficients are from least squares regression.

2.2. Mercapturates of DON

In plants, GSH conjugates of xenobiotics are thought to be deposited in the vacuoles [13], where they can be degraded enzymatically to CysGly and Cys conjugates [14]. However, the metabolic fate of the resulting cysteine conjugates is less well established. In plants, the resulting Cys-conjugates can be *N*-acetylated, or *N*-malonylated followed by decarboxylation, to form *N*-acetylcysteine (NAC) conjugates (mercapturates) [15]; while in mammals they could also be cleaved by cysteine *S*-conjugate β-lyase (usually present in liver and kidneys) to mercaptans [16]. Although NAC biotransformation products are rarely reported in plants, we specifically looked for DON–NAC conjugates in extracts from naturally DON-contaminated grain and the DON-injected wheat by plotting the appropriate LC-HRMS extracted ion chromatograms (Figure 6). We then compared the chromatograms with those from the reaction of DON with NAC (Figure 6). As with DON–GSH, –CysGly, –γ-GluCys and –Cys conjugates, the DON—NAC conjugates afforded both [M + H]$^+$ (*m/z* 460.1641, Δ1.1 ppm) and [M − H]$^-$ (*m/z* 458.1515–458.1517, Δ 5.4–5.8 ppm) ions upon electrospray ionization. However, the signal/noise of extracted ion chromatograms from the naturally-contaminated grain samples was higher in the chromatograms from negative ionization. Where detected, the isomer-profile of the putative DON–NAC conjugates in the grain resembled those observed for the other DON–thiol conjugates in grain, with one major early-eluting conjugate consistent with addition to C-13, and another minor later-eluting conjugate consistent with addition to C-10. DON–NAC conjugates were not detected in the DON-injected wheat sample. The identity of the conjugates was confirmed by comparing product-ion spectra from targeted LC-HRMS/MS of the [M − H]$^-$ ions of the products from reaction of DON with NAC with those of the putative DON–NAC conjugates in a concentrated oat extract (Figure 6). Both the retention time and the HRMS/MS spectrum of [M − H]$^-$ for the putative DON-13-NAC in the oat extract were essentially identical to those of DON-13-NAC in the reference mixture (Figure 6). The base peak in the HRMS/MS spectrum of DON-13-NAC was at *m/z* 299.0955–299.0957 ($C_{14}H_{19}O_5S^-$, $\Delta-1.2$ to -0.6 ppm) (Figure 6). This product-ion has been observed for other DON-13-thiol conjugates and can be attributed to cleavage of the S–C bond on the amino-acid-side of the sulfide bond with concomitant loss of CH_2O from C-6 (Figure 7) [6]. On the other hand, the base peak in the HRMS/MS spectrum of the putative DON-10-NAC conjugate was at *m/z* 162.0217–162.0218 ($C_5H_8O_3NS^-$, $\Delta-8.3$ to -7.6 ppm), corresponding to deprotonated NAC (Figures 6 and 7) produced by retro-Michael addition. Thus, the above data confirm the presence of DON-13-NAC and DON-10-NAC in Norwegian grain.

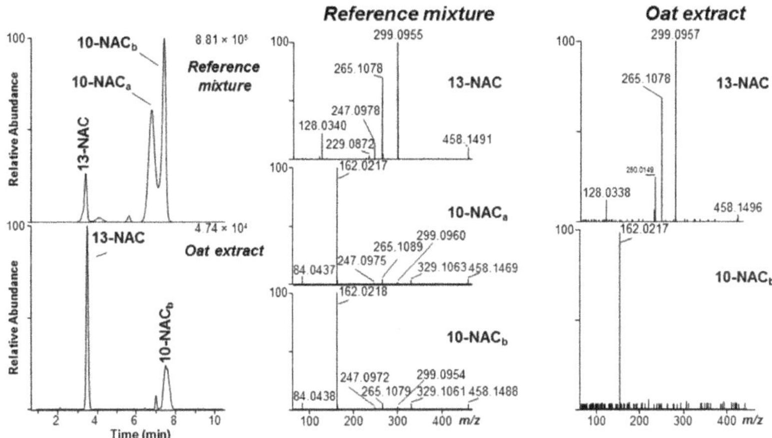

Figure 6. Left, extracted ion LC-HRMS chromatograms ($[M-H]^-$, ±3 ppm) for DON–NAC adducts (*m/z* 458.1490) ("DON mercapturates") in a semi-synthetic reference mixture (upper), and in an extract from ripe naturally-contaminated Norwegian oat grains (lower). Comparison of the HRMS/MS spectra of the $[M-H]^-$ ions (*m/z* 458.15) in the reference mixture with those in the grain extract supported the identity of compounds in oats as being DON-13-NAC (early-eluting conjugate) and DON-10-NAC (later-eluting conjugate). Subscripts (a,b) indicate different diastereoisomers.

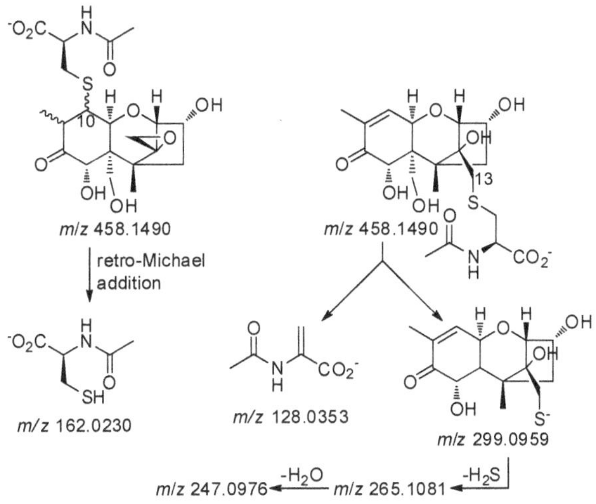

Figure 7. Major product ions observed during fragmentation of $[M-H]^-$ for DON–NAC adducts, and their proposed origins. The negative charge is arbitrarily depicted on the most acidic position in the fragments.

2.3. Separation of DON–Thiol Adducts

In earlier LC-MS studies of thiol-conjugation of DON, it was challenging to separate the isomeric reaction products on an octadecylsilane stationary phase [6,8]. Pentafluorophenylpropyl (PFPP) phases have recently shown to provide unique selectivity, especially for separation of regio- and stereo-isomers [17]. We used a PFPP column with core-shell particles that provided improved separation of the DON–GSH conjugates [7]. Up to three of the four possible 9,10-diastereoisomers of DON-10-Cys and DON-10-NAC could be separated, while only two baseline-separated peaks were obtained for DON-10-GSH, DON-10-CysGly and DON-10-γ-GluCys (Figure 2 and Figure S1). However, we have earlier shown that the broader later-eluting peak of DON-10-GSH contained at least two closely-eluting isomers with small retention time differences but marked differences in positive ion HRMS/MS spectra [7]. Negative ion HRMS/MS spectra did not discriminate between these isomers of DON-10-GSH [7], as was also observed in the present study (Figure 3). Similar broadened peaks were also observed for DON-10-GSH (t_R 5.5 min), DON-10-CysGly (t_R 7.0 min), DON-10-γ-GluCys (t_R 4.7 min), DON-10-Cys (t_R 6.5 min) and DON-10-NAC (t_R 6.7 min) in negative ion LC-MS (Figure 2 and Figure S1).

2.4. DON–Acetates

Several of the grain samples contained both DON-3-*O*-acetate and DON-15-*O*-acetate, with DON-3-*O*-acetate as the dominant analogue (Figure 8). The chromatographic separation of the two isomers has rarely been shown but was recently achieved using a similar HPLC column chemistry (pentafluorophenyl) as was used in this study for separation of DON–thiol isomers [18]. While DON-3-*O*-acetate is a common contaminant of cereal grain in Norway [19], this is the first report of DON-15-*O*-acetate in Norwegian grain and corresponds well with the DON-15-*O*-acetate producing genotypes of *Fusarium graminearum* observed for the first time in Norway in 2006 [20]. The identity of the two DON-acetates was supported by comparison of the LC-HRMS/MS spectra of the $[M + H]^+$ ions in the reference standard and the oat extract (Figure S3).

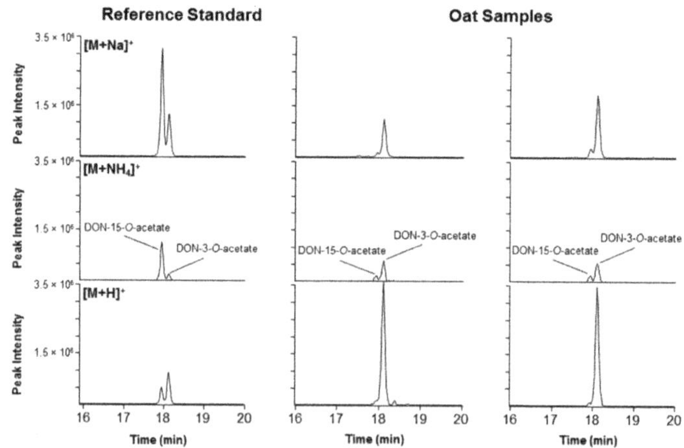

Figure 8. Extracted ion chromatograms (±3 ppm) from LC-HRMS showing the separation and concurrent presence of DON-3-*O*-acetate and DON-15-*O*-acetate in grain. Left, a mixed standard containing 50 ng/mL of both acetates; middle and right, extracts from two naturally-contaminated Norwegian oat samples. The top, middle and lower panels show the extracted ion chromatograms at *m/z* 361.1258 $[M + Na]^+$, 356.1704 $[M + NH_4]^+$, and 339.1438 $[M + H]^+$, respectively. The relatively higher peak intensities of $[M + H]^+$ for DON-3-*O*-acetate in the samples vs. the reference standard could be due to matrix signal enhancement.

3. Conclusions

This is the first report of the natural occurrence of 10- and 13-conjugates of DON with GSH, CysGly, Cys and NAC. The finding of NAC-conjugates is of special interest since such biotransformation products are rarely reported in plants. We did not aim to quantify the concentrations of DON–thiol conjugates. However, our experience regarding relative responses of different types of molecules in the used LC-HRMS instrument suggests that the concentrations of the DON–thiol conjugates in the samples were substantially lower than those of DON itself or DON-3-glucoside. The origin of the DON–NAC conjugates remains to be elucidated, i.e., if they are the result of acetylation of DON–Cys or the product of decarboxylation of a possible DON-malonylcysteine conjugate. Another question that needs to be addressed is the involvement of glutathione *S*-transferases in the conjugation in plants. Our results suggest that at least the formation of DON-13-GSH is merely the result of chemical reaction rather than due to enzyme-catalyzed conjugation. Since DON-13-GSH is likely to be nontoxic and its formation irreversible, the identification of cereal genotypes that utilize a detoxification pathway that leads to such a product may prove useful in future breeding strategies aiming to reduce DON accumulation in cereals.. Furthermore, the presence of the sulfur-linked DON conjugates suggests that other plant-derived thiols, such as cysteine-containing proteins, also may bind to DON. The latter suggest a possible use of such conjugates as biomarkers.

4. Materials and Methods

4.1. Chemicals and Reagents

DON (≥98%), GSH (≥98%), Cys, γ-GluCys, CysGly, NAC and Na_2CO_3 (pro analysis) were from Sigma-Aldrich (Steinheim, Germany), and $NaHCO_3$ (pro analysis) was from Fluka (Steinheim, Germany). $NaHCO_3$ and Na_2CO_3 were used to prepare 0.2 M buffer with pH of 10.7 as measured with a Mettler Delta 320 pH meter at ambient temperature. Purified quantitative standards of DON-10-Cys, DON-10-GSH, DON-13-Cys and DON-13-GSH were available from previous work [6,7], while DON-3-*O*-β-D-glucoside, DON-3-*O*-acetate and DON-15-*O*-acetate were from Romer Labs (Tulln, Austria).

4.2. Preparation of Reference Standard Mixtures

DON (100 μg) was dissolved in 1 mL of a freshly prepared solution of GSH, CysGly, γ-GluCys and Cys (10 mM each) in 0.2 M carbonate buffer (pH 10.7). Aliquots (20 μL) were periodically transferred to chromatography vials, diluted to 1 mL with water, and 1.5 μL acetic acid was added to stop the reaction. Separate reference mixtures were prepared by dissolving 100 μg DON in 1 mL 10 mM NAC in 0.2 M carbonate buffer (pH 10.7). Aliquots were prepared in the same way as for the mixture containing the four thiols above.

4.3. Samples and Extraction

Milled samples of naturally DON-contaminated grains of Norwegian spring wheat and oats were obtained from the Norwegian Institute of Bioeconomy Research, NIBIO [21]. The samples were from the 2004 to 2011 growing seasons, and were selected because of their relatively high DON contamination (1100–11,000 μg/kg). The grain samples had been milled after harvest and stored at −20 °C thereafter. The DON concentrations in these samples had been determined by LC-MS/MS in earlier projects [21]. The presence of DON in these samples was verified in all samples (Table S1), but we did not make efforts to re-quantify the toxin in this study. Aliquots of 0.1 g were weighed into 1.5 mL Eppendorf tubes and 1 mL methanol/water (3:1, *v/v*), containing 0.1% formic acid, was added. The samples were vortex-mixed for ca. 10 s, and then placed in an ultrasonic bath for 15 min followed by centrifugation at 15,000 *g* for 5 min. Supernatants were filtered through 0.45 μm PTFE syringe filters (Phenomenex) and transferred to chromatography vials and sealed.

An extract from DON-treated spikelets of spring wheat cultivar "Remus", which is highly susceptible to *Fusarium* head blight, was obtained from the IFA-Tulln, Austria. The plant ears had been treated with 1 mg DON/wheat ear in a greenhouse experiment. Treated spikelets had been sampled and extracted after incubation for 96 h [4]. The extraction protocol was identical to that used for extraction of the naturally-contaminated grain samples.

4.4. LC-HRMS(/MS) Analyses

Separation was achieved using a 150 × 2.1 mm, i.d. 2.6 μm Kinetex F5 column (Phenomenex, Torrance, CA, USA). Injection volumes were 1–5 μL. The mobile phase (250 μL/min) consisted of 5 mM ammonium formate (A), and 5 mM ammonium formate in 95:5 methanol–water (B), in a linear gradient from 3%–40% B over 14.5 min, then to 100% B at 14.7 min (2-min hold), followed by return to 3% B at 16.9 min, and equilibration with 3% B for 3.1 min using a Dionex UltiMate 3000 UPLC pump (Thermo Fischer Scientific, Waltham, MA, USA). The detector was a Q-Exactive Fourier-transform high-resolution mass spectrometer (Thermo Fischer Scientific) equipped with a heated electrospray ionization interface. The HRMS was run in positive and negative ion full-scan mode using fast polarity switching (i.e., alternating positive and negative ion scans), in the mass range *m/z* 150–1200 (Table 1). The mass resolution was set to 70,000 at *m/z* 200. The spray voltage was 4 kV, the transfer capillary temperature was 250 °C, and the sheath and auxiliary gas flow rates were 35 and 10 units, respectively. Exact values of *m/z* used for extracted ion LC-HRMS chromatograms for DON acetates, thiol conjugates and their sulfoxides (Table 1) were in this study obtained using Fusarium toxin mass calculator version 9 [22], while Xcalibur 2.3 was used to calculate the errors in the accurate masses.

Table 1. Exact values of *m/z* for ions of DON and its derivatives used in this study [22].

Compound	Ion Species ESI$^+$	Calculated Mass (*m/z*)	Ion Species ESI$^-$	Calculated Mass (*m/z*)
Deoxynivalenol (DON)	-	-	$[M + formate]^-$	341.1242
3/15-*O*-acetyl-DON	$[M + H]^+$	339.1438	-	-
	$[M + NH_4]^+$	356.1704		
	$[M + Na]^+$	361.1258		
DON–GSH	$[M + H]^+$	604.2171	$[M - H]^-$	602.2025
DON–GS(O)H	$[M + H]^+$	620.2120	$[M - H]^-$	618.1974
DON–CysGly	$[M + H]^+$	475.1745	$[M - H]^-$	473.1599
DON–γ-GluCys	$[M + H]^+$	547.1956	$[M - H]^-$	545.1811
DON–Cys	$[M + H]^+$	418.1530	$[M - H]^-$	416.1385
DON–NAC	$[M + H]^+$	460.1636	$[M - H]^-$	458.1490
DON-3-β-D-glucoside	$[M + NH_4]^+$	476.2126	$[M + formate]^-$	503.1770

LC-HRMS/MS was used to acquire high-resolution MS/MS spectra for selected DON–thiol conjugates in a concentrated oat sample and compared to those obtained from semi-synthetic reference compounds, using the same chromatographic conditions and detector as above. Analyses were performed in negative mode, using targeted MS/MS scans. The precursor ion mass for the MS/MS spectra was selected with a quadrupole isolation window of *m/z* 2.0. The normalized collision energy for higher-energy collisional dissociation was set to 30 and the resolution during product ion scanning was set to 17,500 or 35,000.

4.5. Oxidation of DON–GSH to DON–GS(O)H with Hydrogen Peroxide

Hydrogen peroxide (30%, 15 μL) was added to a 100 μL aliquot of a methanol–water extract from oats in a sample vial and sealed. The same amount was also added to a 100 μL aliquot of the acidified reference mixture (t = 1 week) from reaction of DON with GSH, CysGly, γ-GluCys and Cys. Oxidation of DON–GSH to its sulfoxide, DON–GS(O)H, was shown by injections into the LC-HRMS system just before addition of hydrogen peroxide and after a reaction time of 90 min (20 °C). The signal/noise for other DON–thiols in the sample was too low to obtain clear data for products from sulfide oxidation.

Supplementary Materials: The following are available online at www.mdpi.com/2072-6651/8/11/329/s1, Table S1: Samples and individual peak areas, Figure S1: Extracted ion chromatograms for products from reaction of DON with γ-GluCys and NAC, Figure S2: *S*-oxidation of DON–GSH reference mixture, Figure S3: Product ion spectra of DON–acetates in reference standard and an oats sample.

Acknowledgments: This work was part of the project "Mycotoxins and toxigenic fungi in Norwegian pig farming: consequences for animal health and possible intervention strategies" funded by the Research Council of Norway (grant No. 225332) and co-financed by Animalia, Lantmännen Research Foundation and Felleskjøpet Fôrutvikling. The authors B.K. and R.S. also acknowledge financial support received by the Austrian Science Fund FWF project SFB Fusarium (F3706, F3715).

Author Contributions: S.U., C.O.M. and A.S. conceived, designed and performed the experiments, analyzed the data, and wrote the paper; I.S.H. selected and provided naturally contaminated samples, discussed the data, and wrote the paper; and B.K. and R.S. provided the DON-injected wheat sample, discussed the data, and wrote the paper.

Conflicts of Interest: The authors declare no conflict of interest.

Abbreviations

The following abbreviations are used in this manuscript:

Cys	L-cysteine
DON	4-deoxynivalenol
γ-GluCys	γ-L-glutamyl-L-cysteine
CysGly	L-cysteinyl-L-glycine
GSH	L-glutathione
HRMS	high-resolution mass spectrometry
LC	liquid chromatography
MS/MS	tandem mass spectrometry
NAC	*N*-acetylcysteine
PFPP	pentafluorophenylpropyl
t_R	retention time

References

1. Nesic, K.; Ivanovic, S.; Nesic, V. Fusarial toxins: Secondary metabolites of *Fusarium* fungi. *Rev. Environ. Contam. Toxicol.* **2014**, *228*, 101–120. [PubMed]
2. Bonnet, M.S.; Roux, J.; Mounien, L.; Dallaporta, M.; Troadec, J.-D. Advances in deoxynivalenol toxicity mechanisms: The brain as a target. *Toxins* **2012**, *4*, 1120–1138. [CrossRef] [PubMed]
3. Pinton, P.; Oswald, I.P. Effect of deoxynivalenol and other type B trichothecenes on the intestine: A review. *Toxins* **2014**, *6*, 1615–1643. [CrossRef] [PubMed]
4. Kluger, B.; Bueschl, C.; Lemmens, M.; Michlmayr, H.; Malachova, A.; Koutnik, A.; Maloku, I.; Berthiller, F.; Adam, G.; Krska, R.; et al. Biotransformation of the mycotoxin deoxynivalenol in *Fusarium* resistant and susceptible near isogenic wheat lines. *PLoS ONE* **2015**, *10*. [CrossRef] [PubMed]
5. Kluger, B.; Bueschl, C.; Lemmens, M.; Berthiller, F.; Häubl, G.; Jaunecker, G.; Adam, G.; Krska, R.; Schuhmacher, R. Stable isotopic labelling-assisted untargeted metabolic profiling reveals novel conjugates of the mycotoxin deoxynivalenol in wheat. *Anal. Bioanal. Chem.* **2013**, *405*, 5031–5036. [CrossRef] [PubMed]
6. Stanic, A.; Uhlig, S.; Solhaug, A.; Rise, F.; Wilkins, A.L.; Miles, C.O. Preparation and characterization of cysteine adducts of deoxynivalenol. *J. Agric. Food Chem.* **2016**, *64*, 4777–4785. [CrossRef] [PubMed]
7. Stanic, A.; Uhlig, S.; Sandvik, M.; Rise, F.; Wilkins, A.L.; Miles, C.O. Characterization of deoxynivalenol–glutathione conjugates using nuclear magnetic resonance spectroscopy and liquid chromatography–high-resolution mass spectrometry. *J. Agric. Food Chem.* **2016**, *64*, 6903–6910. [CrossRef] [PubMed]
8. Stanic, A.; Uhlig, S.; Solhaug, A.; Rise, F.; Wilkins, A.L.; Miles, C.O. Nucleophilic addition of thiols to deoxynivalenol. *J. Agric. Food Chem.* **2015**, *63*, 7556–7566. [CrossRef] [PubMed]

9. Fruhmann, P.; Weigl-Pollack, T.; Mikula, H.; Wiesenberger, G.; Adam, G.; Varga, E.; Berthiller, F.; Krska, R.; Hametner, C.; Frohlich, J. Methylthiodeoxynivalenol (MTD): Insight into the chemistry, structure and toxicity of thia-Michael adducts of trichothecenes. *Org. Biomol. Chem.* **2014**, *12*, 5144–5150. [CrossRef] [PubMed]
10. Roy, K.-M. Sulfones and Sulfoxides. In *Ullmann's Encyclopedia of Industrial Chemistry*; Wiley-VCH Verlag GmbH & Co. KGaA: Weinheim, Germany, 2000.
11. Sato, K.; Hyodo, M.; Aoki, M.; Zheng, X.-Q.; Noyori, R. Oxidation of sulfides to sulfoxides and sulfones with 30% hydrogen peroxide under organic solvent- and halogen-free conditions. *Tetrahedron* **2001**, *57*, 2469–2476. [CrossRef]
12. Miles, C.O.; Melanson, J.E.; Ballot, A. Sulfide oxidations for LC-MS analysis of methionine-containing microcystins in *Dolichospermum flos-aquae* NIVA-CYA 656. *Environ. Sci. Technol.* **2014**, *48*, 13307–13315. [CrossRef] [PubMed]
13. Coleman, J.; Blake-Kalff, M.; Davies, E. Detoxification of xenobiotics by plants: Chemical modification and vacuolar compartmentation. *Trends Plant Sci.* **1997**, *2*, 144–151. [CrossRef]
14. Noctor, G.; Mhamdi, A.; Chaouch, S.; Han, Y.; Neukermans, J.; Marquez-Garcia, B.; Queval, G.; Foyer, C.H. Glutathione in plants: An integrated overview. *Plant Cell Environ.* **2012**, *35*, 454–484. [CrossRef] [PubMed]
15. Brazier-Hicks, M.; Evans, K.M.; Cunningham, O.D.; Hodgson, D.R.W.; Steel, P.G.; Edwards, R. Catabolism of glutathione conjugates in *Arabidopsis thaliana*: Role in metabolic reactivation of the herbicide safener fenclorim. *J. Biol. Chem.* **2008**, *283*, 21102–21112. [CrossRef] [PubMed]
16. Cooper, A.J.L.; Krasnikov, B.F.; Niatsetskaya, Z.V.; Pinto, J.T.; Callery, P.S.; Villar, M.T.; Artigues, A.; Bruschi, S.A. Cysteine *S*-conjugate β-lyases: Important roles in the metabolism of naturally occurring sulfur and selenium-containing compounds, xenobiotics and anticancer agents. *Amino Acids* **2011**, *41*, 7–27. [CrossRef] [PubMed]
17. Havlíková, L.; Matysová, L.; Hájková, R.; Šatínský, D.; Solich, P. Advantages of pentafluorophenylpropyl stationary phase over conventional C18 stationary phase—Application to analysis of triamcinolone acetonide. *Talanta* **2008**, *76*, 597–601. [CrossRef] [PubMed]
18. Tamura, M.; Mochizuki, N.; Nagatomi, Y.; Harayama, K.; Toriba, A.; Hayakawa, K. A method for simultaneous determination of 20 *Fusarium* toxins in cereals by high-resolution liquid chromatography-orbitrap mass spectrometry with a pentafluorophenyl column. *Toxins* **2015**, *7*, 1664–1682. [CrossRef] [PubMed]
19. Uhlig, S.; Eriksen, G.S.; Hofgaard, I.S.; Krska, R.; Beltrán, E.; Sulyok, M. Faces of a changing climate: Semi-quantitative multi-mycotoxin analysis of grain grown in exceptional climatic conditions in Norway. *Toxins* **2013**, *5*, 1682–1697. [CrossRef] [PubMed]
20. Aamot, H.U.; Ward, T.J.; Brodal, G.; Vrålstad, T.; Larsen, G.B.; Klemsdal, S.S.; Elameen, A.; Uhlig, S.; Hofgaard, I.S. Genetic and phenotypic diversity within the *Fusarium graminearum* species complex in Norway. *Eur. J. Plant Pathol.* **2015**, *142*, 501–519. [CrossRef]
21. Hofgaard, I.S.; Aamot, H.U.; Torp, T.; Jestoi, M.; Lattanzio, V.M.T.; Klemsdal, S.S.; Waalwijk, C.; Van der Lee, T.; Brodal, G. Associations between *Fusarium* species and mycotoxins in oats and spring wheat from farmers' fields in Norway over a six-year period. *World Mycotoxin J.* **2016**, *9*, 365–378. [CrossRef]
22. Miles, C.O. *Fusarium* Toxin Mass Calculator (Version 9 in Excel). 2016. Available online: https://www.researchgate.net/publication/304249256_Fusarium_toxin_mass_calculator_version_9_in_Excel (accessed on 23 June 2016).

© 2016 by the authors. Licensee MDPI, Basel, Switzerland. This article is an open access article distributed under the terms and conditions of the Creative Commons Attribution (CC BY) license (http://creativecommons.org/licenses/by/4.0/).

Article

Effect of Various Compounds Blocking the Colony Pigmentation on the Aflatoxin B1 Production by *Aspergillus flavus*

Vitaly G. Dzhavakhiya, Tatiana M. Voinova, Sofya B. Popletaeva, Natalia V. Statsyuk *, Lyudmila A. Limantseva and Larisa A. Shcherbakova

All-Russian Research Institute of Phytopathology, Bolshie Vyazemy, Moscow 143050, Russia; dzhavakhiya@yahoo.com (V.G.D.); tatiana.voinova@bk.ru (T.M.V.); unavil@yandex.ru (S.B.P.); lutik47@yandex.ru (L.A.L.); larisavniif@yahoo.com (L.A.S.)

* Correspondence: nataafg@gmail.com; Tel.: +7-926-242-7241

Academic Editors: Kris Audenaert, Sarah De Saeger and Siska Croubels

Received: 15 September 2016; Accepted: 24 October 2016; Published: 28 October 2016

Abstract: Aflatoxins and melanins are the products of a polyketide biosynthesis. In this study, the search of potential inhibitors of the aflatoxin B1 (AFB1) biosynthesis was performed among compounds blocking the pigmentation in fungi. Four compounds—three natural (thymol, 3-hydroxybenzaldehyde, compactin) and one synthetic (fluconazole)—were examined for their ability to block the pigmentation and AFB1 production in *Aspergillus flavus*. All compounds inhibited the mycelium pigmentation of a fungus growing on solid medium. At the same time, thymol, fluconazole, and 3-hydroxybenzaldehyde stimulated AFB1 accumulation in culture broth of *A. flavus* under submerged fermentation, whereas the addition of 2.5 μg/mL of compactin resulted in a 50× reduction in AFB1 production. Moreover, compactin also suppressed the sporulation of *A. flavus* on solid medium. In vivo treatment of corn and wheat grain with compactin (50 μg/g of grain) reduced the level of AFB1 accumulation 14 and 15 times, respectively. Further prospects of the compactin study as potential AFB1 inhibitor are discussed.

Keywords: aflatoxin B1; fungal melanins; polyketide biosynthesis; compactin; thymol; fluconazole; 3-hydroxybenzaldehyde; *Aspergillus flavus*

---

1. Introduction

Fungal polyketides represent a large group of biologically active compounds synthesized by enzymes from the polyketide synthase (PKS) family [1]. Aflatoxins are secondary metabolites produced by *Aspergillus flavus* via the polyketide pathway. The ability to produce aflatoxins is probably not related to the essential features of *A. flavus*, but may give some competitive advantages to this fungus compared with other toxin-sensitive microorganisms [2,3]. A highly hepatotoxic and carcinogenic aflatoxin B1 (AFB1) is of greatest concern because it can contaminate a wide range of agricultural products that results in great losses in agricultural income and has a significant negative economic impact. A number of methods are developed for pre- and post-harvest aflatoxin management [4]. Traditional approaches, such as fungicides, have shown very limited success in the AFB1 control [5]. Therefore, the search and development of efficient preparations able to inhibit AFB1 production still remains of great theoretical and practical importance.

Fungal melanins, representing high-molecular hydrophobic pigments, are also synthesized via the PKS-depending pathway [6]. The blocking of the melanin biosynthesis in pathogenic fungi can result in the loss of their pathogenicity and an increased susceptibility of fungi to biotic and abiotic stresses [7–9]. For example, a defect in one of the PKS-encoding genes of *A. flavus* resulting in a formation of non-pigmented sclerotia significantly increased the susceptibility of the fungus to the deleterious effect of UV light and heat; in addition, the melanin-deficient mutant was less resistant to insect predation as compared with the wild-type strain generating melanized sclerotia [10].

Based on the data of experiments with radiolabeled polyketide intermediates, the following simplified scheme of AFB1 biosynthesis was proposed: acetate → hypothetic polyketide intermediates → norsolorinic acid → averantin → averufin → versiconal hemiacetal acetate → versicolorin A → sterigmatocystin → AFB1 [11]. Later, this scheme was significantly enlarged. To date, AFB1 biosynthesis is considered to include 23 enzymatic reactions, and 15 intermediates of this pathway have been identified [12].

Norsolorinic acid is the first stable intermediate in the aflatoxin biosynthesis. A structural similarity between the pigments of *A. flavus* spores and norsolorinic acid made it possible to suppose that these pigments and aflatoxins have common precursors; therefore, aflatoxin and melanin biosynthetic pathways have common initial stages [13]. Due to this fact, the search for compounds able to block the early stages of the polyketide biosynthetic pathway before its branching to the aflatoxin and melanin biosyntheses represented a promising task, since such inhibitors would prevent AFB1 accumulation in treated food and feed products and simultaneously decrease the contamination of these products with *A. flavus* due to the decreased viability of melanin-deficient fungus.

In our earlier studies, we tested various phosphoanalogues of amino acids and their derivatives for their ability to block different stages of the polyketide biosynthetic pathway. As a result, some compounds, which blocked either AFB1 or melanin biosynthesis, have been revealed; in the last case, a simultaneous stimulation of the toxinogenesis was observed [14,15]. However, none of these compounds was able to block early stages of polyketide biosynthesis, i.e., simultaneously inhibit the production of AFB1 and melanin.

This study continued the search for inhibitors of the early stages of toxinogenesis among natural and synthetic compounds, which, as we found earlier, are able to block the pigmentation of some plant pathogenic fungi. Our earlier study showed that compactin, a natural inhibitor of the sterol biosynthesis, which inhibits HMG-CoA reductase catalyzing the conversion of HMG-CoA into mevalonic acid [16], causes the depigmentation of colonies of several plant pathogenic fungi [17]. Another possible inhibitor of AFB1 biosynthesis could be fluconazole, a synthetic triazole-based fungicide. It is known that the fungicide action of triazoles is determined by their ability to inhibit the biosynthesis of sterols [18]. In the preliminarily study, we showed that this fungicide was able to depigment the colonies of some fungi including *A. flavus*, so we supposed it could also influence on toxinogenesis.

According to one of the existing hypotheses, the evolution of plants, which are infected with toxigenic fungi, could result in their ability to produce specific compounds inhibiting the biosynthesis of toxins [19]. In fact, some plant terpenoids are really able to suppress AFB1 production by *A. flavus* [20]. One of such compounds is thymol, widely used in medicine, veterinary, and plant protection, so we included this compound into this study.

It is known that 2-chloroethyl phosphoric acid inhibits AFB1 biosynthesis due to oxidative stress alleviation [21]. Some other studies showed an important role of oxidative stress at the initial stages of AFB1 biosynthesis, though the mechanisms of this effect and the specific forms of reactive oxygen influencing on the toxinogenesis still remain unclear [22]. The earlier study devoted to the chemosensitization of *Aspergillus* spp. and some other fungi to the action of various antifungal agents showed that sensitizers, which can act as antioxidants preventing the oxidative stress, increased the sensitivity of fungi to industrial fungicides [23]. Among the tested compounds, the most active chemosensitizer was 3-hydroxybenzaldehyde (3-HBA), a compound of a plant origin. Since the antioxidant properties of this compound could probably inhibit the toxigenesis in *A. flavus*, it was also included into the current study.

Thus, based on the above-described considerations, the purpose of this study was the examination of four compounds of different nature (Figure 1) for their effect on both AFB1 and melanin production to find compound(s) able to block the early stages of polyketide biosynthesis, as well as the in vivo evaluation of the AFB1 accumulation in wheat and corn grain infected with toxigenic *A. flavus* and treated with the most efficient compound.

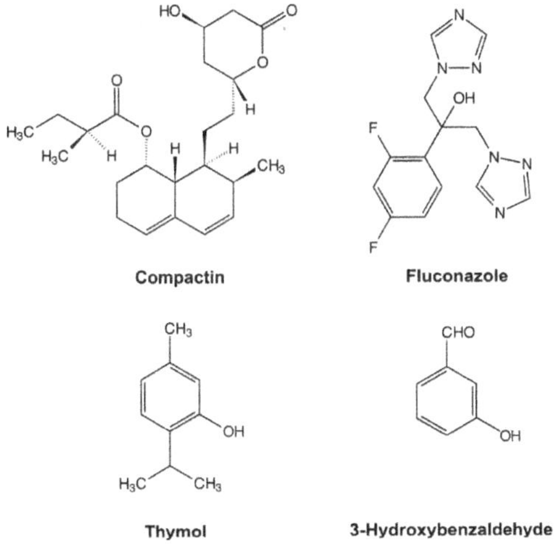

Figure 1. Compounds used in the study.

2. Results

2.1. *Effect of Tested Compounds on the Pigment Production in Aspergillus flavus*

According to the obtained results, all tested compounds were able to suppress the pigment production in *A. flavus* (Figure 2).

The maximum effect was observed for compactin; the mycelium changed its color from green to yellow starting from the compactin concentration equal to 2.5 μg/mL, while no significant growth suppression was revealed (Figure 2, Table 1). Such a picture was observed for either stab-inoculation into the center of a Petri dish or the spreading of spore suspension on the whole agar surface. Taking into account changes in the colony diameter in the case of the stab-inoculation, the compactin concentration providing the fungal growth suppression (25 μg/mL) is ten times greater than the concentration, at which mycelium begins to change its color (2.5 μg/mL).

In the case of a stub-inoculation of control (inhibitor-free) Petri dishes, *A. flavus* grows as coalesced small pigmented colonies, which is caused by abundant sporulation accompanied by a discharge of spores onto nutrient medium with their further germination. In the case of compactin-containing medium, the situation is quite different starting from the compactin concentration equal to 25 μg/mL. After a stub-inoculation, only one discolored colony grows in the center of a Petri dish (Figure 2). This phenomenon was observed for all compactin concentrations causing a full discoloration of mycelium (>25 μg/mL) and is explained by the inhibition of the spore generation on aerial mycelium, since the morphological examination of the mycelium under a light microscope did not reveal any spores in colonies grown on compactin-containing medium. In the case of the compactin concentration range 1–10 μg/mL, characterized by incomplete mycelium discoloration, the microscoping of colonies showed the presence of conidia and spores, but their color was clarified as compared with the control.

In the case of the *A. flavus* cultivation on the liquid compactin-containing medium, it was difficult to evaluate the effect of compactin on the mycelium pigmentation, since the fungal growth was significantly suppressed, when the compactin concentration exceeded 1 μg/mL. The dry weight of mycelium in the control variant and at the compactin concentrations of 1, 2.5, and 5 μg/mL was

1.53 ± 0.04, 1.33 ± 0.34, 0.10 ± 0.04, and 0.12 ± 0.03 g, respectively; a weak discoloration was observed at the compactin concentration of 1 μg/mL (Figure 3).

Figure 2. Effect of various compounds on the colony growth and pigmentation of *Aspergillus flavus*. (**A**) 3-hydroxybenzaldehyde; (**B**) thymol; (**C**) fluconazole; (**D**) compactin.

Table 1. Effect of various compactin content in agar medium on the growth of *Aspergillus flavus* colonies and the color of aerial mycelium.

Compactin Concentration, μg/mL	Colony Diameter, mm	Aerial Mycelium Color
0 (control)	90	Green
1	90	Green
2.5	90	Bright yellow/white
5	89	Bright yellow/white
10	88	Bright yellow/white
25	56	White with yellow center
50	45	White with yellow center

Figure 3. Effect of various compactin concentrations on the mycelium growth of *Aspergillus flavus* on liquid medium.

2.2. Effect of Tested Compounds on the Aflatoxin Production by Aspergillus flavus

The effect of the tested compounds on AFB1 production by *A. flavus* grown in liquid medium is shown in Figure 4. According to the obtained data, fluconazole fungicide added to the liquid medium at the concentrations of 50–100 µg/mL several times increased AFB1 production by *A. flavus*. An increased AFB1 production was also observed in the case of thymol and 3-HBA. Thus, the effect of these three compounds was similar to that observed earlier for *N*-hydroxyputrescine and (1-aminoethyl)phosphonic acid [24].

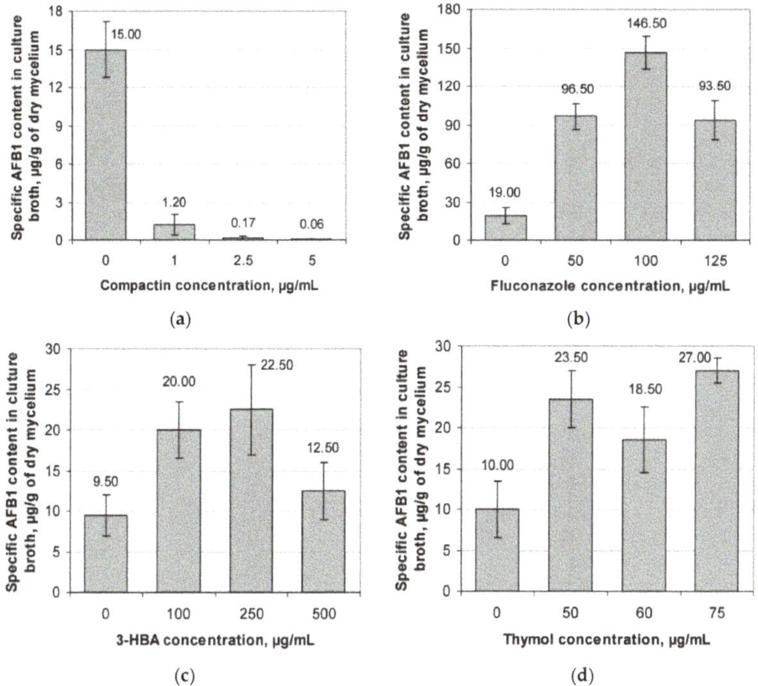

Figure 4. Effect of the tested compounds on the aflatoxin B1 production by *Aspergillus flavus*: (**a**) compactin; (**b**) fluconazole; (**c**) 3-hydroxybenzaldehyde; (**d**) thymol.

Unlike other compounds, compactin significantly reduced the AFB1 accumulation in culture broth. The calculated coefficient of the correlation between the compactin concentration and specific AFB1 content in culture broth was –0.646, which, according to the Chaddock scale, indicates a significant negative correlation between the mentioned parameters. Therefore, a conclusion can be made that compactin inhibits AFB1 production in *A. flavus* more actively than it suppresses the growth of mycelium. The experimental data confirm this conclusion: for example, the presence of 1 µg/mL of compactin almost did not suppress the fungal growth, while significantly reducing the toxin production (Figures 3 and 4).

2.3. In Vivo Evaluation of the Aflatoxin B1 Content in Treated Grain Samples

The data on the effect of the preliminary treatment of corn and wheat grain with compactin on the AFB1 accumulation in grain infected with *A. flavus* are shown in Figure 5.

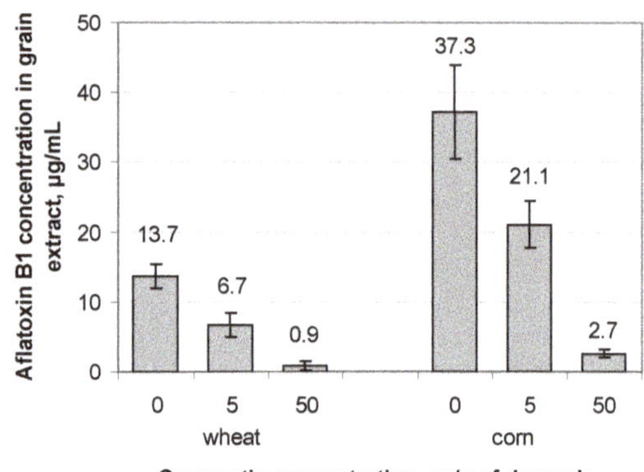

Figure 5. Aflatoxin B1 content in extracts from wheat and corn grain treated with compactin and artificially infected with toxigenic *Aspergillus flavus*.

In both cases, AFB1 concentration in extracts obtained from compactin-treated grain was significantly lower than that in the control reducing up to 6%–7% for the maximum applied compactin concentration (50 µg/g of dry grain). The calculated correlation coefficients between the applied compactin concentration and the extracted AFB1 were –0.93 and –0.90 for corn and wheat, respectively.

A difference between the growth of the fungus on treated and untreated grain was clearly visible (Figure 6), especially in the case of corn. The surface of untreated grains was covered by green colonies, whereas, in the case of the treated grain, less developed white mycelium was observed, which did not contain mature pigmented conidia.

Figure 6. Development of *Aspergillus flavus* on grain treated with compactin (5 µg/g of dry grain): (**a**) control and (**b**) treated wheat grain; (**c**) control and (**d**) treated corn grain.

3. Discussion

To date, there have been many studies describing various natural and synthetic compounds that have been shown to be specific inhibitors of AFB1 production. The majority of these compounds are of plant origin [20,25]. There are also some microbial metabolites, such as the melanin pigment from *Streptomyces torulosis* [26], some other metabolites from *Streptomyces* sp. [27], and respiration inhibitors of a fungal origin, characterized by the ability to significantly reduce AFB1 production by *A. flavus* without any significant effect on the fungal colony growth [28]. It is also known that some synthetic inhibitors of the pentaketide biosynthesis of melanin, such as the tricyclazole fungicide, are also able to inhibit AFB1 biosynthesis presumably via the inhibition of reductase involved in the versicolorin A conversion to dimethyl-sterigmatostycin at the late stages of AFB1 biosynthesis [29].

In this and previous studies, we examined a hypothesis about the possibility to reveal effective inhibitors of the early stages of AFB1 biosynthesis among compounds able to block the pigmentation of the fungus. As far as the authors know, compounds able to simultaneously inhibit both melanin and AFB1 production have only been described in a few publications [29,30], but no any special study of melanogenesis inhibitors was carried out.

Three of four compounds tested in this study suppressed the biosynthesis of the fungal pigment and, at the same time, stimulated the toxinogenesis. We also revealed several such compounds in our previous studies [14,24]. This effect can be explained by the common initial biosynthetic stages of both metabolites first hypothesized by Brown and Salvo [13]. At the later stages, the polyketide biosynthetic chain may branch in different pathways resulting in the production of melanin and AFB1. In this case, the observed stimulation of AFB1 production may be explained in the following way: these compounds block the melanin production at the stages located after the branching point that promotes the accumulation of preceding intermediates, which are also AFB1 precursors. An increased precursor accumulation provides increased AFB1 production.

The last tested compound, compactin, was able to suppress both pigment and toxin formation. One of the possible explanations for this is that it blocks the polyketide biosynthetic chain prior to the branching point; another possibility is that compactin is able to simultaneously block both melanin and AFB1 biosynthetic chains after the branching point. A more detailed study is required to determine if any of these variants is correct.

All inhibitors of the colony pigmentation tested in our current and previous studies showed either a stimulating or an inhibiting effect on AFB1 biosynthesis. Therefore, our data probably confirms the interrelation between the aflatoxinogenesis and the melanin biosynthesis in *A. flavus* occurring via the DOPA pathway [6] and can be used for the further elucidation of the early stages of toxinogenesis. The revealed interrelation can probably be used for practical purposes: the examination of potential AFB1 inhibitors for their ability to block melanin production can serve as a preliminary screening stage able to reveal highly effective toxin inhibitors, such as compactin. In addition, efficient AFB1 inhibitors could be probably revealed among the known inhibitors of the melanin biosynthesis.

The analysis of existing scientific and patent publications showed a lack of any information about the ability of compactin or other statins to inhibit the biosynthesis of mycotoxins. Therefore, we probably revealed a new group of AFB1 inhibitors that opens wide prospects for future studies. Among the planned theoretical tasks, we can mention the determination of the exact point of compactin action in the polyketide biosynthetic pathway using known intermediates and the study of the compactin effect on other mycotoxins of the polyketide origin, such as zearalenone. The authors also plan to continue the study of other statins for their potential as AFB1 inhibitors and have already obtained promising results for lovastatin [24]. Note that, according to the obtained data, compactin is able to inhibit the toxinogenesis at very low concentrations. Finally, an in vivo study of the compactin ability to efficiently reduce AFB1 accumulation in wheat and corn grain contaminated with toxigenic *A. flavus* demonstrated a potential for the practical application of this compound. Since the treatment of feed using such concentrations of active substances does not require high-purity compactin, the technology of its isolation and purification from culture broth can be simplified to reduce the cost of the final product. According to our calculations, the cost of such partially purified compactin obtained using our overproducing strain should be about 50 USD/kg. Taking into account the working compactin dose of 0.1 mg/kg of grain, the final cost of the grain treatment with compactin will make about 5 USD/ton.

In our further work, we plan to study some practical aspects of the possible statin application as AFB1 inhibitors for feed treatment, such as the evaluation of the duration of their inhibiting effect and toxicity assay towards birds and mammals.

4. Materials and Methods

4.1. Strains, Chemicals, and Culture Conditions

Aspergillus flavus AF11 (VKM F-27) was used as a producer of AFB1 throughout the study. The strain was purchased in the All-Russian Collection of Microorganisms (Skryabin Institute of Biochemistry and Physiology of Microorganisms, Pushchino, Russia). AF11 was maintained on agar medium of the following composition (g/L): agar, 20; yeast extract, 5; glucose, 20. In addition, a microelement stock solution ($FeSO_4 \cdot 7H_2O$ (24 mg/mL), $MnCl_2 \cdot 4H_2O$ (42 mg/mL), $CuCl_2 \cdot 2H_2O$ (12 mg/mL), $ZnSO_4 \cdot 7H_2O$ (84 mg/mL)) was added at a ratio of 100 μL per 1 L of the agar medium. The same medium was used to observe mycelium color changes.

A spore suspension prepared from a 10-day-old culture at a concentration of 1×10^7 spores/mL was used as the inoculum. Spore suspension (1 mL) was inoculated into 250 mL shaken flasks containing 50 mL of liquid Payne–Hagler medium [31]. Before the inoculation, the medium was supplemented with solutions of tested compounds sterilized by a filtration through 0.22 μm Millipore membrane filters (MilliporeSigma, St. Louis, MO, USA). In the case of control flasks, the medium did not contain potential inhibitors. The culture was incubated at 27 °C for 170 h under aeration conditions on a New Brunswick™ Excella E25/25R incubation shaker (New Brunswick Scientific, Edison, NJ, USA) at 200 rpm and a 5 cm orbit.

Thymol, fluconazole, and 3-HBA were purchased from Medsi LLC (Moscow, Russia), Makiz-Pharma LLC (Moscow, Russia), and Sigma-Aldrich (St. Louis, MO, USA), respectively.

Compactin was obtained using an overproducing *Penicillum citrinum* 18–12 strain developed by a multi-step induced mutagenesis. The cultivation of this strain and the compactin isolation and purification were carried out as described earlier [32]. Since compactin has poor water solubility, for our purposes, it was converted from the lactone form to the more soluble salt form. To do this, the known amount of compactin was dissolved in ethanol (the maximum concentration did not exceed 100 mg/mL), heated to a boil, then mixed with the equal volume of the equimolar NaOH solution and cooled to the room temperature. Then, the solution was diluted to the required concentration; pH of the final solution should be about 7.0.

The concentration range of the tested substances included minimal working concentrations providing a significant effect on the melanin production.

4.2. Analysis of Aflatoxin B1

After the completion of cultivation, culture broth of each experimental or control flask was centrifuged at 8000 *g* to separate the mycelia. The mycelia were washed with 3 mL of distilled water twice and dried at room temperature up to a constant weight.

The AFB1 content in the supernatant of culture broth was determined by HPLC as described earlier [33]. Aflatoxin B1 purchased from Sigma-Aldrich (St. Louis, MO, USA) was used as a reference sample.

To exclude the possible effect of fungal growth inhibition by the tested compounds, their inhibiting effect was assessed by the specific AFB1 content in culture broth (AFB1 content in culture broth normalized to the dry weight of mycelium). The effect of tested compounds on AFB1 biosynthesis was determined by a comparison between experimental and control flasks.

4.3. Evaluation of the Aflatoxin B1 Content in Treated Grain Samples

To assess the ability of the most efficient tested compound to prevent in vivo AFB1 accumulation in grain, wheat (var. Zlata) and corn (var. Ross 140 SB) grain samples were used as the model objects. Twenty grams of wheat or corn grains were placed into 250 mL shaken flasks containing 10 or 20 mL of distilled water, respectively. Flasks were autoclaved for 1 h under a pressure of 0.5 atmosphere. Then, 1 mL of a sterile compactin solution (0.1 mg/mL or 1 mg/mL) or sterile distilled water (control) was added into the flasks. After the mixing of the flask content, each flask was inoculated by 1 mL of the spore suspension of *A. flavus* containing 1×10^7 spores/mL. After the 8-day incubation at 26 °C, 50 mL of chloroform was added to each flask. Flasks were incubated for 3 h at 26 °C on a New Brunswick™ Excella E25/25R incubation shaker at 250 rpm, their content was then centrifuged for 30 min at 8000 *g* at room temperature to separate grain from the organic phase, and 200 µL of the chloroform extract was sampled from each flask. After the chloroform evaporation, the residue was dissolved in 200 µL of methanol. The AFB1 concentration was further determined by HPLC as described above.

4.4. Data Analysis

Measurements were analyzed using a STATISTICA v. 6.1 software (StatSoft Inc., Tulsa, OK, USA) and depicted as their mean (over at least three independent experiments) ± standard deviation. Differences were considered significant when $p < 0.05$.

Acknowledgments: This work was financially supported by the Russian Science Foundation (project No. 14-16-00150).

Author Contributions: V.G.D., T.M.V., and L.A.L. conceived and designed the experiments. T.M.V., S.B.P., and L.A.L. performed the experiments. V.G.D., T.M.V., N.V.S., and L.A.S. analyzed the data and wrote the paper.

Conflicts of Interest: The authors declare no conflict of interest.

References

1. Crawford, J.M.; Townsend, C.A. New insights into the formation of fungal aromatic polyketides. *Nat. Rev. Microbiol.* **2010**, *8*, 879–889. [CrossRef] [PubMed]
2. Cotty, P.J. Virulence and cultural characteristics of two *Aspergillus flavus* strains pathogenic on cotton. *Phytopathology* **1989**, *79*, 808–814. [CrossRef]
3. Cary, J.W.; Ehrlich, K.C. Aflatoxigenicity in *Aspergillus*: Molecular genetics, phylogenetic relationships and evolutionary implications. *Mycopathologia* **2006**, *162*, 167–177. [CrossRef] [PubMed]
4. Miller, J.D.; Schaafsma, A.W.; Bhatnagar, D.; Bondy, G.; Carbone, I.; Harris, L.J.; Harrison, G.; Munkvold, G.P.; Oswald, I.P.; Pestka, J.J.; et al. Mycotoxins that affect the North American agri-food sector: State of the art and directions for the future. *World Mycotoxin J.* **2014**, *7*, 63–82. [CrossRef]
5. Abbas, H.K.; Wilkinson, J.R.; Zablotowicz, R.M.; Accinelli, C.; Abel, C.A.; Bruns, H.A.; Weaver, M.A. Ecology of *Aspergillus flavus*, regulation of aflatoxin production, and management strategies to reduce aflatoxin contamination of corn. *Toxin Rev.* **2009**, *28*, 142–153. [CrossRef]
6. Pal, A.K.; Gajjar, D.U.; Vasavada, A.R. DOPA and DHN pathway orchestrate melanin synthesis in *Aspergillus* species. *Med. Mycol.* **2014**, *52*, 10–18. [PubMed]
7. Dzhavakhiya, V.G.; Aver'yanov, A.A.; Minaev, V.I.; Ermolinskiy, B.S.; Voinova, T.M.; Lapikova, V.P.; Petelina, G.G.; Vavilova, N.A. Structure and function of cell wall melanin of the micromycete *Pyricularia oryzae* Cav., the rice blast pathogen. *Zh. Obshch. Biol.* **1990**, *51*, 528–535.
8. Hamada, T.; Asanagi, M.; Satozawa, T.; Araki, N.; Banba, S.; Higashimura, N.; Akase, T.; Hirase, K. Action mechanism of the novel rice blast fungicide tolprocarb distinct from that of conventional melanin biosynthesis inhibitors. *J. Pestic. Sci.* **2014**, *39*, 152–158. [CrossRef]
9. Takagaki, M. Melanin biosynthesis inhibitors. In *Fungicide Resistance in Plant Pathogens*; Takagaki, M., Ishii, H., Hollomon, D.W., Eds.; Springer: Tokyo, Japan, 2015; pp. 175–180.

10. Cary, J.W.; Harris-Coward, P.Y.; Ehrlich, K.C.; Di Mavungu, J.D.; Malysheva, S.V.; De Saeger, S.; Dowd, P.F.; Shantappa, S.; Martens, S.L.; Calvo, A.M. Functional characterization of a veA-dependent polyketide synthase gene in *Aspergillus flavus* necessary for the synthesis of asparasone, a sclerotium-specific pigment. *Fungal Genet. Biol.* **2014**, *64*, 25–35. [CrossRef] [PubMed]
11. Bennett, J.W.; Christensen, S.B. New perspectives on aflatoxin biosynthesis. *Adv. Appl. Microbiol.* **1983**, *29*, 53–92. [PubMed]
12. Yu, J.; Chang, P.-K.; Ehrlich, K.C.; Cary, J.W.; Bhatnagar, D.; Cleveland, T.E.; Payne, G.A.; Linz, J.E.; Woloshuk, C.P.; Bennet, J.W. Clustered pathway genes in aflatoxin biosynthesis. *Appl. Environ. Microbiol.* **2004**, *70*, 1253–1262. [CrossRef] [PubMed]
13. Brown, D.W.; Salvo, J.J. Isolation and characterization of sexual spore pigments from *Aspergillus nidulans*. *Appl. Environ. Microbiol.* **1994**, *60*, 979–983. [PubMed]
14. Dzhavakhiya, V.; Statsyuk, N.; Shcherbakova, L. Blocking of some stages of melanogenesis can enhance aflatoxin B1 production in *Aspergillus flavus*. In Proceedings of Conference abstracts of the 37th Mycotoxin Workshop, Bratislava, Slovakia, 1–3 June 2015; Slovak University of Technology: Bratislava, Slovakia, 2015; p. 81.
15. Khomutov, R.M.; Dzhavakhiya, V.G.; Khurs, E.N.; Osipova, T.I.; Shcherbakova, L.A.; Zhemchuzhina, N.S.; Mikituk, O.D.; Nazarova, T.A. Chemical regulation of mycotoxin biosynthesis. *Dokl. Biochem. Biophys.* **2011**, *436*, 25–28. [CrossRef] [PubMed]
16. Ukraintseva, S.N.; Pridannikov, M.V.; Dzhavakhiya, V.G. Compactin: A potential biopesticide. *Zas. Karantin. Rast.* **2008**, *2*, 63. (In Russian)
17. Endo, A. The discovery and development of HMG-CoA reductase inhibitors. *Lipid Res.* **1978**, *33*, 1569–1582. [CrossRef] [PubMed]
18. Odds, F.C.; Brown, A.J.P.; Gow, N.A.R. Antifungal agents: Mechanisms of action. *Trends Microbiol.* **2003**, *11*, 272–279. [CrossRef]
19. Gonsales, E.; Felicio, J.D.; Pinto, M.M. Biflavonoids inhibit the production of aflatoxin by *Aspergillus flavus*. *Braz. J. Med. Biol. Res.* **2001**, *34*, 1453–1456.
20. Holmes, R.A.; Boston, R.S.; Payne, G.A. Diverse inhibitors of aflatoxin biosynthesis. *Appl. Microbiol. Biotechnol.* **2008**, *78*, 559–572. [CrossRef] [PubMed]
21. Huang, J.Q.; Jiang, H.F.; Zhou, Y.Q.; Lei, Y.; Wang, S.-Y.; Liao, B.-S. Ethylene inhibited aflatoxin biosynthesis in due to oxidative stress alleviation and related to glutathione redox state changes in *Aspergillus flavus*. *Int. J. Food Microbiol.* **2009**, *130*, 17–21. [CrossRef] [PubMed]
22. Grintzalis, K.; Vernardis, S.I.; Klapa, M.I.; Georgiou, C.D. Role of oxidative stress in sclerotial differentiation and aflatoxin B1 biosynthesis in *Aspergillus flavus*. *Appl. Environ. Microbiol.* **2014**, *80*, 5561–5571. [CrossRef] [PubMed]
23. Campbell, B.C.; Chan, K.L.; Kim, J.H. Chemosensitization as a means to augment commercial antifungal agents. *Front. Microbiol.* **2012**, *3*. [CrossRef] [PubMed]
24. Dzhavakhiya, V.G.; Voinova, T.M.; Popletaeva, S.B.; Statsyuk, N.V.; Mikityuk, O.D.; Nazarova, T.A.; Shcherbakova, L.A. Some natural and synthetic compounds inhibiting the biosynthesis of aflatoxin B1 and melanin in *Aspergillus flavus*. *Agric. Biol.* **2016**, *51*, 533–542. [CrossRef]
25. Zhou, W.; Hu, L.B.; Zhao, Y.; Wang, M.Y.; Zhang, H.; Mo, H.Z. Inhibition of fungal aflatoxin B1 biosynthesis by diverse botanically-derived polyphenols. *Trop. J. Pharm. Res.* **2015**, *14*, 605–609. [CrossRef]
26. Shaaban, M.T.; El-Sabbagh, S.M.M.; Alam, A. Studies on an actinomycete producing a melanin pigment inhibiting aflatoxin B1 production by *Aspergillus flavus*. *Life Sci. J.* **2013**, *10*, 1437–1448.
27. Sakuda, S. Mycotoxin production inhibitors from natural products. *JSM Mycotoxins* **2010**, *60*, 79–86. [CrossRef]
28. Sakuda, S.; Prabowo, D.F.; Takagi, K.; Shiomi, K.; Mori, M.; Omura, S.; Nagasawa, H. Inhibitory effects of respiration inhibitors on aflatoxin production. *Toxins* **2014**, *6*, 1193–1200. [CrossRef] [PubMed]
29. Wheeler, M.H.; Bhatnagar, D.; Rojas, M.G. Chlobenthiazone and tricyclazole inhibition of aflatoxin biosynthesis by *Aspergillus flavus*. *Pest. Biochem. Physiol.* **1989**, *35*, 315–323. [CrossRef]
30. Okamoto, S.; Sakurada, M.; Kubo, Y.; Tsuji, G.; Fujii, I.; Ebizuka, Y.; Ono, M.; Nagasawa, H.; Sakuda, S. Inhibitory effect of aflastatin A on melanin biosynthesis by *Colletotrichum lagenarium*. *Microbiology* **2001**, *147*, 2623–2628. [CrossRef] [PubMed]

31. Payne, G.A.; Hagler, W.M. Effect of specific amino acids on growth and aflatoxin production by *Aspergillius parasiticus* and *Aspergillius flavus* in defined media. *Appl. Environ. Microbiol.* **1983**, *46*, 805–812. [PubMed]
32. Ukraintseva, S.N.; Voinova, T.M.; Dzhavakhiya, V.G. Obtaining the highly productive mutants *Penicillium citrinum* producing compactin and optimization of fermentation process in shaken flasks. In *Biotechnology in Biology and Medicine*; Egorov, A.M., Zaikov, G., Eds.; Nova Science Publishers: New York, NY, USA, 2006; pp. 233–241.
33. Shcherbakova, L.A.; Statsyuk, N.V.; Mikityuk, O.D.; Nazarova, N.A.; Dzhavakhiya, V.G. Aflatoxin B1 degradation by metabolites of *Phoma glomerata* PG41 isolated from natural substrate colonized by aflatoxigenic *Aspergillus flavus*. *Jundishapur J. Microbiol.* **2015**, *8*. [CrossRef] [PubMed]

© 2016 by the authors. Licensee MDPI, Basel, Switzerland. This article is an open access article distributed under the terms and conditions of the Creative Commons Attribution (CC BY) license (http://creativecommons.org/licenses/by/4.0/).

Article

Effects of Milk Yield, Feed Composition, and Feed Contamination with Aflatoxin B1 on the Aflatoxin M1 Concentration in Dairy Cows' Milk Investigated Using Monte Carlo Simulation Modelling

H. J. van der Fels-Klerx * and Louise Camenzuli

RIKILT Wageningen University & Research, P.O. Box 230, Wageningen 6700 AE, The Netherlands; louise.camenzuli@wur.nl

* Correspondence: ine.vanderfels@wur.nl; Tel.: +31-0-317-481963; Fax: +31-0-317-417717

Academic Editors: Sarah De Saeger, Siska Croubels and Kris Audenaert
Received: 5 September 2016; Accepted: 30 September 2016; Published: 9 October 2016

Abstract: This study investigated the presence of aflatoxin M1 (AfM1) in dairy cows' milk, given predefined scenarios for milk production, compound feed (CF) contamination with aflatoxin B1 (AfB1), and inclusion rates of ingredients, using Monte Carlo simulation modelling. The model simulated a typical dairy farm in the Netherlands. Six different scenarios were considered, based on two lactation and three CF composition scenarios. AfB1 contamination of the CF was based on results from the Dutch national monitoring programme for AfB1 in feed materials from 2000 until 2010. Monitoring data from feed materials used in CF production for dairy cattle in the Netherlands were used. Additionally, AfB1 contamination data from an incident in maize in 2013 were used. In each scenario, five different transfer equations of AfB1 from feed to AfM1 in the milk were used, and 1000 iterations were run for each scenario. The results showed that under these six scenarios, the weekly farm concentration of AfM1 in milk was above the EC threshold in less than 1% of the iterations, with all five transfer equations considered. However, this increased substantially in weeks when concentrations from the contaminated maize batch were included, and up to 28.5% of the iterations exceeded the EC threshold. It was also observed that an increase in the milk production had a minimal effect on the exceedance of the AfM1 threshold due to an apparent dilution effect. Feeding regimes, including the composition of CF and feeding roughages of dairy cows, should be carefully considered based on the potential AfM1 contamination of the farm's milk.

Keywords: transfer; aflatoxins; dairy chain; maize; contamination

1. Introduction

Aflatoxin B1 (AfB1) is a genotoxic and carcinogenic mycotoxin that is produced by fungi, in particular *Aspergillus flavus* and *Aspergillus parasiticus*. AfB1 can be metabolized to aflatoxin M1 (AfM1) by cows, sheep, and goats. Like AfB1, AfM1 is also considered to be genotoxic and carcinogenic to animals and humans [1]. Elevated concentrations of AfB1 in feed result in elevated levels of AfM1 in milk and milk products. Due to their toxic effects on human and animal health, the presence of AfB1 and AfM1 in foodstuffs is strictly regulated within the European Union (EU). Regulation (EC) No 1881/2006 sets the maximum levels for aflatoxin M1 in dairy products, i.e., 0.05 μg/kg for raw milk, heat-treated milk, and milk for the manufacture of milk-based products. In addition, a maximum level for AfB1 in all feed materials was set at 0.02 mg/kg, as well as for compound feed for cattle, sheep, and goats, with the exception of dairy cattle, dairy sheep, and dairy goats at 0.005 mg/kg (Directive 2002/32/EC, Consolidated version 27 February 2015), all relative to a feed with a moisture content of 12%.

Historically, aflatoxins have been mainly found in products originating from countries with tropical weather conditions favourable for the growth of *Aspergillus* spp., like India, Brazil, and Colombia. However, during the last decade, serious contaminations of maize with aflatoxins have been reported in southern Europe. These include maize grown in 2003 in Italy [2] and maize from the 2012 harvest from the Balkan region [3]. Since climate change is increasingly affecting the formation of mycotoxins in Europe [4], these recent incidents are most likely not the last cases of aflatoxins in maize for European farmers.

Maize is a commonly used ingredient for feeding dairy cows. Feed manufacturers produce feed by mixing and grinding maize and other feed ingredients. In recent decades, the maize consumption of dairy cows has increased due to the prices of raw materials, compound feed composition, and the increase in the amount of concentrated feeds in diets. For example, in the Netherlands, the inclusion rates of maize in compound feed for dairy cows has increased from approximately 1% to 18% in the period 2010–2013 [5]. Additionally, the milk production of dairy cows is steadily increasing, which will coincide with an increased intake of feed, particularly compound feed, by cows.

The abovementioned developments have possibly increased the total exposure of dairy cows to aflatoxins, which might in turn lead to a higher probability of dairy cows' milk to be contaminated by AfM1.

The aim of this study was to estimate the AfM1 contamination in dairy cows' milk, using transfer modelling under multiple scenarios of compound feed composition, feed contamination with AfB1, feed consumption, and milk yield. The inclusion rate of maize in compound feed and the contamination of maize by AfB1 were modelled by Monte Carlo simulations, and hence the AfM1 concentration in dairy cows' milk with the intake distribution could be investigated. Monte Carlo simulation is a computerized mathematical technique that is often used for quantitative analyses and decision-making. The strength of this simulation technique is that it provides the decision-maker with a range of possible outcomes together with their probabilities.

2. Results

2.1. Transfer of AfB1 in Feed to AfM1 in Milk

The transfer of AfB1 in feed to AfM1 in milk was modelled for a typical Dutch dairy farm using transfer modelling combined with Monte Carlo modelling (1000 iterations) under six different scenarios. Namely, three compound feed (CF) composition scenarios, under two different milk yield scenarios (normal and extreme lactation). Five different transfer equations (Materials & Methods Table 4) obtained from literature were used to model the transfer from AfB1 in feed to AfM1 in milk. AfB1 contamination in feed ingredients was modelled in line with the results from the Dutch monitoring data, except for weeks 25 and 26, during which data from a contaminated maize batch were used [3].

Model output is weekly resolved, and for each week, the percentage of simulations (from the 1000 iterations) which resulted in an exceedance of the EC limit of 0.05 μg/kg for AfM1 in milk was calculated. For each scenario, the maximum of these weekly percentage exceedances are shown in Table 1. The week with the maximum percentage exceedance coincided with the use of highly contaminated maize in compound feed. In Table 1, the results of the week with the highest percentage exceedance rate without using the contaminated maize batch are also shown (in italics). Less than 1% of all the weekly simulations were above the EC limit for AfM1 in milk, when only monitoring data were used.

Table 1. Maximum weekly percentage * of simulations above the threshold of AfM1 in milk from the whole farm. Numbers in italics represent the maximum percent of simulations in all weeks excluding weeks 25 and 26 (when contaminated maize was used).

CF Composition Scenario	Milk Yield Scenario	Transfer Model									
		Masoero et al. [6]		Veldman et al. [7]		Britzi et al. [8]		Van Eijkeren et al. [9]		Pettersson from EFSA Opinion [10]	
1	normal	4.9	*0.0*	16.5	*0.3*	7.0	*0.1*	6.0	*0.0*	12.5	*0.2*
	extreme	4.8	*0.0*	16.3	*0.3*	8.9	*0.1*	4.7	*0.1*	12.3	*0.1*
2	normal	11.2	*0.0*	28.3	*0.5*	15.0	*0.1*	13.7	*0.1*	23.0	*0.2*
	extreme	11.9	*0.1*	28.5	*0.3*	17.3	*0.1*	11.2	*0.1*	22.8	*0.2*
3	normal	8.6	*0.3*	20.9	*0.6*	11.6	*0.3*	10.3	*0.3*	16.6	*0.5*
	extreme	7.5	*0.1*	18.9	*0.3*	11.2	*0.2*	7.2	*0.2*	14.8	*0.2*

* calculated as (the number of simulations in which the farm milk concentration is above 0.05 μg/kg) /1000 simulations × 100.

The transfer equation from Veldman et al. [7] resulted in the highest percentage of simulations above the EC limit, with an exceedance in 28.3% of the weekly simulations. The use of a low-protein compound feed (CF composition Scenario 2) resulted in the highest percentage of simulations above the EC limit for all transfer equations. This is in line with the fact that the low-protein compound feed has a high maize inclusion rate. With most of the transfer models, however, no clear differences could be observed in the amount of simulations above the EC threshold between the two lactation scenarios. In the extreme lactation scenario, all cows start lactating at the same time, as opposed to having different cows starting their lactation cycle on different weeks (normal lactation). Under the extreme lactation scenario, even when the cows consume highly contaminated feed in the same weeks (weeks 25 & 26) during their lactation peak, a higher transfer rate of AfM1 has not resulted in an increased exceedance rate due to the high volume of milk produced in the farm, and even a decrease when using the model of Van Eijkeren et al. [9]. In our model, the farm weekly milk production in week 25 amounts to 12,000 kg under the normal lactation milk yield scenario. The extreme lactation scenario resulted in 25% higher milk yield in the same week. Hence, the reason for the similar number of simulations above the EC threshold for both lactation scenarios is probably due to dilution. This dilution effect is most clear for the scenarios using the transfer equation provided in the EFSA opinion [10] from Pettersson [11], where the maximum weekly percentage of simulations above the EC limit is lower under the extreme lactation scenarios, than under the normal lactation scenarios. In the equation from Pettersson [11], the concentration of AfM1 in milk is only dependent on the total AfB1 intake (Materials & Methods Table 4). In the remaining scenarios, the transfer is dependent both on the total AfB1 intake and on the milk yield, and hence this dilution effect is less obvious due to the interaction between the intake and milk yield. Consequently, when comparing the percent of simulations above the EC threshold of a scenario under normal lactation and the corresponding scenario under extreme lactation, the difference is minimal. This implies that a higher milk production has an overall minimal effect on the concentration of AfM1 in the milk of the farm. When using the model from Van Eijkeren et al. [9], the dilution effect is such that with the extreme lactation the probability of AfM1 exceedance is even lower than with the normal lactation.

For each weekly model output, the mean concentration of AfM1 in the milk produced at the farm over 1000 iterations was also calculated. The maximum values for all these weekly mean concentrations are presented in Table 2. For all the scenarios modelled, the maximum weekly mean falls within the EC limit for AfM1 in milk. In some of the considered scenarios, this maximum of the weekly mean concentrations can reach up to 0.04 μg/kg (Table 2). Concentrations of AfM1 in milk as high as 0.32 μg/kg were modelled, which is 6.4 times the EC limit. Notably, this high concentration was calculated under CF Scenario 3, using the transfer equation from Veldman et al. [7]; it also coincided with the weeks when contaminated maize was used.

Table 2. Maximum of weekly mean AfM1 concentrations (μg/kg) in milk from the whole farm (over all iterations) *.

CF Composition Scenario	Milk Yield Scenario	Transfer Model				
		Masoero et al. [6]	Veldman et al. [7]	Britzi et al. [8]	Van Eijkeren et al. [9]	Pettersson [11] from EFSA Opinion [10]
1	normal	0.015	0.028	0.018	0.017	0.029
	extreme	0.015	0.028	0.020	0.015	0.029
2	normal	0.022	0.040	0.026	0.025	0.037
	extreme	0.022	0.041	0.029	0.022	0.037
3	normal	0.018	0.033	0.021	0.020	0.032
	extreme	0.017	0.031	0.022	0.017	0.031

* Including the contaminated batch in weeks 25 & 26. Please note that the maximum of the highest of the weekly mean concentrations is always seen with the contaminated batch.

2.2. *Effect of Milk Yield and Feed Intake*

Additional scenarios other than the standard scenarios were modelled in order to investigate the interaction between high/low yielding cows and high/low feed intake. The yearly milk yield per cow as modelled under the standard scenarios (Section 2.1) was equivalent to 9111 kg, with high and low yielding cows producing 11,845 and 6378 kg per year, respectively. Similar to the standard scenarios, each week, the percentage of simulations above the EC limit of 0.05 μg/kg for AfM1 in milk was calculated. The maximum values of these weekly percentages for each additional scenario are shown in Table 3. In comparison with Table 2 (specifically CF composition 1 under normal lactation), the high feed (HF) scenarios resulted in a higher weekly percent of simulations exceeding the EC limit. This is true for both the high milk yield (HY) and the low milk yield (LY) cows; however, the effect is larger for the high milk yield scenarios. The low feed (LF) scenarios resulted in a lower percent of simulations exceeding the EC limit, even for simulations with a high milk yield (HY). Additionally, the maximum weekly percent of simulations above the EC limit also coincided with the weeks when contaminated maize was used. If these weeks are excluded, the maximum percent of simulations above the limit is drastically reduced (percent shown in italics).

Table 3. Maximum weekly percentage * of simulations above the AfM1 threshold in milk from the whole farm (HF = High Feed; LY = Low Yield). Numbers in italics represent the maximum percent of simulations in all weeks excluding weeks 25 and 26 (when contaminated maize was used).

Feed and Yield Scenario	Masoero et al. [6]		Veldman et al. [7]		Britzi et al. [8]		Van Eijkeren et al. [9]		Pettersson [11] from EFSA Opinion [10]	
HF_HY	8.7	*0.1*	23.4	*0.5*	14.5	*0.3*	6.9	*0.0*	16.7	*0.3*
HF_LY	6.1	*0.1*	22.2	*0.4*	10.0	*0.1*	15.0	*0.3*	17.2	*0.3*
LF_HY	1.8	*0.0*	11.3	*0.1*	6.1	*0.0*	1.0	*0.0*	7.3	*0.1*
LF_LY	0.6	*0.0*	10.6	*0.1*	2.7	*0.0*	5.4	*0.0*	8.0	*0.1*
contaminated silage	8.6	*0.1*	33.0	*0.6*	13.6	*0.3*	12.0	*0.2*	23.8	*0.4*

* calculated as (the number of simulations in which the farm milk concentration is above 0.05 μg/kg) /1000 simulations × 100.

The mean AfM1 concentration as modelled for the whole farm was calculated every week as the mean of the 1000 iterations. The maximum of these weekly mean AfM1 concentrations in milk followed a similar trend when compared to the standard feed/standard milk yield scenario. The maximum of the weekly mean concentrations in all four scenarios did not exceed the EC limit; however, the highest farm AfM1 concentration modelled was 0.22 μg/kg under the high feed and low yield scenario (HF_LY) when using the equation by Veldman et al. [7] during the weeks with contaminated maize.

The final scenario modelled included corn silage as an additional source of AfB1 in the daily diet. The AfB1 contamination in corn silage was set at 1 μg/kg, being the limit of quantification.

When compared to the corresponding scenario without the inclusion of contaminated silage (CF composition 1 under normal lactation), the percentage of simulations above the EC limit for AfM1 in milk is approximately doubled. Nevertheless, excluding weeks 25 and 26, the percent of simulations above the AfM1 limit is still less than 1%. This shows that this additional contamination at a low level does not contribute significantly to raising the daily AfB1 intake above the threshold required to exceed the AfM1 threshold in milk.

2.3. Effect of the Variation in the Transfer Equations

Under the same milk yield scenario, and the same compound composition, the equation used for the transfer of AfB1 in feed to AfM1 in milk played an important role. The exceedance percentage was up to six times higher, depending on which transfer equation is used. The equations established by Masoero et al. [6], Veldman et al. [7], Britzi et al. [8], and Van Eijkeren et al. [9] relate the concentration of AfM1 in milk to the daily milk yield and to the daily intake of AfB1. The equation from Pettersson in 1998 [11]—presented in the EFSA opinion in 2004 [10] on AfB1 in animal feed—related the AfM1 in milk to the daily intake of AfB1, irrespective of the daily milk yield. The transfer rate in the equations from Masoero et al. [6] and Van Eijkeren et al. [9] depends similarly on milk yield, and hence similar results were obtained when using these two transfer equations. On the other hand, the equation from Veldman et al. [7] results in an overall higher transfer rate. Hence, the modelled concentration of AfM1 in milk when using the equation from Veldman et al. [7] will always be the highest. The equation from Britzi et al. [8] has the highest dependence of transfer rate on milk yield. Hence, the modelled concentration of AfM1 in milk will vary greatly depending on the phase in the lactation cycle. On the contrary, in the model described by Van Eijkeren et al. [9] the dependence is very small.

3. Discussion

This study aimed at investigating the potential exceedance of the maximum limit for AfM1 concentrations in milk, under different scenarios for AfB1 contamination of feed ingredients, inclusion rates of ingredients in compound feed, milk production, and transfer rates. Monte Carlo simulation modelling was used to investigate the AfM1 contamination of milk for the wide range of situations that realistically occur in practice. The results show that milk production of the farm and milk yield appeared to have minimal effects on the outcomes; a higher milk production did not result in increased concentrations of AfM1 in the farm milk, due to a potential dilution effect. While, according to transfer equations, a higher milk yield will result in a higher transfer rate, the concentration in the milk produced on the farm changes only slightly due to an increased milk production.

The Monte Carlo approach was applied to simulate the AfB1 contamination in each individual feed ingredient, using monitored mean and standard deviation AfB1 concentrations in the respective feed ingredient monitored over a 10-year period. Thus, the monitoring results were used to fit a distribution for AfB1 in each feed ingredient, and this distribution was in turn used as an input in the Monte Carlo simulation. The same approach was used to simulate the compound feed composition for Scenario 3, provided a known minimum and maximum percentage inclusion for each ingredient. This approach allowed us to simulate 1000 different contamination/composition pairs for the daily feed intake for AfB1, changing every two weeks. With 10-year monitoring data of AfB1 contamination in feed ingredients and the guidelines available for feed composition, less than 0.6% of all the simulations for individual two-week periods were above the EC limit for AfM1 concentration in milk, with a maximum concentration of 0.08 μg/kg (over all five transfer equations). This resulted from either a high inclusion rate of maize with an AfB1 concentration below the EC limit in compound feed, or from a low inclusion rate of maize above the EC limit for AfB1 in the compound feed. Nevertheless, it can be concluded that given the current practices, the probability of exceeding the EC limit of AfM1 in dairy milk is very low. In our model we also tested the effect of corn silage contaminated with AfB1 on the AfM1 content in milk. The contamination of corn silage was set to the limit of quantification of

AfB1. With an inclusion rate of corn silage in the dairy cow diet of 27% and a low AfB1 contamination level, a minimal change in the AfM1 content in milk was observed.

Our model concludes that under current inclusion rates of ingredients in compound feed and current AfB1 levels in feed ingredients, the probability of exceeding this EC limit is very low, and only a highly contaminated batch would be cause for concern. When considering the same lactation scenario and the same compound feed composition scenario, an increase in the milk yield and the daily feed intake by a factor of 1.3 resulted in a maximum increase of 0.2% in the probability of exceeding the EC limit in milk. Hence, using the current guidelines for compound feed composition, and the current limits on AfB1 in feed materials, an increase in milk yield does not appear to increase the probability of exceeding the limit of AfM1 in milk.

In 2013, a contaminated shipment of maize intended for feed materials was imported into the Netherlands [3]. The batch had a mean AfB1 concentration of 50.2 (±36.1) μg/kg, which is much higher than the EC legal limit for using maize as an ingredient (being 0.02 mg/kg at a moisture content of 12%). This batch was, however, not found to be contaminated during regular monitoring and was used for the production of a compound feed for dairy cattle. Such a batch was included in our transfer model, also using Monte Carlo simulations for the simulation of the AfB1 contamination of the feed ingredients and for the inclusion rate of each ingredient for a two-week period (weeks 25 and 26). Under normal lactation, 5%–28% of the simulations exceeded the EC limit for AfM1 in milk in the respective weeks. However, the probability, when including this contaminated feed in our model, was considerably dependent on the transfer equation used.

The transfer rate used to set the EC legislative limit in feed for dairy cows was 1%–2%. However, several studies found higher transfer rates of aflatoxins in cows with a higher milk yield [6,7,12] and in early/mid-lactating cows [7,8,13–15]. This was incorporated in our model through a 45-week lactation cycle, with milk yield varying through the cycle. The transfer rate varied, possibly due to differences between the metabolisms of the cows, the milk yield, and the source of contamination. In fact, the source of the contamination, milk yield, and cow breed in the studies available varied. Veldman et al. [7] used contaminated groundnut meal, Britzi et al. [8] and Masoero et al. [6] used contaminated corn meal, and the model by Van Eijkeren et al. [9] was fitted to data by Frobish et al. [16] using contaminated cottonseed. Concerning breed, Britzi et al. [8] carried out their study on Israeli Holstein cows (high-yield cows with an average yield of 11,400 kg $_{\text{milk}}$/cow), Masoero et al. [6] carried out their study on Holstein cows, and Veldman et al. [7] did not specify which cows were used. The maximum transfer rate from Veldman et al. [7] and Britzi et al. [8] was about 6%; however, in the model set up by Van Eijkeren et al. [9], the maximum transfer rate was 3.2%. When the model set up by Van Eijkeren et al. [9] was applied to the results from Masoero et al. [6] and Veldman et al. [7], the model did not agree with the data. The rate of exceedance of the EC threshold varied considerably depending on the equation used. However, it is unclear which of these equations is most suitable within our model, and hence all should be considered.

4. Conclusions

Given the ranges in the available transfer equations, only when a highly contaminated batch was included in the model was there a high possibility of exceeding the EC threshold of AfM1 in milk. Additionally, this depended on the inclusion rate of maize in the compound feed. Regarding the increased use of maize in feed for dairy cows, and the increasing milk yield, we hypothesised higher concentrations of AfM1 in dairy cows' milk. This study showed that under current practise and for the current limits on feed ingredients, an increased milk yield alone will not affect the contamination rate of milk with AfM1. However, to some extent, an increased use of maize in compound feed, combined with a higher contamination of maize with AfB1, will indeed increase the probability of exceedance of the EC limit for AfM1 in milk. Therefore, composition of compound feeds for dairy cows should be carefully performed, and should include information on the potential of AfB1 contamination of the ingredients used, as was shown by Van der Fels-Klerx and Bouzembrak [17], so as to comply with the EC limit for AfB1 in the final feeds.

5. Materials and Methods

5.1. Model

A simulation model was developed in order to estimate the distribution of AfM1 concentrations in dairy milk in a typical farm in the Netherlands. The model was developed in MATLAB R2015b. The model outline, given below, is based on a Monte Carlo simulation (1000 iterations) of the daily intake of aflatoxin B1 from compound feed. Monte Carlo simulation was used to assess the wide variety of possibilities that realistically occur in each of these steps, i.e., in compound feed composition, contamination of the feed ingredients with AfB1, milk production, and transfer of AFB1 to AFM1 in milk.

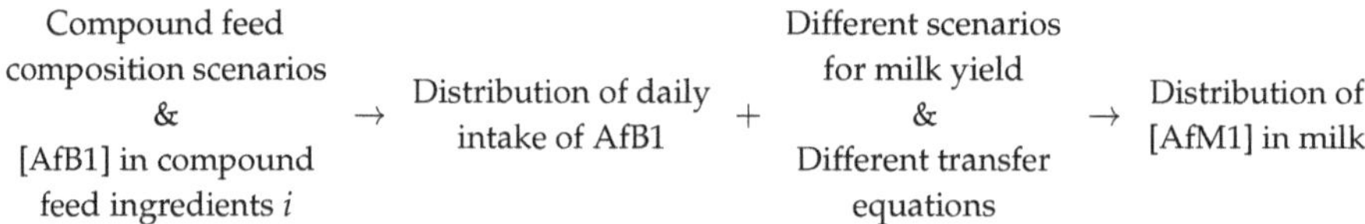

In the model, it is assumed that all the cows in the farm are housed indoors throughout the year. The modelled farm was set up based on the survey by Driehuis et al. [18] among 24 dairy farms across the Netherlands. An average herd size of 69 cows was assumed, with an average total daily feed intake (and standard deviation) of 18.7 (1.3) kg $_{DM}$/cow, of which 4.3 (0.2) kg $_{DM}$/cow is compound feed.

The milk yield per cow was modelled according to an incomplete gamma model developed by Wood [19] and discussed in Olori et al. [20]. The lactation period per cow was set to 45 weeks, followed by a four-week dry period. The milk yield for the whole farm was modelled in two different milk yield scenarios, namely (1) an individual cow started a new lactation cycle each week (normal lactation scenario); and (2) as an extreme scenario, all cows started the lactation cycle on the same day (extreme lactation scenario).

The concentration of AfM1 in milk ($[AfM1]_{milk}$) can be modelled through several equations; Table 4 presents the equations published in the scientific literature. Masoero et al. [6],Veldman et al. [7], and Britzi et al. [8] experimentally related the concentration of AfM1 in milk to the total intake of AfB1 (total intake$_{AfB1}$) and the daily milk yield (Y_{milk}). Van Eijkeren et al. [9] estimated the concentration of AfM1 at steady state, also from the total intake of AfB1 and the daily milk yield. The EFSA opinion from 2004 [10] on AfB1 in animal feed uses the equation from Pettersson [11], which is based on a collection of different experimental studies.

Table 4. Equations used for modelling the transfer of AfB1 in feed to AfM1 in dairy milk.

Equation	Source
$[AfM1]_{milk}\ (\mu g_{AfM1}/kg_{milk}) = \text{Total intake}_{AfB1} \cdot Y_{milk}^{-1} \cdot [Y_{milk} \cdot 0.13 - 0.26]/100$	[6]
$[AfM1]_{milk}\ (\mu g_{AfM1}/kg_{milk}) = \text{Total intake}_{AfB1} \cdot Y_{milk}^{-1} \cdot [Y_{milk} \cdot 0.077 - 0.326]/100$	[7]
$[AfM1]_{milk}\ (\mu g_{AfM1}/kg_{milk}) = \text{Total intake}_{AfB1} \cdot Y_{milk}^{-1} \cdot [0.5154 \cdot e^{Y_{milk} \cdot 0.0521}]/100$	[8]
$[AfM1]_{milk}\ (\mu g_{AfM1}/kg_{milk}) = \text{Total intake}_{AfB1} \cdot [Y_{milk} \cdot 0.032] \cdot (17 + Y_{milk})^{-1}$	[9]
$[AfM1]_{milk}\ (ng_{AfM1}/kg_{milk}) = \text{Total intake}_{AfB1} \cdot 0.787 + 10.95$	[10]

5.2. Input Data

5.2.1. AfB1 Contamination in Compound Feed

Data on aflatoxin concentrations in maize and other feed ingredients used in compound feed production for dairy cows, over the years 2000–2010, were extracted from the national monitoring database named KAP (Quality of Agricultural Products) in the Netherlands. These included 3427 records in total. All measurements below the limit of detection (LOD) of 1 μg/kg are recorded as zero in the database. A summary table of the data is presented in Table 5. In addition to the information from the KAP database, data from a recent incident in 2013 [3] of maize from the Balkan region, being contaminated above the maximum level of 20 μg/kg, were also considered. These data were from a shipment of maize dated from March 2013, intended for feed production in the Netherlands, as described by De Rijk et al. [3].

Table 5. Summary data for AfB1 concentration (μg/kg) in individual feed ingredients.

Ingredient	# of Records	Min. [AfB1]	Max. [AfB1]	Mean [AfB1]	Std. Deviation
Wheat [1]	346	0	1.4	0.019	0.148
Barley [1]	155	0	1.5	0.016	0.144
Corn [1]	768	0	115	0.653	5.52
Triticale [1]	24	0	1.9	0.079	0.388
Rye [1]	23	0	0	0	0
Soybean meal [1]	751	0	5.00	0.0395	0.269
Sunflower scrap [1]	136	0	7.5	1.09	1.72
Palm kernel [1]	484	0	26.0	0.515	3.08
Rapeseed scrap [1]	71	0	0	0	0
Corn gluten feed [1]	517	0	42.0	0.461	2.03
Flour [1]	2	0	0	0	0
Citrus pulp [1]	114	0	1.5	0.022	0.168
Dried beet pulp [1]	34	0	0	0	0
Molasses [1]	2	0	0	0	0
Contaminated maize [2]	72	6.2	168	50.2	36.1

[1] Data from the KAP database, [2] Data from De Rijk et al. [3].

5.2.2. Compound Feed (CF) Composition

Data available for compound feed composition in the Netherlands for lactating dairy cows, including the recipe and the rate to which individual ingredients are used in the compound feed, are presented in Table 6 [21]. These include the mean rate of use of each ingredient in 2014, for each of high- and low-protein compound feed for dairy cattle, as well as the range (minimum–maximum) to which each ingredient is used. Based on these data, three different scenarios for compound feed composition were defined. In the high- and low-protein scenarios, the fixed percentage of each ingredient was used in each simulation, whereas in the third scenario, the minimum and maximum ranges were used to simulate 1000 different compound feed compositions, with composition percentage within the suggested range.

Table 6. Feed composition (%) under a high-protein (Scenario 1) and a low-protein diet (Scenario 2). The general guidelines for feed composition are provided in Scenario 3 as the minimum and maximum percentages for each feed material.

Feed Ingredients	High-Protein	Low-Protein	Minimum	Maximum
CF composition scenario	1	2	3	
Total grains			*20*	*60*
Wheat	5	6	0	35
Barley	0	2.39	0	0
Corn	10.24	15.06	0	35
Triticale	1.2	1.9	0	15
Rye	0	1	0	15
Soybean meal	14.96	0.23	0	30
Sunflower seed meal	4.5	3.83	0	25

Table 6. *Cont.*

Feed Ingredients	High-Protein	Low-Protein	Minimum	Maximum
Palm kernel	15.01	15	0	20
Rapeseed meal	7.94	5.54	0	30
Corn gluten feed	3.67	1	0	30
Flour	0	0.04	0	20
Dried beet pulp	0.08	7.86	0	40
Citrus pulp	0	3.37	0	25
Molasses	1.5	1.55	0	10

The data from Tables 4 and 5 were combined to calculate the total intake of AfB1 from compound feed (CF) composed of ingredients i, using the equation below. It is assumed that a compound feed consists of 85% dry matter (DM) [22].

$$\text{Total daily intake}_{\text{AfB1}} = \frac{\text{daily feed intake}\cdot\%\text{ CF in feed}}{0.85}\cdot\left(\sum_{n=1}^{i}[\text{AfB1}]_i\cdot\%\ i\text{ in CF}\right).$$

Additionally, it is assumed that a fresh batch of compound feed is used every two weeks and that all cows eat the same amount of feed, meaning that over a two-week period the total daily intake of AfB1 is constant for all cows. Figure 1 shows an example of the total daily intake for the 1000 iterations, changing every two weeks.

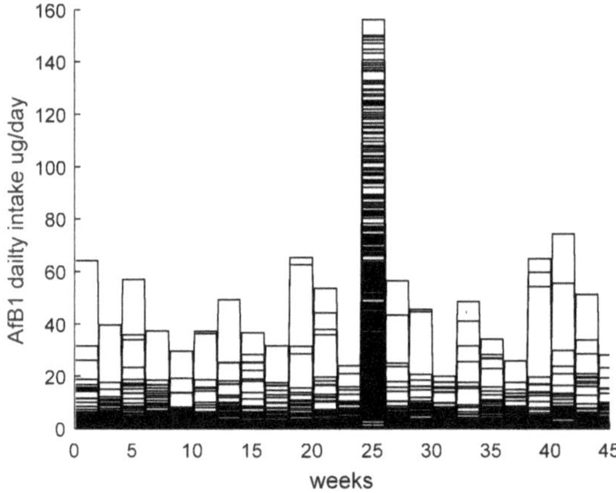

Figure 1. Daily intake of AfB1 from dairy cows' compound feed over for 1000 simulations over the whole lactation period.

5.3. Scenarios

The Monte Carlo simulation model was run for each of the six different basis scenarios, with two lactation scenarios by three compound feed composition scenarios, using each of the five different transfer equations (Table 4). In each scenario, 1000 iterations were used. AfB1 contamination data from KAP were used for all weeks, except for weeks 25 and 26, when data from the contaminated maize batch was used. Under the extreme lactation scenario, all cows have maximum milk yield during weeks 25 and 26, hence coinciding with the fortnight of contaminated maize.

In order to investigate the interaction between high/low yielding cows and high/low feed intake, additional scenarios were also modelled. High and low yielding cows were modelled as 1.3 and 0.7 times, respectively, of the milk yield used in the scenarios above. In accordance with the average total feed intake (and standard deviation) presented by Driehuis et al. [18], total daily feed intake for high and low feeding cows is modelled as 23 (high feed, HF) and 14.5 $kg_{\cdot DM}$/cow (low feed, LF), respectively. These four feed/milk yield pairs were modelled with compound feed Scenario 1 (high-protein diet), and the normal lactation scenario. The final scenario modelled, also the worst case scenario, assumed contaminated corn silage at the limit of quantification (LOQ) of AfB1. Information from the KAP database shows that the contamination of corn silage in the Netherlands from 2001 until 2010 is below the LOQ, and hence setting the contamination at the LOQ represents the worst case scenario under current conditions. The percentage of corn silage in the total daily feed intake was set at 27% [18], with a contamination of AfB1 of 1 μg/kg and an assumed dry matter content of 30% [22].

Acknowledgments: The authors acknowledge financial contribution of the Netherlands Ministry of Economic Affairs through the Topsector project AF-14225, and the contribution of the private partners in this project. Leo van Raamsdonk and Ron Hoogenboom, RIKILT Wageningen University and Research, are thanked for their suggestions made to this manuscript.

Author Contributions: H.J.v.d.F.-K. initiated the study; H.J.v.d.F.-K. and L.C. designed the study; L.C. did the programing and running of the models; H.J.v.d.F.-K. and L.C. evaluated the results, and wrote the paper.

Conflicts of Interest: The authors declare no conflict of interest.

References

1. International Agency for Research on Cancer. *IARC Monographs on the Evaluation on Carcinogenic Risks to Humans: Aflatoxin M1*; International Agency for Research on Cancer: Lyon, France, 1993; Volume 56.
2. Piva, G.; Battilani, P.; Pietri, A. Emerging issues in Southern Europe: Aflatoxins in Italy. In *The Mycotoxin Factbook*; Barug, D., Bhatnagar, D., van Egmond, H.P., van der Kamp, J.W., van Osenbruggen, W.A., Visconti, A., Eds.; Wageningen Academic Publishers: Wageningen, The Netherlands, 2006; pp. 139–153.
3. De Rijk, T.; van Egmond, H.; van der Fels-Klerx, H.J.; Herbes, R.; de Nijs, M.; Samson, R.; Slate, A.; van der Spiegel, M. A study of the 2013 Western European issue of aflatoxin contamination of maize from the Balkan area. *World Mycotoxin J.* **2015**, *8*, 641–651. [CrossRef]
4. Battilani, P.; Toscano, P.; van der Fels-Klerx, H.J.; Moretti, A.; Camardo Leggieri, M.; Brera, C.; Rortais, A.; Goumperis, T.; Robinson, T. Aflatoxin B1 contamination in maize in Europe increases due to climate change. *Sci. Rep.* **2016**, *6*. [CrossRef] [PubMed]
5. Remmelink, G.; Wageningen UR Livestock Research, Lelystad, The Netherlands. Use of maize in feeds for cows in The Netherlands in the period 2010–2013. Personal communication, 2013.
6. Masoero, F.; Gallo, A.; Moschini, M.; Piva, G.; Diaz, D. Carry-over of aflatoxin from feed to milk in dairy cows with low or high somatic cell counts. *Animal* **2007**, *1*, 1344–1350. [CrossRef] [PubMed]
7. Veldman, A.; Meijs, J.A.C.; Borggreve, G.J.; Heeres-van der Tol, J.J. Carry-over of aflatoxin from cows' food to milk. *Anim. Sci.* **1992**, *55*, 163–168. [CrossRef]
8. Britzi, M.; Friedman, S.; Miron, J.; Solomon, R.; Cuneah, O.; Shimshoni, J.; Soback, S.; Ashkenazi, R.; Armer, S.; Shlosberg, A. Carry-over of aflatoxin B1 to aflatoxin M1 in high yielding Israeli cows in mid- and late-lactation. *Toxins* **2013**, *5*, 173–183. [CrossRef] [PubMed]
9. Van Eijkeren, J.C.H.; Bakker, M.I.; Zeilmaker, M.J. A simple steady-state model for carry-over of aflatoxins from feed to cow's milk. *Food Addit. Contam.* **2006**, *23*, 833–838. [CrossRef] [PubMed]
10. European Food Safety Authority. Opinion of the scientific panel on contaminants in the food chain on a request from the Commission related to aflatoxin B1 as undesirable substance in animal feed. *EFSA J.* **2004**, *39*, 1–27.
11. Pettersson, H. Concerning Swedish derogation on aflatoxin. Complement to the Memo of 97-03-03 on "Carry-over of aflatoxin from feedingstuffs to milk". Department of Animal Nutrition and Management, Swedish University of Agricultural Sciences: Uppsala, Sweden, 1998.
12. Japanese Food Safety Commission. *Mycotoxin Evaluation Report Aflatoxin M1 in Milk and Aflatoxin B1 in Feed*; Food Safety Commission: Tokyo, Japan, 2013.
13. Diaz, D.; Hagler, W., Jr.; Blackwelder, J.; Eve, J.; Hopkins, B.; Anderson, K.; Jones, F.; Whitlow, L. Aflatoxin Binders II: Reduction of aflatoxin M1 in milk by sequestering agents of cows consuming aflatoxin in feed. *Mycopathologia* **2004**, *157*, 233–241. [CrossRef] [PubMed]

14. Sumantri, I.; Murti, T.W.; van der Poel, A.F.B.; Boehm, J.; Agus, A. Carry-over of aflatoxin B1 feed into aflatoxin M1 milk in dairy cows treated with natural sources of aflatoxin and bentonite. *J. Indones. Trop. Anim. Agric.* **2012**, *37*, 271. [CrossRef]
15. Kutz, R.E.; Sampson, J.D.; Pompeu, L.B.; Ledoux, D.R.; Spain, J.N.; Vázquez-Añón, M.; Rottinghaus, G.E. Efficacy of Solis, NovasilPlus, and MTB-100 to reduce aflatoxin M1 levels in milk of early to mid lactation dairy cows fed aflatoxin B1. *J. Dairy Sci.* **2009**, *92*, 3959–3963. [CrossRef] [PubMed]
16. Frobish, R.A.; Bradley, B.D.; Wagner, D.D.; Long-Bradley, P.E.; Hairston, H. Aflatoxin residues in milk of dairy cows after ingestion of naturally contaminated grain. *J. Food Prot.* **1986**, *49*, 781–785.
17. Van der Fels-Klerx, H.J.; Bouzembrak, Y. Modelling approach to limit aflatoxin B1 contamination in dairy cattle compound feed. *World Mycotoxin J.* **2016**, *9*, 455–464. [CrossRef]
18. Driehuis, F.; Spanjer, M.C.; Scholten, J.M.; te Giffel, M.C. Occurrence of mycotoxins in feedstuffs of dairy cows and estimation of total dietary intakes. *J. Dairy Sci.* **2008**, *91*, 4261–4271. [CrossRef] [PubMed]
19. Wood, P.D.P. Algebraic model of the lactation curve in cattle. *Nature* **1967**, *216*, 164–165. [CrossRef]
20. Olori, V.E.; Brotherstone, S.; Hill, W.G.; McGuirk, B.J. Fit of standard models of the lactation curve to weekly records of milk production of cows in a single herd. *Livest. Prod. Sci.* **1999**, *58*, 55–63. [CrossRef]
21. Van der Fels-Klerx, H.J.; Adamse, P.; de Nijs, M.; de Jong, J.; Bikker, P. A model for risk-based monitoring of contaminants in feed ingredients. *Food Control* **2016**, in press. [CrossRef]
22. The Professionnal Nutrient Managament Group (The Industry). Tried and Tested: Feed Planning for Cattle and Sheep. Available online: http://www.nutrientmanagement.org/assets/12028 (accessed on 21 September 2015).

© 2016 by the authors. Licensee MDPI, Basel, Switzerland. This article is an open access article distributed under the terms and conditions of the Creative Commons Attribution (CC BY) license (http://creativecommons.org/licenses/by/4.0/).

Article

Aflatoxin B_1 and M_1 Degradation by Lac2 from *Pleurotus pulmonarius* and Redox Mediators

Martina Loi [1,2], Francesca Fanelli [1], Paolo Zucca [3], Vania C. Liuzzi [1], Laura Quintieri [1], Maria T. Cimmarusti [1,2], Linda Monaci [1], Miriam Haidukowski [1], Antonio F. Logrieco [1], Enrico Sanjust [3] and Giuseppina Mulè [1,*]

1 Institute of Sciences of Food Production, National Research Council of Italy (CNR), via Amendola 122/O, Bari 70126, Italy; martina.loi@ispa.cnr.it (M.L.); francesca.fanelli@ispa.cnr.it (F.F.); vania.liuzzi@ispa.cnr.it (V.C.L.); laura.quintieri@ispa.cnr.it (L.Q.); teresa.cimmarusti@ispa.cnr.it (M.T.C.); linda.monaci@ispa.cnr.it (L.M.); miriam.haidukowski@ispa.cnr.it (M.H.); antonio.logrieco@ispa.cnr.it (A.F.L.)

2 Department of Economics, University of Foggia, via Napoli 25, Foggia 71122, Italy

3 Department of Biomedical Sciences, University of Cagliari, Cittadella Universitaria, Complesso Universitario, SP Monserrato-Sestu Km 0.700, Monserrato 09042, Italy; pzucca@unica.it (P.Z.); sanjust@unica.it (E.S.)

* Correspondence: giuseppina.mule@ispa.cnr.it; Tel./Fax: +39-080-5929317

Academic Editors: Sarah De Saeger, Siska Croubels and Kris Audenaert
Received: 12 July 2016; Accepted: 15 August 2016; Published: 23 August 2016

Abstract: Laccases (LCs) are multicopper oxidases that find application as versatile biocatalysts for the green bioremediation of environmental pollutants and xenobiotics. In this study we elucidate the degrading activity of Lac2 pure enzyme form *Pleurotus pulmonarius* towards aflatoxin B_1 (AFB_1) and M_1 (AFM_1). LC enzyme was purified using three chromatographic steps and identified as Lac2 through zymogram and LC-MS/MS. The degradation assays were performed in vitro at 25 °C for 72 h in buffer solution. AFB_1 degradation by Lac2 direct oxidation was 23%. Toxin degradation was also investigated in the presence of three redox mediators, (2,2′-azino-bis-[3-ethylbenzothiazoline-6-sulfonic acid]) (ABTS) and two naturally-occurring phenols, acetosyringone (AS) and syringaldehyde (SA). The direct effect of the enzyme and the mediated action of Lac2 with redox mediators univocally proved the correlation between Lac2 activity and aflatoxins degradation. The degradation of AFB_1 was enhanced by the addition of all mediators at 10 mM, with AS being the most effective (90% of degradation). AFM_1 was completely degraded by Lac2 with all mediators at 10 mM. The novelty of this study relies on the identification of a pure enzyme as capable of degrading AFB_1 and, for the first time, AFM_1, and on the evidence that the mechanism of an effective degradation occurs via the mediation of natural phenolic compounds. These results opened new perspective for Lac2 application in the food and feed supply chains as a biotransforming agent of AFB_1 and AFM_1.

Keywords: laccase; *Pleurotus*; mycotoxins; aflatoxin B_1; aflatoxin M_1; biodegradation; redox mediators

---

1. Introduction

Laccases (LCs, benzenediol: oxygen oxidoreductase, EC 1.10.3.2) are multicopper oxidases widely distributed in plants, bacteria, insects, and fungi [1]. Among fungi, white rot basidiomycetes, such as *Pleurotus* spp. are the most efficient producers of LCs [2]. Laccases typically contain four cupric ions, classified within three distinct spectroscopic types, T1, T2, and T3 [3]. They are essential for the one-electron oxidation of a reducing substrate and for the reoxidation of the enzyme by means of molecular oxygen, which is in turn reduced to water. Some "white" or "yellow" fungal laccases have been described [4,5], lacking the T1 cupric ion, which confers the blue color to the enzyme.

Pleurotus pulmonarius Fr. (Quél.), or Indian oyster mushroom, is an edible mushroom known for its medicinal properties and biotechnological potential [6]. It produces several LC isoforms, which are encoded by complex multi-gene families. LCs have different substrate specificity, catalytic properties, regulatory mechanisms and localization. Their synthesis and secretion depends upon nutrient levels, culture conditions, developmental stage, and can be increased by the addition of a wide range of inducers to cultural media [7].

LC catalyzes the oxidation of phenols, aromatic amines, and other non-phenolic compounds, while reducing molecular oxygen to water; LC activity can be further extended to non-phenolic substrates by the use of synthetic or natural redox mediators [8]. The mediators, after being oxidized by LC, diffuse out of the active site and oxidize recalcitrant compounds which possess high redox potential or high molecular weight. Being structurally diverse, different mediators may act on chemically-unrelated compounds, widening LC substrate range [9].

Several compounds have been used as redox mediators in the laccase mediator system (LMS). Synthetic mediators such as 2,2-azino-bis-[3-ethylbenzo-thiazolin-sulfonate] (ABTS), 2,2,6,6-tetramethylpiperidine-*N*-oxyl (TEMPO), and 1-hydroxybenzotriazole (HBT), have been widely used in many biocatalytic processes [10,11]. However, their use is limited due to their high cost, toxicity, and the high mediator-substrate molar ratio needed.

Being a green catalyst with a broad range of substrates, LC has been industrially applied since the early 1990s in chemical synthesis, the food industry, and bioremediation [12].

In addition, LC and LC-like activities in crude fungal extracts have been positively correlated with mycotoxin degradation [13–16], though neither the mechanism of action, nor the degradation products, have been yet fully elucidated.

Mycotoxins are secondary metabolites mainly produced by *Aspergillus*, *Penicillium*, and *Fusarium* spp., which display toxic, carcinogenic, teratogenic, and mutagenic activity towards humans and animals, and contaminate staple food commodities worldwide. Aflatoxins are mycotoxins produced by *Aspergillus* spp., and aflatoxin B_1 (AFB_1) is the most toxic: it has been classified by the International Agency for Research on Cancer (IARC) as Group 1, carcinogenic to humans, and it is known for its teratogenic, hepatotoxic, and immunosuppressive effects on humans and animals [17].

Due to their stability, which confers resistance to physical and chemical treatments of food processing, aflatoxins can persist and are usually found in cereal-based and animal products.

Aflatoxin M_1 (AFM_1) is the animal catabolic product of AFB_1 and contaminates milk and dairy products. AFM_1 is classified in group 2B by the IARC due to its demonstrated hepatotoxic and carcinogenic effect on animals, although its toxicity is one order of magnitude lower than AFB_1 [17].

Due to aflatoxin contamination, every year billions of dollars are lost along the food and feed supply chain worldwide [18], constituting a huge economic problem and a public health concern.

In the current study we purified a LC isoform from *P. pulmonarius*, a well characterized LC producer and a source of LC isozymes with recognized biotechnological potential in the field of bioremediation. We tested the degrading activity of purified LC towards AFB_1 and AFM_1, elucidating the effect of direct and mediated oxidation using a model synthetic mediator, ABTS, and two naturally-occurring phenols, acetosyringone (AS) and syringaldehyde (SA).

2. Results

2.1. LC Production and Purification

After 24 days of static incubation, the total activity (enzymatic activity, EU) of the crude extract (4 L) was 17,120 EU with specific activity of 6 U/mg. The purification steps for LC are detailed in Table 1. The apparent increase in total activity after Ca-phosphate gel might be due either to inhibition effect or interference with ABTS analysis by dark brown pigments (probably arising from oxidation/polymerization of ferulic acid; these pigments are nearly totally removed during this first purification step).

Table 1. Summary of Lac2 purification from *P. pulmonarius* culture filtrate.

Purification Step	Total Volume (mL)	Total Activity (EU)	Total Protein (mg)	Specific Activity (U/mg)	Purification Fold
Crude extract	4000	17,120	2800.00	6	1
Ca-phosphate gel	500	33,400	95.00	351	59
DEAE cellulose	50	23,785	11.50	2068	344
DEAE FF	10	3837	1.20	3224	538
Superdex	14	2053	0.19	10,920	1820

As preliminary steps, the batch treatment with calcium phosphate gel and the first anion exchange chromatography on diethylamino ethyl (DEAE) cellulose were performed to remove the majority of the brown pigments and contaminating proteins and resulted in the greatest increase in specific activity, from 6 to 2068 U/mg. In the last two chromatographic steps LC activity was further purified from remaining impurities and from the other contaminant proteins. This strategy resulted in 1820-fold purification with 12% final yield of laccase enzyme with respect to the crude extract.

2.2. Zymography

Figure 1, panel A shows the zymogram of extracellular LC activity by using ABTS as the substrate; protein bands exhibiting activity in zymogram were also compared with the related electrophoretic pattern stained with Coomassie stain (Figure 1, panel B and C).

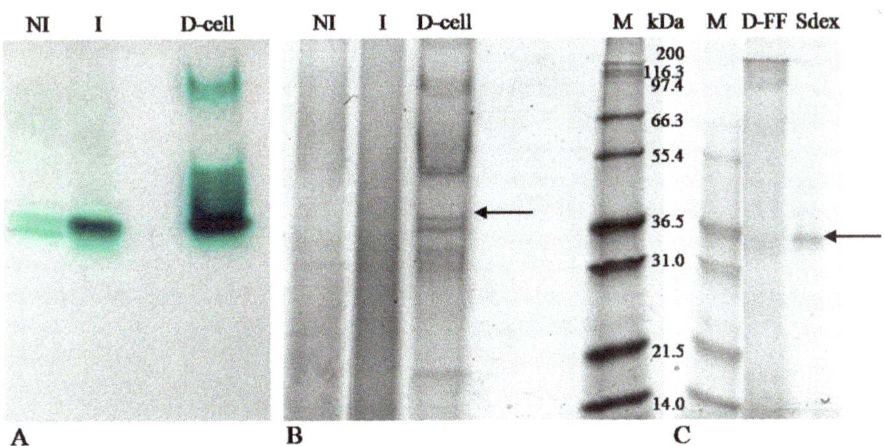

Figure 1. Zymography (**A**) using ABTS as the substrate and SDS PAGE (**B** and **C**) of *P. pulmonarius* LC preparations. NI-not induced; I-induced; D-cell-sample after DEAE cellulose, D-FF-sample after DEAE FF, Sdex-sample after Superdex, M-Marker. The arrows indicate Lac2 bands.

In particular, the not induced (NI) sample putatively produced two laccase isoforms corresponding to a molecular weight ranging from 31 to 36.5 kDa in the electrophoretic profile stained with Coomassie (Figure 1, panel B). The induced (I) sample showed an analogous zymogram pattern with the exception of the band showing a more intense activity than the NI sample due to LC induction by ferulic acid. However, the zymogram was more sensitive than Coomassie staining, as previously reported [19]. LC activity with ABTS was still clearly detectable when as low as 0.3 μg of proteins were loaded on SDS-PAGE (data not shown). After the final purification step with Superdex, Lac2 appeared as one single band in SDS PAGE (Figure 1, panel C).

2.3. Laccase Identification by MS/MS

Table 2 summarizes the identification of the band digested from the polyacrylamide gel with the highest score. After protein digestion, the resulting peptide mixture was analyzed in data-dependent MS/MS acquisition mode, The acquired MS data were processed by Proteome Discoverer software (version 1.4, 2012, Thermo Scientific, San José, CA, USA) and searched against a customized DB containing all *Pleurotus* spp. protein sequences present in UniProt. As a result, seven unique peptides were detected and assigned to Lac2 and Lac4 of *P. pulmonarius* and *P. sajor-caju*, respectively. In Table 2 the peptide sequences identified by MS and matching with part of Lac2 and Lac4 sequences in the Uniprot DB are reported. Considering that *P. pulmonarius* and *P. sajor-caju* are synonymous [20], and that the strain used in this study belonged to *P. pulmonarius* species, we identified our protein as Lac2 of *P. pulmonarius*.

Table 2. Overview of protein assignments referred to the analysis of the excised band obtained by the Sequest scoring algorithm interrogating a customized *Pleurotus* database imported by UniProt.

Assigned Protein	Accession Number	Protein Coverage	No. of Identified Peptides	Sequences	Confidence Level	m/z (Da)
Laccase 2 *Pleurotus pulmonarius*	Q2VT18	21.24%	7	YSFVLTADQTPDNYWIR	High	1045.07686
				YAGGPTSPLAVINVESTKR	High	980.65433
				SAGSTTYNFDTPAR	High	744.42450
				GDNFQLNVVNQLSDTTMLK	High	1069.20405
Laccase 4 *Pleurotus sajor-caju*	Q7Z8S3			SAGSTTYNFDTPARR	High	822.55662
				ANPNLGSTGFAGGINSAILR	High	644.12372
				SVPITGPTPATASIPGVLVQGNK	High	735.54871
				GDNFQLNVVNQLSDTTMLK	High	718.38048

2.4. In Vitro Degradation of AFB_1 and AFM_1 with Laccase and Redox Mediators

Degradation results are shown in Figure 2, while examples of HPLC chromatograms of AFB_1 and AFM_1 degradation are shown in Figure 3. Direct oxidation of AFB_1 by means of Lac2 alone accounted for 23% degradation. The addition of a redox mediator resulted in a very effective degradation of the toxin. The lowest concentrations of ABTS and AS (1 mM) were able to double the degradation percentage compared to Lac2 alone (45% and 42%, respectively) while, in the case of SA, the presence of mediator at 1 mM lowered the degradation percentage (13%). Absolute values of aflatoxin concentrations are shown in Table 3.

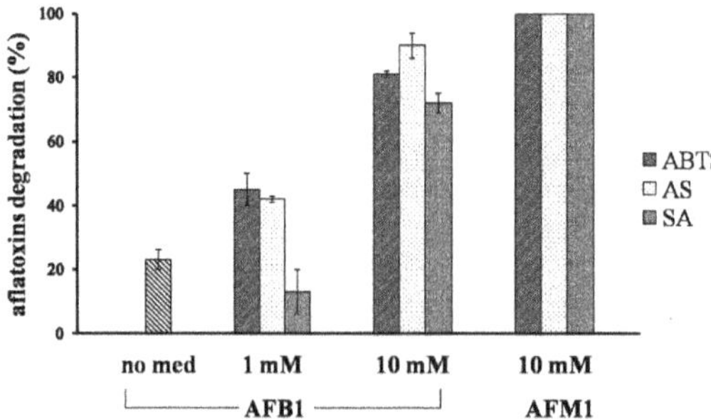

Figure 2. AFB_1 and AFM_1 degradation (%) after three days of incubation at 25 °C, performed by Lac2 and the respective redox mediator in buffered solution (1 mM sodium acetate pH 5). ABTS-[2,2′-azino-bis-(3-ethylbenzothiazoline-6-sulfonic acid)]; AS-acetosyringone, SA-syringaldehyde. Values are the mean of three replicates and the error bars represent the standard error measured between independent replicates.

Figure 3. *Cont.*

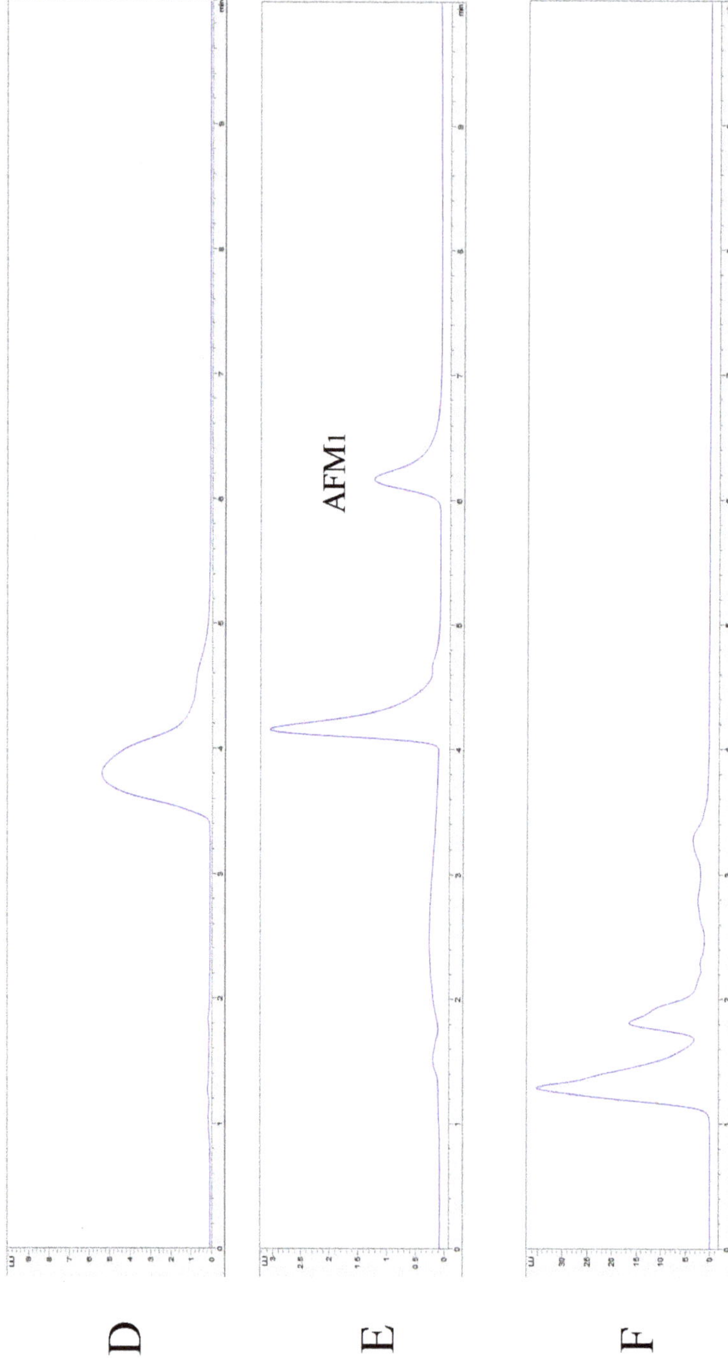

Figure 3. Examples of HPLC chromatograms of AFB_1 (**A**—negative control, **B**—positive control, **C**—sample after degradation by Lac2) and AFM_1 (**D**—negative control, **E**—positive control, **F**—sample after degradation by Lac2).

Table 3. Absolute values of aflatoxin concentrations (ng/mL) after LC treatment and statistical analysis of aflatoxin degradation by Lac2 and redox mediators.

Sample	AFB_1		AFM_1
	med 1 mM	med 10 mM	med 10 mM
Positive control	923 ± 33 *	923 ± 33 *	53 ± 7 *
No med	710 ± 27 *	710 ± 27 *	n.t.
ABTS	508 ± 46 *	175 ± 5 *	0 ± 0 *
AS	535 ± 9 *	92 ± 27 *	0 ± 0 *
SA	803 ± 120	258 ± 16 *	0 ± 0 *

n.t. = not tested; Comparisons between controls and treated samples were performed using a *t*-test. A p value < 0.001 was considered statistically significant (*).

At 10 mM each mediator further enhanced AFB_1 degradation which reached 90% for AS, 81% for ABTS and 72% for SA. With regards to AFM_1, Lac2 proved to completely degrade the toxin with all mediators added at 10 mM since after the reaction no AFM_1 peak was detected by HPLC analysis.

3. Discussion

In this work we purified and identified a Lac2 isoform from a strain of *P. pulmonarius* and evaluated the ability of the pure enzyme to degrade AFB_1 and AFM_1 either by direct or mediated oxidation with three different redox mediators, ABTS, AS, and SA.

Mycotoxin degradation by fungi and bacteria is a widely investigated topic, especially in the last 10–15 years [21–23], but a deep understanding of which enzyme is directly responsible for mycotoxin degradation and the mechanism involved, is still lacking. Many bioremediation applications exploit the fungus as a whole-cell biocatalyst, or its secretome, thus involving the concerted activity of several enzymatic systems, including laccases, extracellular peroxidases (Lignin peroxidase-LiP, EC 1.11.1.14, manganese peroxidase-MnP, EC 1.11.1.13, and versatile peroxidase-VP, EC 1.11.1.16) and oxidases that generate the H_2O_2 needed for peroxidase activity (tyrosinase-EC 1.14.18.1 and aryl-alcohol oxidase-EC 1.1.3.7). Furthermore, low molecular weight compounds that act as mediators might be present in the culture media. This limits the discrimination between the direct action of the enzyme and the mediated one, which is a crucial point to develop industrial or biotechnological applications. Even commercial fungal preparations may contain contaminant proteins: Margot and colleagues [24] recently verified that the most used commercial laccase from Sigma (Milan, Italy, Ref. 38429) actually contains a mixture of different proteins, from 17 to ~80 kDa, and different LC isoforms.

Although aflatoxin degradation by fungal laccase enzymes has already been reported [15,25,26], to date no direct and unambiguous correlation between laccase and aflatoxin degradation has been described, since cultured filtrates or LC commercial preparation were tested in the degradation assays.

In our study the LC isozyme responsible for aflatoxins degradation was identified as Lac2. This isozyme has been extensively biochemically characterized in a previous work by Zucca and colleagues [27]. They reported a copper content of 3.8 cupric ions per protein molecule and a sugar content of 6.7% ± 0.3% (expressed as glucose equivalents), measured the enzyme activity at different pH values, and its stability at different temperatures.

Lac2 production was induced by low molecular weight compounds; among various putative inducers, ferulic acid proved to be by far the most effective [27]; according to this, the induced sample in zymography showed one band with a much more intense activity than the related NI sample. In order to remove contaminants and to purify laccase, a preparative chromatography was performed. As expected, this procedure increased the activity of bands detected in the NI and I samples (Figure 1, panel A) and resulted in a laccase band with an apparent molecular weight close to 35 kDa, as determined by the comparison with the lanes stained with Coomassie (Figure 1, panel B and C). Similar results were described by Diaz [28], who detected four laccase isoenzymes with molecular weights of 65, 47, 38, and 29 kDa.

The predicted molecular weight of Lac2 is 56.6 kDa, in agreement with the previous estimation by SDS PAGE and RP-HPLC–electrospray ionization-MS which assigned Lac2 a molecular weight of 55–61 kDa [27]. However, in this study Lac2 apparent molecular weight, estimated by SDS PAGE, was approximately 35 kDa; a possible explanation of this divergence is that under non-reducing conditions LC shows increased mobility and a lower apparent molecular weight due to the extensive glycosylation, as measured for this LC isozyme by Zucca et al. [27]. Moreover, glycosylation has been reported to be responsible for unconventional electrophoretic behavior under non-reducing or native conditions [29] and to influence SDS/protein interaction; it may facilitate LC migration through the gel net, making the LC external structure more flexible and elastic.

The limited reactiveness of Lac2 alone towards AFB_1 might be explained by the high electrochemical oxidation potential, high ionization potential, or steric hindrance that prevents the substrate from being oxidized or enter the active site of LC [30]. Those limitations can be overcome by the use of redox mediators, which are effective LC substrates and, in turn, oxidize recalcitrant compounds.

Effective degradation of aflatoxins proceeds via LMS, both with ABTS and natural mediators, despite their proposed different mode of action. ABTS mediation is reported to occur via an electron transfer (ET) route, while phenoxy radicals mediate via hydrogen atom abstraction (HAT), at least when working at acidic or neutral pH values; only under alkaline conditions, where the mediators are in their anionic form, the HAT mechanism turns into an ET one [8,9,31].

ABTS was the first used artificial mediator [32]. It is soluble in water and, upon one-electron oxidation, produces a stable radical cation with a high absorbance at 420 nm. Thanks to these features ABTS has been successively used as an oxidation mediator towards polycyclic aromatic hydrocarbons, in organic synthesis, and in the treatment of textile wastewater [33–35]. ABTS is a model compound for LMS and an efficient mediator for AFB_1 and AFM_1 (81% and 100% degradation, respectively). However, it is a synthetic mediator and, as such, its high cost and concerns related to its potential toxicity have restricted industrial implementation of this LMS even in non-food applications, and raised the need for safe, cost-effective, and readily available mediators.

With this aim, we investigated the role of two naturally-occurring compounds, AS and SA, which are both 2,6-dimethoxy-substituted phenols derived from syringyl lignin units. They were described as the fastest and most efficient laccase mediators for the degradation of industrial dyes, sulfonamide antibiotics, and for the removal of lignin from paper pulps [36–38]. In addition to being natural compounds, they are inexpensive, safe, and can be as effective as the synthetic ones.

Both AS and SA were demonstrated to be effective mediators for AFB_1 degradation at 10 mM, although SA was less efficient than AS and ineffective at 1 mM. Most probably, the SA-derived radical undergoes an internal redox reaction leading to syringic acid and, therefore, wasting a significant fraction of the reactive radical. Such a decrease of the reactive radical did not occur in the case of AS, explaining the noticeable effectiveness of the compound.

Methoxy substitutions in syringyl-type compounds decrease the redox potential and increase electron density at the phenoxy group. Those compounds are readily oxidized by LC and generate relatively stable radicals since the substitutions in the phenol ring impose steric hindrances for the polymerization via radical coupling [9,39]. The substituent in the para position also influences the phenoxy radical stability, since electron donor groups at the para-position stabilize the phenoxy radicals, while electron-withdrawing substituents lead to a decreased radical stability [40]. Accordingly, AS harbors the weakest electron acceptor group and generates a more stable radical than SA.

With respect to AFM_1, Lac2 was able to degrade it completely with all mediators tested at 10 mM with no differences emerging among ABTS, AS, and SA. The decontamination of AFM_1 in buffered solution, in model and real food matrices, was previously investigated using lactic acid bacteria [41]. Nevertheless, the decontamination was a result of a reversible binding to the carbohydrates and peptidoglycan of the bacterial cell wall surface and not a biological degradation. To our knowledge, the present study reports for the first time the effective degradation of AFB_1 and AFM_1 by means of Lac2 from *P. pulmonarius*.

The degradation products of aflatoxins have not been identified yet. Depending on the degrading agent, aflatoxins can be degraded by several mechanisms, such as epoxidation, hydroxylation, dehydrogenation, and reduction. A wide range of putative potential reaction products obtained accounts for the difficulties in the development of sensitive identification methods, as well as for the limited data on their toxicological characterization. Considering the degradation products deriving from laccase treatment, only a study conducted by Alberts and colleagues [15] reported a reduced mutagenicity (using Ames test) of the degradation products of AFB_1.

According to our results we hypothesize that LMS acts on the lactone ring, which is responsible for fluorescence properties: cleavage of the lactone ring results in a non-fluorescent compound that has greatly reduced biological activity [42,43]. However, the ring cleavage of lactones is a hydrolysis, rather than an oxidation. Therefore, as laccases cannot catalyze lactone hydrolysis, fluorescence quenching should be the consequence of a deeper modification of the coumarin-like core of the toxins, which is responsible for their fluorescence. In fact, oxidative demethoxylations of simple aromatics by means of fungal laccases are well known [44,45]; methanol is released and the aromatic ring is changed into its quinonoid counterpart, which rearranges with the result of an irreversible and deep degradation. Hypothesizing that this mechanism is also valid for substituted aromatics, like aflatoxins, the reversibility of the reaction is unlikely.

This forecast is confirmed by our observations reported here. Additionally, as expected, the use of redox mediators strongly enhances the degrading ability of Lac2.

Laccase has been increasingly applied in food industry in the last 30 years. The demonstrated biodegrading activity towards mycotoxins, the green catalysis, and the use of natural mediators support a potential and feasible application in food and feed. A mandatory requisite for feed application is that products of mycotoxin degradation have to be stable and non-toxic. The development of these applications has to overcome the cost of production of the enzyme, the optimization of the degradation reaction, as well as the gap of knowledge related to the degradation products and their toxicity as required by the EU commission [46].

Pleurotus spp. ligninolytic system is a source of biotechnologically important enzymes which play an essential role in green bioremediation; so far, *P. pulmonarius* LCs have been extensively used in the removal of industrial dyes and the treatment of lignocellulosic waste.

This is the first time that *P. pulmonarius* Lac2 was unambiguously identified as capable of degrading AFB_1 and AFM_1 in the presence of natural redox mediators. Although further studies are needed to optimize the degradation assay, this study clearly illustrates the potentiality of Lac2 for its use as a biotransformation agent.

4. Materials and Methods

4.1. Organism, Culture Conditions, LC Induction, and Production

The *Pleurotus pulmonarius* strain ACR-16 of Cattedra di Chimica Biologica Collection, department of Biomedical Sciences, University of Cagliari (), syn. *P. sajor-caju*, [20], was maintained as ITEM17144 in the Agri-Food Toxigenic Fungi Culture Collection of the Institute of Sciences of Food Production, CNR, Bari [47]. ITEM17144 was routinely grown on malt extract agar plates (MEA, Oxoid) at 25 °C.

For LC production, ITEM 17144 was grown in liquid medium (2% *w/v* malt extract, 0.5% *w/v* yeast extract, 10 mM of potassium phosphate buffer pH 6, 0.1 mM of $CuSO_4$) supplemented with 10 mM ferulic acid as a laccase inductor for 24 days, in darkness at 25 °C, in static conditions (relative humidity 70%).

4.2. Chemicals and Reagents and Standards Preparation

Chemicals for gel electrophoresis including Bio-safe Coomassie stain and Bradford reagent were supplied by Bio-Rad Laboratories (BioRad, Milan, Italy).

Acetonitrile (ACN) (LC–MS grade), formic acid, acetic acid, ammonium bicarbonate, trizma^{TM} base, tween 20, hydrochloric acid, trifluoroacetic acid (TFA), iodoacetamide (IAA), dithiothreitol (DTT), 2-azino-di-[3-ethylbenzo-thiazolin-sulphonate] (ABTS), acetosyringone (AS), syringaldehyde (SA), AFB_1 and AFM_1 standards (purity >99%) were obtained from Sigma-Aldrich (Milan, Italy). Trypsin (proteomic grade) was purchased from Promega (Milan, Italy). Regenerate cellulose syringe filters, 0.2 μm (size 4 mm) were obtained from Sartorius Italy S.r.l. (Muggiò, Italy)

Mycotoxin stock solution of AFB_1 (10 μg/mL) was prepared by dissolving solid commercial toxins in toluene:ACN (9:1, *v*/*v*) (HPLC grade). The exact concentration of aflatoxin B_1 was determined according to Association of Official Analytical Chemists (AOAC) Official Method 971.22 [48]. Aliquots of the stock solution were transferred to 4 mL amber silanized glass vials and evaporated to dryness under a stream of nitrogen at 50 °C. The residue was dissolved with water:methanol (60:40, *v*/*v*) to obtain calibrant standard solutions at 0.4, 1.2, 2.0, 4.0, 5.0, and 10.0 ng/mL. Standard solutions were stored at −20 °C and warmed to room temperature before use.

Mycotoxin stock solution of AFM_1 (10 μg/mL) was prepared by dissolving solid commercial toxins in ACN (HPLC grade). The exact concentration of standard aflatoxin solution was determined according to AOAC official method 2000.08 [49]. Aliquots of the stock solution were transferred to 4 mL amber glass vials and evaporated to dryness under a stream of nitrogen at 50 °C. The residue was dissolved with water:ACN (75:25, *v*/*v*) to obtain calibrant standard solutions at 1.0, 2.5, 5.0, 7.5, and 10.0 ng/mL. Standard solutions were stored at −20 °C and warmed to room temperature before use.

4.3. LC Purification

LC purification was performed according to Zucca et al. [27] with slight modifications. Briefly, after 24 days of incubation, the culture medium was collected, diafiltered, and concentrated in 50 mM potassium phosphate buffer and 50 mM 6-aminohexanoic acid (protease inhibitor) using a Vivaflow 200 apparatus (Vivascience AG, Hannover, Germany) equipped with a Hydrosart membrane module (nominal MW cut-off 10,000 Da) and a Masterflex L/S system pump (Cole-Parmer, Vernon Hills, IL, USA) at 4 °C. The enzyme solution was added with NaCl to a final concentration of 0.25 M and gently stirred with freshly prepared calcium phosphate gel at 4 °C for 30 min. The slurry was centrifuged at 8000× *g* for 30 min, the supernatant recovered and diafiltered as previously described.

The resulting solution was adjusted to 0.2 M NaCl and loaded onto a DEAE-cellulose column (15 cm × 5 cm), which was pre-equilibrated with 50 mM potassium phosphate buffer pH 6 and 0.2 M NaCl. Bounded brown pigments were separated from unbound LC, which was eluted with the same buffer and desalted by dialysis against 50 mM potassium phosphate buffer prior to a second ion exchange chromatography.

Desalted LC fractions were loaded onto a Hiprep 16/10 DEAE FF assembled on an Akta Prime FPLC (Amersham Bioscience, Milan, Italy) equipped with a UV detector for protein absorbance monitoring at 280 nm. Column equilibration was performed with 50 mM potassium phosphate buffer pH 6 at a constant flow of 5 mL/min. Unbound proteins were washed out while LC was eluted with a linear gradient of 50 mM potassium phosphate buffer pH 6 containing 0.5 M NaCl in 40 min.

LC-rich fractions were pooled, concentrated, dialyzed, separated by size exclusion chromatography with a HiLoad 16/60 Superdex 75 column (GE Healthcare, Milan, Italy) assembled on the Akta Prime FPLC; equilibration and run were performed with 50 mM potassium phosphate buffer pH 6 at constant flow of 0.4 mL/min. Proteins were eluted and stored at −20 °C until use.

Protein content was determined using the Coomassie Brilliant Blue G250 method [50], the standard curve was performed using bovine serum albumin (BSA, 1–0.025 mg·mL^{-1}).

4.4. LC Spectrophotometric Activity Assay

Laccase activity was photometrically measured (Ultraspec 3100pro, Amersham Pharmacia Biotech Italia, Cologno Monzese, Italy). The reaction was performed in 100 mM sodium acetate pH 4.5, 2 mM ABTS and an appropriate amount of enzyme solution in a final volume of 1 mL. The oxidation of ABTS was determined after 10 min by photometric assay at 420 nm (ε_{420} = 36,000 $M^{-1}\cdot cm^{-1}$). One unit was defined as the amount of enzyme which oxidized 1 μmol of substrate per min [51].

4.5. Zymography

The activities of the crude extract, both in presence (Induced, I) and absence (Not Induced, NI) of ferulic acid, as well as DEAE cellulose fractions were detected by zymograms, as previously reported [52].

Amounts of 5 μg (on average) both for I sample and the DEAE cellulose fraction, and 2 μg of NI samples, were dissolved in denaturant, non-reducing sample buffer (62.5 mM Tris-HCl pH 6.8, 25% glycerol, 2% SDS, and 0.01% Bromophenol Blue); all samples were loaded on two sodium dodecyl sulfate-polyacrylamide gels electrophoresis (SDS-PAGE, 12% T, 3% C), performed according to Laemmli [53]. The unstained molecular weight marker M12 (2.5–200 KDa, Life Technology, Waltham, MA, USA) was used as the reference. Electrophoretic separation was performed in a Miniprotean System (Biorad, Segrate, Italy) filled with running buffer composed of 25 mM Tris and 0.19 M glycine at 100 V for 15 min and 150 V for 1 h.

After the run, the gel was divided into two segments: one segment was washed with distilled water (four washes of 15 min and one of 30 min) at room temperature in order to remove SDS and then incubated with a solution of 5 mM ABTS in 50 mM potassium phosphate buffer pH 6; LC activity was revealed within six min. The remaining segment was fixed with 40% ethanol, 10% acetic acid 50% H_2O for 30 min and then stained with Bio-safe Coomassie stain (Bio-Rad), following the manufacturer's instructions.

Destained gels were digitally acquired by an Image Scanner III (GE Healthcare, Pittsburgh, PA, USA). The experiment was performed in two replicates. The putative laccase band corresponding to the induced isoform, detected in the electrophoretic pattern of DEAE sample stained with Coomassie, was excised and analysed by mass spectrometry.

4.6. LC-MS/MS Analysis

The excised gel band shown in Figure 1 was cut into small pieces and placed in 1.5 mL vials for in-gel digestion, with trypsin chosen as the proteolytic enzyme. Protein digestion was accomplished according to the manufacturer's instructions with slight modifications. Firstly, gel slices were destained by adding 200 μL of 100 mM NH_4HCO in 50% ACN and kept at 37 °C for 45 min; this step was repeated until electrophoresis dye was removed. Gel slices were then dehydrated in 100 μL of ACN and dried in a Speed Vac. Then, 130 μL of 10 mM DTT solution (prepared in 25 mM NH_4HCO_3) were added to the batch and incubated in shaking conditions for 1 h at 57 °C. Thirty microliters of 55 mM IAA solution (prepared in 25 mM NH_4HCO_3) were added and the mixture was incubated in darkness for 30 min at room temperature. After incubation, 0.3 μg of trypsin were added to the batch and incubated under shaking conditions overnight at 37 °C. The sample was then incubated with 150 μL of MilliQ water for 10 min, with frequent vortex mixing. Liquid was then removed and saved in a new microcentrifuge tube (Sigma, Milan, Italy). The extraction of gel slices digest was performed by adding 50 μL of 50% can, 5% TFA, 45% H_2O solution. The extract was incubated under shaking conditions at room temperature for 1 h and centrifuged to recover the supernatant fraction. The procedure was repeated twice. Both supernatants were collected, mixed, and added with the aliquot obtained by sample incubation with water, and evaporated in the speed-vac in order to concentrate the sample. The final pellet was re-suspended in 50 μL of H_2O:ACN (90:10 + 0.1% of formic acid) and filtered through regenerate cellulose filters (0.22 μm) before injection into the HPLC-MS system.

For the HPLC-MS/MS analysis a system consisting of UHPLC pump coupled through an ESI interface with a dual pressure linear ion trap mass spectrometer VelosPro^{TM} (Thermo Scientific, San José, CA, USA) was used. Peptide separation was performed on an Aeris peptide 3.6 μm XB-C-18 analytical column (150 × 2.10 mm, 3.6 μm, 100 Å, Phenomenex (Torrance, CA, USA); the injection volume was 20 μL. The following linear elution gradient was used for the analytical separation: solvent B was varied from 5%–60% in 55 min; then was increased up to 90% in 1 min and this ratio was maintained constant for the following 15 min. The precent of B was suddenly decreased at 5% and kept stable for 15 min for column reconditioning. The two reserves used were: A = H_2O + 0.1% formic acid and B = ACN + 0.1% formic acid; flow rate was set at 200 μL/min.

MS system was operated in Data Dependent^{TM} Acquisition mode (DDA) by selecting the option *N*th *order double play* mode. In particular, two events were set for this experiment: (1) full MS in the range 400–2000 *m*/*z*, four microscans; and (2) full MS/MS DDA of the 20 most abundant ions in the MS spectrum using a normalized collision energy at 35%. Dependent settings for the DDA were set as reported [54].

4.7. Bioinformatic Analysis

Raw data obtained from LC-MS/MS acquisitions were searched against a customized database (DB = approximately 12,900 entries) containing amino acid sequences referred to all *Pleurotus* spp., downloaded from the largest UniProt DB available online [55] Protein identification was performed by the commercial software Proteome Discoverer^{TM} based on Sequest™ (version 1.4, Thermo-Fisher-Scientific, San José, CA, USA, 2012) scoring algorithm. Software results were filtered post-acquisition by peptide mass deviation (300 ppm), by setting $n = 3$ as minimum number of peptides for protein identification and peptide confidence medium (meaning better than 5% of confidence level).

4.8. In Vitro Degradation of AFB_1 and AFM_1 with LC and Redox Mediators

Degradation assays were performed in 500 μL of reaction volume of 1 mM sodium acetate buffer pH 5 with 1 μg/mL of AFB_1. 2.5 units of LC were added to each reaction.

Alternatively, ABTS, AS, or SA were independently tested as redox mediators at 1 mM and 10 mM.

With respect to AFM_1, degradation assays were performed by incubating 0.05 μg/mL of AFM_1, 2.5 units of LC and ABTS, AS, or SA as redox mediators at 10 mM. In control samples, the enzymatic solution was replaced by an equal volume of buffer. Reactions were incubated at 25 °C for three days in the dark. Each experiment was performed in triplicate.

4.9. Chemical Analyses

AFB_1 analyses were performed with a HPLC Agilent 1260 Series (Agilent Technology, Santa Clara, CA, USA) with post column photochemical derivatization (UVE™, LCTech GmbH, Dorfen, Germany). The analytical column was a Luna PFP (150 × 4.6 mm, 3 μm) (Phenomenex, Torrance, CA, USA) preceded by a SecurityGuard™ (PFP, 4 × 3.0 mm, Phenomenex).

Samples containing AFB_1 were filtered using RC 0.20 μm filters (Grace) and 100 μL of volume was injected into the HPLC apparatus with a full loop injection system. The fluorometric detector was set at wavelengths of 365 nm (excitation) and 435 nm (emission). The mobile phase consisted of a mixture of H_2O:ACN (70:30, *v*/*v*) and the flow rate was 1.0 mL/min. The temperature of the column was maintained at 40 °C. AFB_1 was quantified by measuring peak areas at the retention time of aflatoxin standard solutions (Sigma-Aldrich, Milan, Italy) and comparing these areas with the relevant calibration curve at 0.4–10.0 ng/mL. With this mobile phase, the retention time was about 14.5 min. The limit of quantification (LOQ) was 0.4 ng/mL, while the LOD of the method was 0.2 ng/mL based on a signal to noise ratio of 3:1.

AFM_1 analyses were performed with a HPLC Agilent 1260 Series with a fluorometric detector (Santa Clara, CA, USA). The column used was a Zorbax SB-C18 (150 × 4.6 mm i.d., 5 μm Agilent, (Santa Clara, CA, USA) with a security guard (4 × 3.0 mm).

AFM_1 levels in samples were determined by HPLC/FLD method. The solutions were filtered using RC 0.20 μm filters (Grace, Taipei, China); 50 μL were injected into the HPLC apparatus with a full loop injection system. The fluorometric detector was set at wavelengths of 365 nm (excitation) and 450 nm (emission). The mobile phase consisted of a mixture of H_2O:ACN (75:25, *v*/*v*) and the flow rate was 1 mL/min. The temperature of the column was maintained at 30 °C. AFM_1 was quantified by measuring peak areas at the retention time of aflatoxin standard solutions and comparing these areas with the relevant calibration curve at 1.0–10.0 ng/mL. With this mobile phase, the retention time of AFM_1 was about 6 min. The LOQ was 1.0 ng/mL, while the limit of detection (LOD) of the method was 0.12 ng/mL, based on a signal to noise ratio of 3:1.

When needed, controls and samples were diluted to fit the calibration ranges of the corresponding HPLC methods. Degradation percentages were calculated as follows:

$$\%\, aflatoxin\ degradation = \frac{aflatoxin_{sample}}{aflatoxin_{control}} \times 100 \tag{1}$$

Acknowledgments: This work was financially supported by H2020-E.U.3.2-678781-MycoKey-Integrated and innovative key actions for mycotoxin management in the food and feed chain.

Author Contributions: M.L., F.F. and G.M. conceived and designed the experiments. M.L. and V.C.L. performed the laccase purification under the supervision of P.Z. and E.S. and performed the degradation assays. M.T.C. and M.H. performed the chemical analysis. M.L. and L.Q. performed the zymogram and the SDS-PAGE, while L.M. performed the LC-MS/MS analysis. All authors reviewed the paper. M.L., F.F. and A.F.L. wrote the paper and G.M. was responsible for the submission.

Conflicts of Interest: The authors declare no conflict of interest.

Abbreviations

ABTS	[2,2′-azino-bis-(3-ethylbenzothiazoline-6-sulfonic acid)]
CAN	Acetonitrile
AFB_1	aflatoxin B_1
AFM_1	aflatoxin M_1
AS	Acetosyringone
DEAE	diethylamino ethyl
ET	electron transfer
EU	enzymatic unit
FLD	fluorescent detector
HAT	hydrogen atom abstraction
HBT	Hydoxybenzotriazole
Lac2	laccase 2
LC	laccase
LMS	laccase mediator system
LOD	limit of detection
LOQ	limit of quantification
SA	Syrinagaldehyde
TEMPO	2,2,6,6-tetramethylpiperidine-*N*-oxyl

References

1. Claus, H. Laccases: Structure, reactions, distribution. *Micron* **2004**, *35*, 93–96. [CrossRef] [PubMed]
2. Osma, J.F.; Toca-Herrera, J.L.; Rodrıguez, S. Uses of Laccases in the Food Industry. *Enzym. Res.* **2010**. [CrossRef] [PubMed]

3. Pardo, I.; Camarero, S. Laccase engineering by rational and evolutionary design. *Cell. Mol. Life Sci.* **2015**, *72*, 897–910. [CrossRef] [PubMed]
4. Leontievky, A.; Myasoedova, N.; Pozdnyakova, N.; Golovleva, L. "Yellow" laccase of *Panus tigrinus* oxidizes non-phenolic substrates without redox mediators. *FEBS Lett.* **2007**, *413*, 446–448. [CrossRef]
5. Palmieri, G.; Giardina, P.; Bianco, C.; Scaloni, A.; Capasso, A.; Sannia, G. A novel white laccase from *Pleurotus ostreatus*. *J. Biol. Chem.* **1997**, *272*, 31301–31307. [CrossRef] [PubMed]
6. Khatun, S.; Islam, S.; Cakilcioglu, U.; Guler, P.; Chatterjee, N.C. Nutritional qualities and antioxidant activity of three edible oyster mushrooms (*Pluerotus* spp.). *NJAS-Wagening. J. Life Sci.* **2015**, *73*, 1–5. [CrossRef]
7. Munoz, C.; Guillen, F.; Martınez, A.T.; Martınez, M.J. Induction and Characterization of Laccase in the Ligninolytic Fungus *Pleurotus eryngii*. *Curr. Microbiol.* **1997**, *34*, 1–5. [PubMed]
8. Zucca, P.; Cocco, G.; Sollai, F.; Sanjust, E. Fungal laccases as tools for biodegradation of industrial dyes. *Biocatalysis* **2015**, *1*, 82–108. [CrossRef]
9. Baiocco, P.; Barreca, A.M.; Fabbrini, M.; Galli, C.; Gentili, P. Promoting laccase activity towards non-phenolic substrates: A mechanistic investigation with some laccase-mediator systems. *Org. Biomol. Chem.* **2003**, *1*, 191–197. [CrossRef] [PubMed]
10. Camarero, S.; Ibarra, D.; Martınez, M.J.; Martınez, A.T. Lignin-Derived Compounds as Efficient Laccase Mediators for Decolorization of Different Types of Recalcitrant Dyes. *Appl. Environ. Microbiol.* **2005**, *71*, 1775–1784. [CrossRef] [PubMed]
11. Moldes, D.; Díaz, M.; Tzanov, T.; Vidal, T. Comparative study of the efficiency of synthetic and natural mediators in laccase-assisted bleaching of eucalyptus kraft pulp. *Bioresour. Technol.* **2008**, *17*, 7959–7965. [CrossRef] [PubMed]
12. Pezzella, C.; Guarino, L.; Piscitelli, A. How to enjoy laccases. *Cell. Mol. Life Sci.* **2005**, *72*, 923–940. [CrossRef] [PubMed]
13. Doyle, M.P.; Marth, E.H. Peroxidase activity in mycelia of *Aspergillus parasiticus* that degrade aflatoxin. *Eur. J. Appl. Microbiol. Biotechnol.* **1979**, *7*, 211–217. [CrossRef]
14. Engelhardt, G. Degradation of Ochratoxin A and B by the White Rot Fungus *Pleurotus ostreatus*. *Mycotoxin Res.* **2002**, *18*, 37–43. [CrossRef] [PubMed]
15. Alberts, J.F.; Gelderblom, W.C.A.; Botha, A.; van Zyl, W.H. Degradation of aflatoxin B_1 by fungal laccase enzymes. *Int. J. Food Microbiol.* **2009**, *135*, 47–52. [CrossRef] [PubMed]
16. Banu, I.; Lupu, A.; Aprodu, I. Degradation of Zearalenone by Laccase enzyme. *Sci. Study Res.* **2013**, *14*, 79–84.
17. Aflatoxins. *IARC Monographs*; International Agency for the Research on Cancer: Lyon, France, 2012; Volume 100F.
18. Wu, F. Global impacts of aflatoxin in maize: Trade and human health. *World Mycotoxin J.* **2015**, *8*, 137–142. [CrossRef]
19. Dong, J.L.; Zhang, Y.W.; Zhang, R.H.; Huang, W.Z.; Zhang, Y.Z. Influence of culture conditions on laccase production and isozyme patterns in the white-rot fungus *Trametes gallica*. *J. Basic Microbiol.* **2005**, *45*, 190–198. [CrossRef] [PubMed]
20. Pegler, D.N. The classification of the genus *Lentinus* Fr. (*Basidiomycota*). *Kavaka* **1975**, *3*, 11–20.
21. Wu, Q.; Jezkova, A.; Yuan, Z.; Pavlikova, L.; Dohnal, V.; Kuca, K. Biological degradation of aflatoxins. *Drug Metab. Rev.* **2009**, *41*, 1–7. [CrossRef] [PubMed]
22. Zhao, L.H.; Guan, S.; Gao, X.; Ma, Q.G.; Lei, Y.; Bai, X.M.; Ji, C. Preparation, purification and characteristics of an aflatoxin degradation enzyme from *Myxococcus fulvus* ANSM068. *J. Appl. Microbiol.* **2011**, *110*, 147–155. [CrossRef] [PubMed]
23. Adebo, O.; Njobeh, P.B.; Gbashi, S.; Nwinyi, O.C.; Mavumengwana, V. Review on Microbial Degradation of Aflatoxins. *Crit. Rev. Food Sci. Nutr.* **2015**. [CrossRef] [PubMed]
24. Margot, J.; Bennati-Granier, C.; Maillard, J.; Blánquez, P.; Barry, D.A.; Holliger, C. Bacterial versus fungal laccase: Potential for micropollutant degradation. *AMB Express* **2013**, *3*, 63–77. [CrossRef] [PubMed]
25. Scarpari, M.; Bello, C.; Pietricola, C.; Zaccaria, M.; Bertocchi, L.; Angelucci, A.; Ricciardi, M.R.; Scala, V.; Parroni, A.; Fabbri, A.A.; et al. Aflatoxin Control in Maize by *Trametes versicolor*. *Toxins* **2014**, *6*, 3426–3437. [CrossRef] [PubMed]

26. Zeinvand-Lorestani, H.; Sabzevari, O.; Setayesh, N.; Amini, M.; Nili-Ahmadabadi, A.; Faramarzi, M.A. Comparative study of in vitro prooxidative properties and genotoxicity induced by aflatoxin B_1 and its laccase-mediated detoxification products. *Chemosphere* **2015**, *135*, 1–6. [PubMed]
27. Zucca, P.; Rescigno, A.; Olianas, A.; Maccioni, S.; Sollai, F.; Sanjust, E. Induction, purification, and characterization of a laccase isozyme from *Pleurotus sajor-caju* and the potential in decolorization of textile dyes. *J. Mol. Catal. B Enzym.* **2011**, *68*, 216–222.
28. Díaz, R.; Téllez-Téllez, M.; Sánchez, C.; Bibbins-Martínez, M.D.; Díaz-Godínez, G.; Soriano-Santos, J. Influence of initial pH of the growing medium on the activity, production and genes expression profiles of laccase of *Pleurotus ostreatus* in submerged fermentation. *Electron. J. Biotechnol.* **2013**, *16*. [CrossRef]
29. Perry, C.R.; Smith, M.; Britnell, C.H.; Wood, D.H.; Thursto, C.F. Identification of two laccase genes in the cultivated mushroom *Agaricus bisporus*. *Microbiology* **1993**, *139*, 1209–1218. [CrossRef] [PubMed]
30. Tadesse, M.A.; D'Annibale, A.; Galli, C.; Gentili, P.; Sergi, F. An assessment of the relative contributions of redox and steric issues to laccase specificity towards putative substrates. *Org. Biomol. Chem.* **2008**, *6*, 868–878. [CrossRef] [PubMed]
31. Cañas, A.; Camarero, S. Laccases and their natural mediators: Biotechnological tools for sustainable eco-friendly processes. *Biotechnol. Adv.* **2010**, *28*, 694–705. [CrossRef] [PubMed]
32. Bourbonnais, R.; Paice, M.G. Oxidation of non-phenolic substrates: An expanded role for lactase in lignin biodegradation. *FEBS Lett.* **1990**, *267*, 99–102. [CrossRef]
33. Solis-Oba, M.; Almendariz, J.; Viniegra-Gonzalez, G. Biotechnological treatment for colorless denim and textile wastewater. *Rev. Int. Contam. Ambient.* **2008**, *24*, 5–11.
34. Wells, A.; Teria, M.; Eve, T. Green oxidations with laccase-mediator systems. *Biochem. Soc. Trans.* **2006**, *34*, 304–308. [CrossRef] [PubMed]
35. Collins, P.J.; Kotterman, M.; Field, J.A.; Dobson, A. Oxidation of Anthracene and Benzo [a] pyrene by Laccases from *Trametes versicolor*. *Appl. Environ. Microbiol.* **1996**, *62*, 4563–4567. [PubMed]
36. Shi, L.; Ma, F.; Han, Y.; Zhang, X.; Yu, H. Removal of sulfonamide antibiotics by oriented immobilized laccase on Fe_3O_4 nanoparticles with natural mediators. *J. Hazard. Mater.* **2014**, *279*, 203–211. [CrossRef] [PubMed]
37. Dubé, E.; Shareck, F.; Hurtubise, Y.; Beauregard, M.; Daneault, C. Decolourization of recalcitrant dyes with a laccase from *Streptomyces coelicolor* under alkaline conditions. *J. Ind. Microbiol. Biotechnol.* **2008**, *35*, 1123–1129. [CrossRef] [PubMed]
38. Camarero, S.; Ibarra, D.; Martınez, A.T.; Romero, J.; Gutierrez, A.; del Rio, J.C. Paper pulp delignification using laccase and natural mediators. *Enzym. Microb. Technol.* **2007**, *40*, 1264–1271. [CrossRef]
39. Campos, R.; Kandelbauer, A.; Robra, K.H.; Cavaco-Paulo, A.; Guébitz, G.M. Indigo degradation with purified laccases from *Trametes hirsuta* and *Sclerotium rolfsii*. *J. Biotechnol.* **2001**, *89*, 131–139. [CrossRef]
40. Rosado, T.; Bernardo, P.; Koci, K.; Coelho, A.V.; Robalo, M.P.; Martins, L.O. Methyl syringate: An efficient phenolic mediator for bacterial and fungal laccases. *Bioresour. Technol.* **2012**, *124*, 371–378. [CrossRef] [PubMed]
41. Elsanhoty, R.M.; Salam, S.A.; Ramadan, M.F.; Badr, F.H. Detoxification of aflatoxin M_1 in yoghurt using probiotics and lactic acid bacteria. *Food Control* **2014**, *43*, 129–134. [CrossRef]
42. Vazquez, I.N.; Albores, A.M.; Martınez, E.M.; Miranda, R.; Castro, M. Role of Lactone Ring in Structural, Electronic, and Reactivity Properties of Aflatoxin B_1: A Theoretical Study. *Arch. Environ. Contam. Toxicol.* **2010**, *59*, 393–406. [CrossRef] [PubMed]
43. Lee, S.; Dunn, J.J.; DeLucca, A.J.; Ciegler, A. Role of lactone ring of aflatoxin B_1 in toxicity and mutagenicity. *Experientia* **1981**, *37*, 16–17. [CrossRef] [PubMed]
44. Kersten, P.; Kalyanaraman, B.; Hammel, K.E.; Reinhammar, B.; Kirk, T.K. Comparison of lignin peroxidase, horseradish peroxidase and laccase in the oxidation of methoxybenzenes. *Biochem. J.* **1990**, *268*, 475–480. [CrossRef] [PubMed]
45. Ander, P.; Eriksson, K.E.; Yu, H. Vanillic acid metabolism by *Sporotrichum pulverulentum*: Evidence for demethoxylation before ring-cleavage. *Arch. Clin. Microbiol.* **1983**, *136*, 1–6. [CrossRef]
46. Commission regulation (EU) 2015/786 of 19 May 2015 defining acceptability criteria for detoxification processes applied to products intended for animal feed as provided for in Directive 2002/32/EC of the European Parliament and of the Council. *Off. J. Eur. Union* **2015**, *58*, 10–14.
47. ITEM Collection. Available online: http://www.ispa.cnr.it/Collection (accessed on 15 June 2016).

48. Association of Official Analytical Chemists (AOAC) Official Method 971.2988 (2000). Available online: http://www.aoacofficialmethod.org/index.php?main_page=product_info&cPath=1&products_id=625 (accessed on 16 August 2016).
49. Association of Official Analytical Chemists (AOAC) Official Method 2008.08-2008. Available online: http://www.aoacofficialmethod.org/index.php?main_page=product_info&cPath=1&products_id=2816 (accessed on 16 August 2016).
50. Bradford, M.M. A Rapid and Sensitive Method for the Quantitaton of Microgram Quantities of Protein Utilizing the Principle of Protein-Dye Binding. *Anal. Biochem.* **1976**, *72*, 248–254. [CrossRef]
51. Bleve, G.; Mita, G.; Rampino, P.; Perrotta, C.; Villanova, L.; Grieco, F. Molecular cloning and heterologous expression of a laccase gene from *Pleurotus eryngii* in free and immobilized Saccharomyces cerevisiae cells. *Appl. Microbiol. Biotechnol.* **2008**, *79*, 731–741. [CrossRef] [PubMed]
52. Téllez-Téllez, M.; Sánchez, C.; Loera, O.; Díaz-Godínez, G. Differential patterns of constitutive intracellular laccases of the vegetative phase of *Pleurotus* species. *Biotechnol. Lett.* **2005**, *27*, 1391–1394. [CrossRef] [PubMed]
53. Laemmli, U.K. Cleavage of structural proteins during the assembly of the head of bacteriophage T4. *Nature* **1970**, *227*, 680–685. [CrossRef] [PubMed]
54. Monaci, L.; Pilolli, R.; De Angelis, E.; Godula, M.; Visconti, A. Multi-allergen detection in food by micro high performance liquid chromatography coupled to a dual cell linear ion trap mass spectrometry. *J. Chromatogr. A* **2014**, *1358*, 36–144. [CrossRef] [PubMed]
55. Uniprot Database. Available online: http://www.uniprot.org (accessed on 13 June 2016).

© 2016 by the authors. Licensee MDPI, Basel, Switzerland. This article is an open access article distributed under the terms and conditions of the Creative Commons Attribution (CC BY) license (http://creativecommons.org/licenses/by/4.0/).

Article

Essential Oils Modulate Gene Expression and Ochratoxin A Production in *Aspergillus carbonarius*

Rachelle El Khoury [1,2], Ali Atoui [3,*], Carol Verheecke [2], Richard Maroun [1], Andre El Khoury [1] and Florence Mathieu [2]

[1] Laboratoire de Mycologie et Sécurité Alimentaire (LMSA), Centre d'analyse et de Recherche (CAR), Campus des Sciences et Technologie, Université Saint-Joseph, Mkalles-Beyrouth 1107-2050, Lebanon; rachelle.khouryel@net.usj.edu.lb (R.E.K.); richard.maroun@usj.edu.lb (R.M.); andre.khoury@usj.edu.lb (A.E.K.)

[2] Laboratoire de Génie Chimique, Université de Toulouse, CNRS, INPT, UPS, Toulouse 31326, France; carol.verheecke@gmail.com (C.V.); florence.mathieu@ensat.fr (F.M.)

[3] Laboratory of Microbiology, Department of Natural Sciences and Earth, Faculty of Sciences I, Lebanese University, Hadath Campus, Beirut P.O Box 11-8281, Lebanon

* Correspondence: a.atoui@cnrs.edu.lb; Tel.: +961-1-450-811

Academic Editor: Antonio Moretti
Received: 21 June 2016; Accepted: 9 August 2016; Published: 19 August 2016

Abstract: Ochratoxin A (OTA) is a mycotoxin, mainly produced on grapes by *Aspergillus carbonarius*, that causes massive health problems for humans. This study aims to reduce the occurrence of OTA by using the ten following essential oils (E.Os): fennel, cardamom, anise, chamomile, celery, cinnamon, thyme, taramira, oregano and rosemary at 1 μL/mL and 5 μL/mL for each E.O. As a matter of fact, their effects on the OTA production and the growth of *A. carbonarius* S402 cultures were evaluated, after four days at 28 °C on a Synthetic Grape Medium (SGM). Results showed that *A. carbonarius* growth was reduced up to 100%, when cultured with the E.Os of cinnamon, taramira, and oregano at both concentrations and the thyme at 5 μL/mL. As for the other six E.Os, their effect on *A. carbonarius* growth was insignificant, but highly important on the OTA production. Interestingly, the fennel E.O at 5 μL/mL reduced the OTA production up to 88.9% compared to the control, with only 13.8% of fungal growth reduction. We further investigated the effect of these E.Os on the expression levels of the genes responsible for the OTA biosynthesis (*acOTApks* and *acOTAnrps* along with the *acpks* gene) as well as the two regulatory genes *laeA* and *vea*, using the quantitative Reverse Transcription-Polymerase Chain Reaction (qRT-PCR) method. The results revealed that these six E.Os reduced the expression of the five studied genes, where the *ackps* was downregulated by 99.2% (the highest downregulation in this study) with 5 μL/mL of fennel E.O. As for the *acOTApks*, *acOTAnrps*, *veA* and *laeA*, their reduction levels ranged between 10% and 96% depending on the nature of the E.O and its concentration in the medium.

Keywords: *Aspergillus carbonarius*; Ochratoxin A; gene expression; essential oils

1. Introduction

Mycotoxins are secondary metabolites produced by a wide range of filamentous fungi which have adverse effects on humans, animals and crops that result in illnesses and economic losses [1]. The most known mycotoxins are the aflatoxins, patulin, citrinin, fumonisin B1, zearalenone and the Ochratoxin A (OTA). In fact, OTA is produced by different species of *Aspergillus* and *Penicillium* such as *A. ochraceus*, *A. carbonarius*, *P. verrucosum* and *P. nordicum*. However, *A. carbonarius* is considered to be the main producer of OTA on coffee and grapes [2]. OTA has nephrotoxic, hepatotoxic, teratogenic, and immunotoxic effects *on* several animal species [3] and was classified by the International Agency for Research on Cancer (IARC) in 1993, as a possible human carcinogen (group 2B) [4]. Additionally, it was associated with the acute renal failure in humans [5] and liver cancer in animals after a long-term exposure with the molecule [6].

Chemical and physical treatments are considered to be inefficient at removing the OTA from grapes and wine without altering their organoleptic properties [7]. Hence, the awareness of the hazardous effects of chemical preservatives has led the scientific community to search for naturally occurring molecules that are able to reduce OTA production without altering the fungal growth, thus preventing its replacement by other undesired mycotoxigenic fungi.

Essential oils (E.Os), herbs and spices antimicrobial properties have been acknowledged and used since ancient times for therapy and food preservation. Thus many studies were carried out on their ability to reduce mycelial growth and/or mycotoxins production by different mycotoxigenic fungi. Soliman and Badeaa [8] found that the E.Os of the thyme, cinnamon, marigold, spearmint, basil and quyssum had inhibitory activities against the growth of *A. flavus*, *A. parasiticus*, *A. ochraceus* and *Fusarium verticillioides*. Similar results were obtained by Basílico and Basílico [9], who found that the mint and oregano E.Os also had a complete inhibitory effect on the growth of *A. ochraceus*.

To date, the biosynthetic pathway of OTA in *A. carbonarius* has not yet been completely elucidated and only few genes have been discovered. However, it was confirmed by several studies that the OTA biosynthesis pathway contains a polykétide synthase (PKS) and a non ribosomal peptide synthase (NRPS) family. A study conducted by Gallo et al. [10] identified and characterized an *acpks* gene in *A. carbonarius* that encodes a conserved ketosynthase and acyl transferase domains, and found that the OTA production strongly depended on the expression levels of the *acpks* gene, proving the implication of this gene in the OTA biosynthesis. Those same authors have identified the *acOTApks* gene that has a different function from *acpks*, the previously described gene. *AcOTApks* encodes an AcOTApks protein that belongs to the (HR)-PKS family, and contains a putative methyltransferase domain likely responsible for the addition of the methyl group to the OTA polyketide structure [11]. Another characterized OTA biosynthesis gene in *A. carbonarius*is the *acOTAnrps* which was proven to be essential for OTA, OTα and OTB biosynthesis and is located about 900 nt upstream of a *pks* gene [12]. Furthermore, the OTA biosynthesis in *A. carbonarius* is regulated by two known genes: *laeA* and *veA*. Their deletion caused a drastic decrease in the OTA production and a dowregulation in the expression of the *nrps* gene [13].

Until now, the effect of E.Os on the expression of the genes responsible for OTA production in *A. carbonarius* is not yet evaluated. Therefore, the aim of this study is to evaluate the effect of 10 different E.Os (anise, rosemary, thyme, oregano, fennel, cardamom, chamomile, cinnamon, taramira and celery) at concentrations of 1 µL/mL and 5 µL/mL, on the OTA produced by *A. carbonarius* (the main contaminant of grapes), and their influence on the expression levels of *acpks*, *acOTApks*, *acOTAnrps*, *laeA* and *veA* genes involved in the OTA biosynthesis, in order to choose the E.O(s) whose effects are only limited to OTA reduction without affecting the growth of *A. carbonarius*.

2. Results

2.1. Essential Oils Effect on Growth of A. carbonarius *S402 and Its OTA Production*

The effect of the ten E.Os used in this study at 1 µL/mL and 5 µL/mL, on the *A. carbonarius* dry weight, its radial growth and OTA production are shown in Table 1.

Table 1. Dry weight (g) and diameter (cm) of *Aspergillus carbonarius* S402, the OTA concentration (ng/mL*g) after four days of culture at 28 °C on Synthetic Grape Medium(SGM) supplemented with the ten essential oils at 1 µL/mL and 5 µL/mL, compared to a control.

Essential Oils	Dry Weight (g)		Radial Growth (cm)		OTA Concentration (ng/mL*g of Dry Weight)	
Control	0.36 ± 0.05 [a]		5.1 ± 0.01 [c]		50.58 ± 0.79	
	5 µL/mL	5 µL/mL	1 µL/mL	5 µL/mL	1 µL/mL	5 µL/mL
Anise	0.38 ± 0.01 [a]	0.3 ± 0.01 [b]	4.5 ± 0.01 [c]	3.5 ± 0.05 [c,d]	16.66 ± 0.55 *	10.36 ± 0.27 *
Cardamom	0.36 ± 0.005 [a]	0.31 ± 0.005 [b]	4.5 ± 0.01 [c]	4.4 ± 1 [c]	13.15 ± 0.20 *	8.37 ± 0.15 *
Celery	0.36 ± 0.05 [a]	0.34 ± 0.05 [a]	4.1 ± 0.05 [c]	3.8 ± 0.00 [d]	16.43 ± 0.24 *	15.60 ± 0.24 *
Chamomile	0.36 ± 0.005 [a]	0.31 ± 0.005 [b]	5.1 ± 0.05 [c]	4.5 ± 0.01 [c]	16.54 ± 0.26 *	10.66 ± 0.19 *
Cinnamon	No growth	No growth	N.D	N.D	N.D	N.D
Fennel	0.32 ± 0.05 [b]	0.31 ± 0.005 [b]	4.1 ± 0.05 [c]	3.4 ± 0.01 [c]	6.8 ± 0.12 *	5.6 ± 0.10 *
Oregano	No growth	No growth	N.D	N.D	N.D	N.D
Rosemary	0.36 ± 0.01 [a]	0.33 ± 0.001 [a]	4.5 ± 0.00 [c]	4.5 ± 0.00 [c]	23.62 ± 0.65 *	11.03 ± 2.80 *
Taramira	No growth	No growth	N.D	N.D	N.D	N.D
Thyme	0.2 ± 0.01 *	No growth	2.1 ± 0.01 *	N.D	3.56 ± 0.17 *	N.D

The Mean of the dry weight (g) and the growth inhibition (%) ± the standard deviation of the triplicates are represented in this table. Statistical differences are indicated as: * = significant difference ($p < 0.01$). Data with the same letters are not significantly different ($p < 0.05$). N.D: Not Detectable.

Our results showed that the radial growth, dry weight and OTA inhibition levels strongly depend on the nature of the E.O and its concentration (Table 1). The E.O of taramira, oregano and cinnamon (1 µL/mL and 5 µL/mL) were able to reduce *A. carbonarius* S402 growth up to 100%, with no visible growth on the SGM medium; therefore, no detectable OTA concentration in the culture medium was found after HPLC (High Performance Liquid Chromatography) analysis. Although the E.O of thyme at 5 µL/mL completely blocked the growth of *A. carbonarius* S402, it only decreased 48.71% of the fungal growth at 1 µL/mL.

E.Os of the fennel, cardamom, chamomile, rosemary, anise and the celery at 1 µL/mL and 5 µL/mL, showed a slight effect on the dry weight and radial growth of *A. carbonarius* S402 but a significant impact on the OTA production. A reduction of 88.9% in OTA was reached when 5 µL/mL of fennel E.O was added to the SGM medium, while it only reduced 13% of the fungal growth. Additionally, this E.O at 1 µL/mL was able to reduce the OTA to 86.6%, with no significant impact on the radial growth (33.3%) and the dry weight (11%). Cardamom, chamomile, rosemary, anise and celery E.Os at 1 µL/mL, also had an important effect in OTA reduction that reached 74.2%, 67.5,% 53.7%, 76.6%, and 68.5% respectively, with growth reduction that ranged between 0% and 11% (cardamom and celery (0%), anise (5%), chamomile and rosemary (11%)). OTA reduction was higher, when the E.O concentrations were at 5 µL/mL reaching 83.5% for cardamom E.O with only 13% of dry weight reduction. The OTA reduction percentages were almost the same for the E.Os of chamomile (79.1%), rosemary (78.3%), and celery (77.1%), with a growth reduction of 13%, 8% and 5%, respectively. Lastly, the E.O of anise reduced the fungal growth to 16% with 83.9% of OTA concentration reduction.

2.2. Microscopic Morphology Evaluation

Scanning Electron Microscopy (SEM) of *A. carbonarius* S402 treated with 1 µL/mL and 5 µL/mL of fennel E.O compared to untreated samples showed that the application of the fennel E.O at 1 µL/mL did not affect the size of the conidial head. For instance, the conidial head of the untreated samples measured 107 µm (Figure 1A), similar to the treated sample with 1 µL/mL (102 µm) (Figure 1C). When the SGM medium was supplemented with 5 µL/mL of fennel E.O, the results were different: the conidial head was smaller and more compact than those in control condition, and measured 84 µm (Figure 1E), i.e., 21.4% lesser than the control.

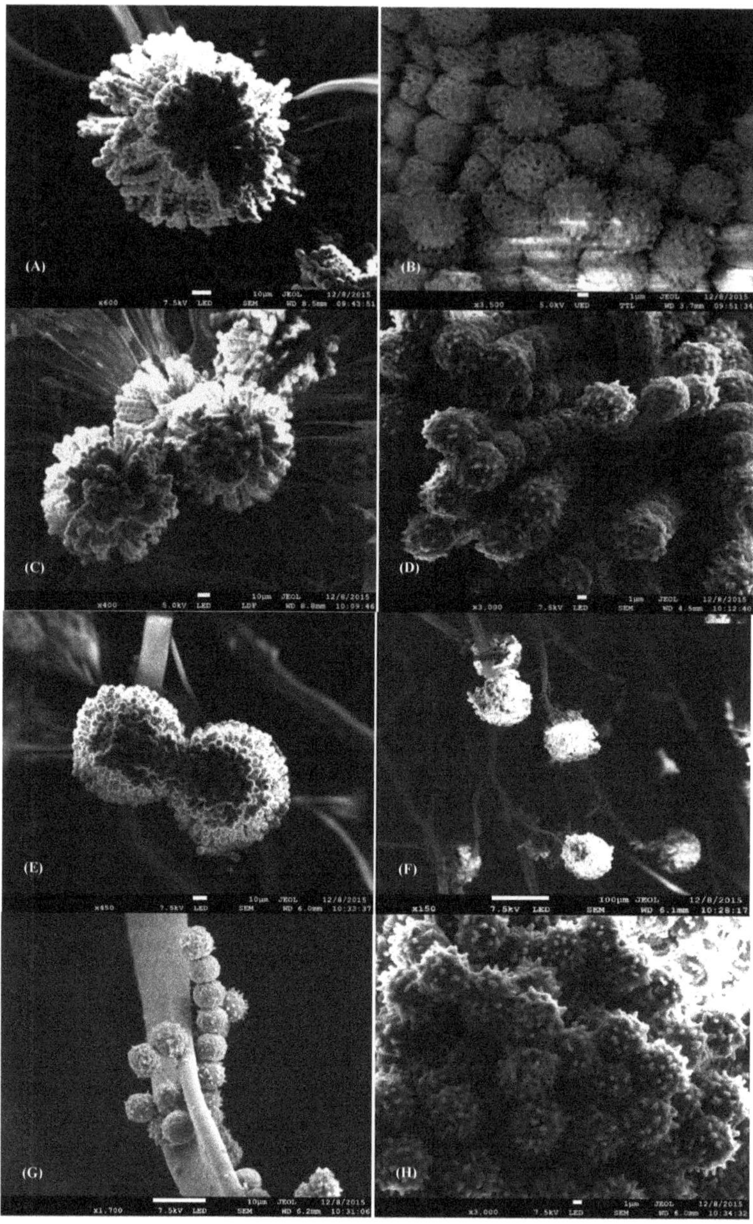

Figure 1. Scanning electron micrographs of the *Aspergillus carbonarius* S402, cultured on Synthetic Grape Medium (SGM) at 28 °C for four days: (**A**) *A. carbonarius* S402 conidial head; (**B**) spores of *A. carbonarius*; (**C**) *A. carbonarius* conidial head cultured with 1 μL/mL fennel oil; (**D**) spores of *A. carbonarius* cultured with 1 μL/mL fennel oil; (**E**,**F**) *A. carbonarius* conidial head cultured with 5 μL/mL fennel oil; (**G**) conidiophores of *A. carbonarius* cultured with 5 μL/mL fennel oil with abnormally placed spores; and (**H**) spores of *A. carbonarius* cultured with 5 μL/mL fennel oil.

As for the spore concentration, a non-significant variation was observed between the untreated *A. carbonarius* and the cultures treated with 1 μL/mL and 5 μL/mL of fennel E.O. In fact, the results showed that the untreated samples had a concentration of 17 $\times$ 16^6 spores/mL, whereas the cultures treated with 1 μL/mL and 5 μL/mL of fennel E.O had a concentrations of 15.8 $\times$ 10^6 spores/mL and 16.1 $\times$ 10^6 spores/mL respectively.

2.3. *Analysis of Genes Expression Involved in OTA Biosynthesis by A. carbonarius*

To better understand the OTA reduction by the E.Os that did not affect the fungal growth, the expressions of two known biosynthetic genes involved in the OTA biosynthesis (*acOTApks* and *acOTAnrps*) as well as the *acpks* gene and the two regulatory genes (*veA* and *laeA*) were evaluated by qRT-PCR.

The expression levels of the target genes were normalized by the expression of the reference gene (β-*tubulin* was chosen as reference gene, as shown in the Supplementary Material, Figure S1) and were expressed relative to the control. Table 2 shows the relative expression values of the target genes of *A. carbonarius* S402 cultured with the six E.Os, along with the OTA reduction.

Results showed that expression level of the target genes in the presence of 1 μL/mL of E.Os varied depending on the nature of the E.O. In fact the E.O of the chamomile, was able to significantly downregulate all the target genes. Others, such as cardamom, rosemary, anise, fennel and celery E.Os, were only able to affect specific genes but had no effect on the others. As a matter of fact, the *acpks* was significantly downregulated when *A. carbonarius* was cultured with the six E.Os at 1 μL/mL. Notably, its expression was reduced by 96% by the E.O of the fennel, 50% and 90% by the celery and rosemary E.Os respectively, accompanied by an OTA reduction that reached 86.9%, 68.8% and 53.7% respectively for each E.O. As for the cardamom, chamomile and anise E.Os, they were able to equally reduce this gene expression by 80% and had approximately the same effect on OTA production, as they reduced it to 74.2% and 76.9% by cardamom and anise E.O, respectively, and to 67.5% by the chamomile E.O.

As for the *acOTApks*, its expression was not downregulated by neither of the six E.Os. In fact, E.O of the rosemary had no significant effect on this gene's expression while the other tested E.Os reduced the *acOTApks* by 99%, 86%, 88%, 81% and 76% for the fennel, cardamom, chamomile, anise and celery E.Os, respectively. While *acOTAnrps* expression was equally reduced by the E.Os of cardamom, chamomile and rosemary by 80%, its expression was not affected by the fennel, celery and anise E.Os.

Regarding the regulatory genes *veA* and *laeA*, they were both not affected by the E.Os of the anise and celery. The *veA* expression was reduced by approximately 90% with the E.Os of the fennel and rosemary, along with 55% and 63% by the E.Os of cardamom and chamomile respectively. As for the *laeA*, it was downregulated by 92%, 71% and 80% with the E.Os of the fennel, chamomile and rosemary respectively. However, the E.O of the cardamom had no significant effect on the *laeA* expression.

At 5 μL/mL, the six E.Os were able to significantly reduce the expressions of the five studied genes. For instance, the *acpks* was highly downregulated by the E.O of the fennel, reaching 99.2% of reduction, followed by the chamomile, rosemary, cardamom and celery E.Os with 98%, 95%, 90%, 85% and 70% reduction, respectively.

As for the *acOTApks*, its expression was reduced up to 99.9% with the fennel E.O, the highest gene downregulation in this study. Moreover, this E.O decreased the OTA production up to 88.6% yet reduced the fungal growth by only 13.6%. The E.Os of the cardamom, rosemary and chamomile also had major effects on the expression of the *acOTApks*. In fact they reduced its expression by 98%, 96% and 95%, respectively, coupled with an OTA reduction by 83.5%, 78.3% and 79.1%, respectively, for each E.O. In addition, celery and anise E.Os downregulated the *acOTApks* by 81%, and 86% respectively as well as the OTA production by 77.1% and 83.9% for each E.O respectively. As for the E.O of fennel, rosemary and anise, they were able to reduce the *acOTAnrps* expression by 98% for both fennel and rosemary, and 99% for anise. Regarding the rest of the E.Os, the cardamom, celery and chamomile, they also reduced this gene expression by 84%, 82% and 52% respectively.

Table 2. Normalized relative Ochratoxin A (OTA) gene expression in *Aspergillus carbonarius* S402 observed after 4 days of culture in presence of six different E.Os at 1 µL/mL and 5 µL/mL, associated with the percentage of OTA reduction levels.

Essential Oils	**Fennel**				**Cardamom**				**Chamomile**			
	1 µL/mL		**5 µL/mL**		**1 µL/mL**		**5 µL/mL**		**1 µL/mL**		**5 µL/mL**	
Genes	Relative expression	% of OTA reduction	Relative expression	% of OTA reduction	Relative expression	% of OTA reduction	Relative expression	% of OTA reduction	Relative expression	% of OTA reduction	Relative expression	% of OTA reduction
acpks	0.04 *		0.008 *		0.24 *		0.15 *		0.2 *		0.02 *	
acOTApks	0.009 *		0.001 *		0.14 *		0.02 *		0.12 *		0.05 *	
acOTAnrps	0.7	86.90%	0.02 *	88.60%	0.2 *	74.20%	0.18 *	83.50%	0.2 *	67.50%	0.16 *	79.10%
veA	0.09 *		0.07 *		0.45 *		0.29 *		0.37 *		0.06 *	
laeA	0.08 *		0.06 *		0.9		0.3 *		0.29 *		0.04 *	

Essential Oils	**Rosemary**				**Anise**				**Celery**			
	1 µL/mL		**5 µL/mL**		**1 µL/mL**		**5 µL/mL**		**1 µL/mL**		**5 µL/mL**	
Genes	Relative expression	% of OTA reduction	Relative expression	% of OTA reduction	Relative expression	% of OTA reduction	Relative expression	% of OTA reduction	Relative expression	% of OTA reduction	% of OTA reduction	Relative expression
acpks	0.1 *		0.05 *		0.2 *		0.1 *		0.5 *		0.3 *	
acOTApks	0.9		0.04 *		0.19 *		0.14 *		0.24 *		0.19 *	
acOTAnrps	0.2 *	53.70%	0.02 *	78.30%	0.8	76.90%	0.01 *	83.90%	0.7	68.50%	0.48 *	77.10%
veA	0.1 *		0.05 *		0.9		0.13 *		0.7		0.12 *	
laeA	0.2 *		0.09 *		0.7		0.2 *		0.8		0.2 *	

Normalized gene expression of the following genes: *acpks, acOTApks, acOTAnrps, veA* and *laeA* along with the OTA reduction in the Synthetic Grape Medium (SGM) after four days of culture at 28 °C. Statistical analysis was made using the GraphPad prism 6 software (version 6.07 for windows, 2015, GraphPad Software Inc., La Jolla, CA, USA). Data with (*) = significant difference ($p < 0.01$).

Finally the *veA* and *laeA* expressions were nearly the same when *A. carbonarius* was cultured with the E.Os of the fennel (93% and 94%), cardamom (71% and 70%), chamomile (94% and 96%), rosemary (95% and 91%), anise (87% and 80%) and lastly celery (88% and 80%), respectively, for each gene.

3. Discussion

Until now, studies have shown that the E.Os can reduce OTA production by reducing fungal growth. Nevertheless, few were able to prove that OTA could also be inhibited without significantly reducing the fungal growth. To that purpose, this work was dedicated to finding an E.O capable of reducing the OTA produced by *A. carbonarius*, without significantly altering its growth, thus preserving the natural balance of the microbial ecosystem and mostly preventing the occurrence of other mycotoxigenic fung.

However, few works were able to prove that the reduction of OTA production could be associated with the downregulation of the expression of genes responsible for the OTA biosynthesis in *A. carbonarius*. In this study, we investigated the effects of ten E.Os on OTA production by *A. carbonarius* S402 and we found that four of these E.Os, taramira, oregano, cinnamon and thyme, completely blocked OTA production by preventing the growth of *A. carbonarius* S402 cultures. These findings are consistent with other studies found in the literature, which tested the effect of some E.Os on the growth of other mycotoxigenic fungi. In fact, Hua et al. [14] found that *A. ochraceus* fungal growth was completely blocked by using the E.O of cinnamon, a natural and synthetic cinnamaldehyde with a concentration of 500 μL/L for the three E.Os. Moreover, the antifungal activities of the cinnamon was studied over the years and proven to be effective against mould growth [8,15–17]. These results regarding the E.O of the cinnamon are consistent with our results, stating that the growth of *A. carbonarius* S402 was completely blocked with the use of this E.O. The oregano E.O also had a great antifungal activity, preventing the growth of *A. niger*, *F. oxysporum* and *Penicillium spp.* in YES broth, at a concentration of 0.1 mL/mL [18]. Additionally, the carvacrol and thymol were identified as the main components of the oregano E.O, responsible for its antifungal activities [19–21]. Similarly, the E.O of the thyme also showed an antifungal effect on the growth of *F. verticillioides*, *A. parasiticus*, *A. flavus* and *A. parasiticus* due to the presence of the thymol, thyme's main component [22–24].

However, the E.O of anise, fennel, cardamom and chamomile did not show a significant antifungal activity against *A. carbonarius* at the concentrations used in this study. The E.O of anise at 1500 μL/L, when used on *A. ochraceus* on a Malt Extract Agar medium (MEA) [14] did not show a significant effect on the growth of the fungus, which similar to our results but contradict those of Soliman and Badeaa [8] who found that at 500 ppm, this E.O blocked the growth of *A. flavus*, *A. ochraceus*, *A. parasiticus* and *F. verticillioides* on PDA medium (Potato Dextrose Agar) after 5–7 days of culture. Furthermore, these same authors found that the E.O of the fennel at 3000 ppm blocked the growth of the four tested fungi, also contradicting with our results, stating that, even at 5 μL/mL (0.5%), this E.O only reduced *A. carbonarius* S402 growth by 13.8% compared to a control. A study conducted by Bansod and Rai [25] showed that the MIC of the fennel E.O against the growth of *A. fumigatus* and *A. niger* was at 2% of this E.O, which means that in order to block the growth of these two fungi, a concentration 4 times higher than ours was needed. The reasons behind these differences could be numerous; first of all, they may be associated with the fact that in our study, we did not extract the E.Os from the plants but instead we bought them from the market; consequently, the extraction methods may vary and alter the antifungal activity of the E.Os. Furthermore, different species of fungi do not have similar spores density or similar growth rates (*A. carbonarius* has a negligible ratio of (spores dry weight)/(biomass dry weight) on DG18 agar medium compared to *A. flavus* and *A. ochraceus*, but it has a significantly higher biomass density than the other two fungi due to its larger conidial size compared to *A. flavus* and *A. ochraceus*, respectively)[26], consequently causing the E.Os to perform differently on different fungal species. Lastly, these contradictions may be associated with the different types of media used in both studies along with the incubation period.

According to Baydar et al. [27] the antifungal properties of E.Os can be attributed to the presence of phenolic compounds such as eugenol, anethole, carvacol, precursors *r*-cimene and *g*-terpinolene and isomers of thymol. Furthermore, they have low molecular weight and lipophilic properties that allows them to easily penetrate cell membrane [28] causing irreversible cell wall damage and cellular organelle as well as affecting the pH homeostasis and equilibrium of inorganic ions [29–31]. Another study on the mode of action of E.Os on fungal growth suggested that *A. conyzoides* E.O was able to cross the plasma membrane of *A. flavus* and interact with the membrane structures of cytoplasmic organelles thus preventing the fungal growth [32].

Regarding the expression levels of the genes involved in the OTA biosynthesis pathway, in presence of E.O, few studies were found in the literature on this topic. Nevertheless, Murthy et al. [33] noticed that the reduction of OTA production was not correlated with the growth reduction of *A. ochraceus* after inoculating the medium with increasing doses of Ajowan Ethanolic Extract (AEE). At 50 and 150 mL/g of AEE, inhibition of OTA production was higher than the reduction of fungal growth. Similarly, Hua, Xing, Selvaraj, Wang, Zhao, Zhou, Liu and Liu [14] also noticed a decreased growth of *A. ochraceus* by 75–150 mg/mL of citral E.O, but the OTA production was completely blocked, therefore suggesting that the reduction was not correlated with the growth reduction, but rather to the suppression of transcription of OTA biosynthesis genes. In fact, they found that the transcription of the *pks* gene (responsible for the synthesis of the polyketide dihydroisocoumarin and involved in the first steps of the OTA biosynthetic pathway) was completely inhibited by 75–150 mg/mL of citral E.O. Likewise; a study conducted by Caceres et al. [34] on the reduction of aflatoxin B1 (AFB1) produced by *A. flavus* showed that all the genes (except for *aflT*) of the AFB1 cluster were drastically downregulated when *A. flavus* was cultured with 0.5 mM of eugenol, resulting an extreme AFB1 reduction with no significant impact on the fungal growth. In the same manner, our study revealed that six of the tested E.Os (anise, fennel, cardamom, chamomile, celery and rosemary) reduced significantly the OTA from the medium without equally reducing the fungal growth. These results suggest that this OTA reduction was not associated with the growth reduction but rather with the repression of certain biosynthetic genes. Fennel, cardamom, chamomile, rosemary, anise and celery E.Os downregulated the *acpks, acOTApks, acOTAnrps, laeA* and *veA* genes, involved in the synthesis of the appropriate enzymes involved in the OTA biosynthesis. A similar study revealed that the expression of the *acOTApks* gene in *A. carbonarius* was also reduced by the addition of 0.065 mg/mL of the hydroxycynnamic acids such as *p*-coumaric and ferulic acids [35]. Our findings suggest that the E.Os used in this study have different ways to reduce OTA from the medium without affecting the fungal growth. Some were able to downregulate the regulatory genes (*laeA* and *veA*) such as the fennel, chamomile and rosemary E.Os at 1 μL/mL and consequently reducing the expression of the genes responsible for the OTA biosynthesis (*acpks, acOTApks* and *acOTAnrps*). Others, such as the anise and celery E.Os at 1 μL/mL did not have significant effect on the expressions of *laeA* and *veA*, but they were able to downregulate directly the *acpks, acOTApks* and *acOTAnrps* genes, suggesting that different types of E.Os has different mode of action on the OTA biosynthesis genes. Similarly, a study was conducted on *F. culmorum* cultured with 0.5 mM of ferulic acid showed that this compound reduced the trichotecene type B production by reducing up to 5.4 times the expression levels of several *tri* genes involved in the biosynthesis pathway [36]. Furthermore, AFB1 production was reduced by adding the caffeic acid in culture with *A. flavus,* resulting in the downregulation of the aflatoxins biosynthesis genes in *A. flavus,* reducing 6.6 fold the *aflD*, 7.1 *aflM*, 9.1 *aflP* and 1.5 *aflS* gene without affecting the fungal growth [37]. Moreover, another study conducted by Yoshinari et al. [38] found that AFG1 production in *A. parasiticus* was reduced by using E.O of the chamomile that inhibited the cytochrome P450 synthesis (CYPA), enzyme implicated in the AFG1 production. Additionally, due to the similarity between the CYPA and TRI4 (an enzyme responsible for trichothecenes early biosynthesis), the E.O of chamomile was also able to decrease the 3-acetyldeoxynivalenol (3-ADON) production in *F. graminearum* by reducing the TRI4 enzyme.

Regarding the microscopic aspects of *A. carbonarius* cultured with 1 µL/mL and 5 µL/mL of fennel E.O, the morphology of the conidial head and the spores remained unchanged, contrary to other results found in the literature. For instance, Hua, Xing, Selvaraj, Wang, Zhao, Zhou, Liu and Liu [14] found that the treatment of *A. ochraceus* with different concentrations of natural cinnamaldehyde, citral and eugenol E.Os showed alteration in the morphology of the hyphæ that appeared collapsed and abnormal, correlated with an OTA reduction. Moreover, the E.O of *Cinnamomum zeylanicum* caused an interesting inhibition of spore germination in *A. flavus*, *A fumigatus* and *A. niger* also causing alterations in the hyphal development, a loss in pigmentation and a lack of sporulation [39]. Another similar study conducted by Sharma and Tripathi [40] showed that the E.O of *Citrus sinensis* at 0.5 µg/mL caused a distortion of the *A. niger* mycelium, squashed and flattened the conidiophores bearing damaged conidial head. In our case, the E.O of the fennel used in this study highly reduced OTA production without altering the growth of *A. carbonarius* S402 or the microscopic aspects of the fungus.

Our findings, along with the data collected from previous studies, indicate that the use of the E.Os on food and crops could be one of the solutions to reduce the OTA from culture media produced by *A. carbonarius*. Some of these E.Os, including anise, celery, cardamom, chamomile, rosemary and fennel, were able to reduce the OTA by repressing the expressions of the genes involved in the OTA biosynthesis without particularly affecting the fungal growth; thus not altering the natural microbial balance.

4. Conclusions

The inhibitory effects of ten E.Os on the OTA production of *A. carbonarius* S402 and their effect on the expression of the genes involved in OTA biosynthesis were evaluated in this study. The E.Os of cinnamon, thyme, oregano and taramira showed an antifungal activity and prevented the growth of *A. carbonarius* S402 cultured on SGM medium. Nevertheless, the objective of our study was to find one or several E.O that has no significant effect on *A. carbonarius* growth, but rather on OTA production. For this aim, we found that the E.O of fennel at 5 µL/mL was the most effective E.O in terms of downregulating the OTA biosynthesis and regulatory genes of *A. carbonarius*, in correlation with a drastic reduction in the OTA production. In addition, this E.O did not alter the morphology of the fungus' conidial heads or the sporulation after four days of culture on SGM medium. Similarly, the E.O of the chamomile, cardamom, celery, rosemary and the anise did not show any antifungal activities and thus did not have a significant effect on the growth of the fungus, but only reduced the OTA production by reducing the genes expressions responsible for this production.

5. Materials and Methods

5.1. Essential Oils

Ten E.Os were chosen in order to evaluate their influence on the growth of *A. carbonarius* and on OTA production: rosemary (*Rosmarinu sofficinalis*), anise (*Pimpinella anisum*), chamomile (*Chamaemelum nobile*), fennel (*Foeniculum vulgare*), thyme (*Thymus vulgaris*), oregano (*Origanum vulgare*), taramira (*Eruca sativ*), cinnamon (*Cinnamomum verum*), cardamom (*Elettariacardamomum*) and celery (*Apium graveolens*). These E.Os were purchased from the local Lebanese market (Tyr, Lebanon), and were used with no additional treatment.

5.2. Strain and Culture Conditions

A. carbonarius strain S402 was provided by the LGC-UMR 5503 (Chemical Engineering Laboratories) located in the Superior National School of Agronomy, in Toulouse France. This strain was sub-cultured in Czapec-Yeast-Agar medium (CYA) as described by El Khoury, Rizk, Lteif, Azouri, Delia and Lebrihi [2] for 4 days at 28 °C. A spore suspension was prepared by adding 8 mL of a sterile Tween 80 (0.005%) solution and scrapping the surface of the cultures with a sterile Pasteur Pipette (Chase Scientific Glass, Inc., Rokwood, TN, USA). The spore count was evaluated by using a

Neubauer haemocytometer (Superior, Marienfeld, Lauda-Konigshofen, Germany), the concentration was adjusted to 10^6 spore/mL and spore suspension was then kept at 4 °C for further use. The culture was then conducted on a Synthetic Grape Medium (SGM), prepared following the instructions described by Bejaoui et al. [41].

Each E.O (at 1 µL/mL and 5 µL/mL) was mixed to 20 mL of the SGM medium before pouring the mix into Petri dishes. In order to compare and evaluate the effect of the E.Os on *A. carbonarius* growth and on OTA production, a control culture was prepared without adding any E.O to the SGM medium. A sterilized transparent cellophane film was placed on the surface of the medium, and then 10^6 spores of *A. carbonarius* were inoculated at the center of the film. The growth of *A. carbonarius* was estimated by the radial growth (cm) and the dry weight (g), both after 4 days of culture. *A. carbonarius* dry weight was evaluated by weighing the mycelium after its desiccation (placing the transparent film carrying the culture at 100 °C for 24 h). As for the radial growth, diameters measurements of *A. carbonarius* cultures were taken after 4 days of incubation at 28 °C.

All assays were carried out in triplicates for each condition.

5.3. Scanning Electron Microscopy (SEM) for Morphologic Study and Spores Count

The fennel E.O at 1 µL/mL and 5 µL/mL was chosen among the ten E.Os to observe its effect on the spores and conidia morphology of *A. carbonarius* in comparison with the control on SGM medium. This E.O was chosen in order to investigate its effect on the fungal sporulation since it had the greatest effect on the OTA production without altering significantly the growth of *A. carbonarius*. After 4 days of culture at 28 °C, the fungus was desiccated at 75 °C for 24 h, prepared for the SEM by coating the samples with 40 nm of gold under vacuum and then viewed with the MEB-FEG (Field Emission Gun); model JSM 7100F (JOEL, Freising, Germany). The spore concentration was assessed using a Neubauer haemocytometer (Superior, Marienfeld, Lauda-Konigshofen, Germany), after taking 4 agar plugs from each condition and mixed with 5 mL of distilled water. The tubes were then shaken vigorously using vortex mixer for 3 min. Samples were made in triplicates and the mean number of spores was considered.

5.4. OTA Extraction and HPLC Analysis

After the incubation period, 3 agar plugs (0.5 cm diameter) were extracted from each SGM medium, placed in 3 mL microtubes and weighed. One milliliter of HPLC grade methanol was added, and the mix was incubated and shaken for 60 min at room temperature. After centrifugation at 13,000 r.p.m (round per minute), the aqueous phase was separated from the debris and diluted with 20 mL phosphate buffer. OTA was then purified using Ochraprep immunoaffinity columns (R-Biopharm, Glasgow, Scotland) by injecting the mix into the columns using a syringe. OTA was eluted by adding 1.5 mL of methanol/acetic acid (98:2, *v/v*) followed by 1.5 mL of distilled water. After filtering with 0.4 µm filters (Sartorius stedim, Biotech), the OTA extract was stored at 4 °C before quantification.

OTA quantification was done with a Water Alliance HPLC system using an Utisphere ODB column, C18 (150 × 4.6 mm, 5 µm, 120 Å) (Interchim, Montluçon, France) at 30 °C. A 30 min isocratic flow was delivered at 49% of eluent A: acidified water (0.2% of acetic acid) and 51% of eluent B: acetonitrile. A flow rate of 1 mL/min was used and 10 µL of extract were injected into the apparatus. OTA was detected by a fluorescent detector at 333/440 nm excitation/emission wavelengths. Peak identity was confirmed by analyzing absorption spectrum with a diode array detector coupled to the system. OTA concentrations were calculated based on a standard calibration curve.

5.5. RNA Extraction, cDNA Synthesis and qRT-PCR

For genes expressions analysis, culture conditions were the same as those described in Section 5.2. After 4 days at 28 °C, the cellophane film was separated from the surface of the SGM medium and the mycelium was then carefully cut out from the cellophane, ground in liquid nitrogen and stored at

−80 °C. Total RNA was extracted from the frozen mycelium using the RNeasy kit (Qiagen, Düsseldorf, Germany) following the manufacturer protocol and then treated with DNase I Amplification Grade (Sigma-Aldrich, St. Louis, MO, USA). The quality and quantity of the total RNA was assessed using the Experion RNA analysis kit (version 3.20, 2015, BioRad, Marnes-la-Coquette, France). The single strand cDNA was synthesized from 1 µg of total RNA using the Advantage RT-for-PCR kit (Clontech, Mountain View, CA, USA) following the manufacturer protocol.

Gene specific primers used are listed in Table 3. The reactions were held using CFX96 Touch Real time PCR detection system Bio-Rad (version 3.0, 2012) using Sso Advanced Universal Sybr Green Supermix (Bio-Rad, Marnes-la-Coquette, France) to evaluate the cDNA amplification, by starting with 100 ng/µL of freshly synthesized cDNA following the manufacturer protocol for the mix preparation and the amplification cycles, with a Tm of 58 °C. The choice of suitable reference genes was conducted by choosing six candidate genes (β-*tubulin, cox5, actin, 18S, calmodulin and gpdA*) used in previous studies, and their stability was tested in the conditions used in our study using the qRT-PCR (version 3.0, 2012, Bio-Rad, Marnes-la-Coquette, France) method and the geNorm software (version 3.0, 2014, Biogazelle, Ghent, Belgium). Their expression stability was assessed with qbase$^+$ biogazelle software (version 3.0, 2014, Biogazelle, Ghent, Belgium) by calculating their geNorm *M* value (*M*) and their coefficient of variation on the normalized relative quantities (*CV*). These values can then be compared against empirically determined thresholds for acceptable stability [42]. The most stable gene is the one with the lowest *M* value and with a *CV* value that does not exceed 0.15 (*M* value of <0.5 and *CV* value of <0.15).

Table 3. List of primers used in this study.

Primer Name	Primer Sequence (5′→3′)	References	Efficiency
acpks-F *acpks-R*	GAGTCTGACCATCGACACGG GGCGACTGTGACACATCCAT	[10]	110.0%
acOTApks-F *acOTApks-R*	CGTGTCCGATACTGTCTGTGA GCATGGAGTCCTCAAGAACC	[11]	102%
acOTAnrps-F *acOTAnrps-R*	ATCCCCGGAATATTGGCACC CCTTCGATCAAGAGCTCCCC	[12]	82.6%
laeA-F *laeA-R*	CACCTATACAACCTCCGAACC GGTTCGGCCAACCGACGACGC	[13]	104.7%
veA-F *veA-R*	TCCCGGTTCTCACAGGCGTA GCTGTCCTTGGTCTCCTCGTA	[13]	101.3%
β-tubulin-F *β-tubulin-R*	CGCATGAACGTCTACTTCAACG AGTTGTTACCAGCACCGGA	[43]	95%
calmodulin-F *calmodulin-R*	CCAGATCACCACCAAGGAGC GTTATCGCGGTCGAAGACCT	[44]	105%
18S-F *18S-R*	GCAAATTACCCAATCCCGAC GAATTGCCGCGGCTGCTG	[44]	98%
cox5-F *cox5-R*	CCCTGTTCTACGTCATTCACTTGTT TCTTCTCGGCCTTGGCATACTC	[45]	87%
actin1-F *actin1-R*	TTGACAATGGTTCGGGTATGTG TTGACAATGGTTCGGGTATGTG	[45]	82%
gpdA-F *gpdA-R*	ACGGCAAGCTCACTGGTATGT CAGCCTTGATGGTCTTCTTGATG	[45]	110%

The PCR efficiencies for each primer pair were determined by a serial cDNA dilution experiments, and were calculated from the slopes of the curves given by the CFX96 Touch Real time PCR detection system software (Bio-Rad).

5.6. Statistical Analysis

Data regarding the dry weight, radial growth, OTA concentrations and genes expressions were analyzed through one-way analysis of variance (ANOVA) and paired *t*-test using GraphPad (version 6.07 for windows, 2015, GraphPad Software Inc., La Jolla, CA, USA) prism 6 software.

Supplementary Materials: The following is available online at www.mdpi.com/2072-6651/8/8/242/s1., Figure S1 Genorm *M* value of the six candidate reference genes: *β-tubulin, calmodulin, cytochrome c oxydase subunit V* (*cox5*), *18S*, *actin* (*actin1*) and *glyceraldehydes 3-phosphate dehydrogenase* (*gdpA*). The most stable gene is the one with the lowest *M* value.

Acknowledgments: The authors would like to thank the Research Council of Saint-Joseph University (Lebanon), and the Polytechnic National Institute of Toulouse (INPT), as well as the French Institute-Campus France and the National Council for Scientific Research, Lebanon (CNRS-L) for the financial support of Rachelle El Khoury.

Author Contributions: Florence Mathieu, Ali Atoui, Andre El Khoury and Richard Maroun conceived, supervised and designed the experiments. Rachelle El Khoury performed the experiments. Rachelle El Khoury, Ali Atoui and Florence Mathieu wrote the paper. Carol Verheecke and Ali Atoui contributed to qRT-PCR performance and data analysis.

Conflicts of Interest: The authors declare no conflict of interest.

Abbreviations

The following abbreviations are used in this manuscript:

A. carbonarius	*Aspergillus carbonarius*
qRT-PCR	Quantitative Reverse Transcription-Polymerase Chain Reaction
OTA	Ochratoxin A
IARC	International Agency for Research on Cancer
E.O	Essential oil
SGM	Synthetic Grape Medium
HPLC	High Performance Liquid Chromatography
mM	Milliomolar
N.D	Not detectable
r.p.m	Round per minute

References

1. Reverberi, M.; Ricelli, A.; Zjalic, S.; Fabbri, A.A.; Fanelli, C. Natural functions of mycotoxins and control of their biosynthesis in fungi. *Appl. Microbiol. Biotechnol.* **2010**, *87*, 899–911. [CrossRef] [PubMed]
2. El Khoury, A.; Rizk, T.; Lteif, R.; Azouri, H.; Delia, M.-L.; Lebrihi, A. Occurrence of ochratoxin A- and aflatoxin B1-producing fungi in lebanese grapes and ochratoxin A content in musts and finished wines during 2004. *J. Agric. Food Chem.* **2006**, *54*, 8977–8982. [CrossRef] [PubMed]
3. Ponsone, M.L.; Chiotta, M.L.; Combina, M.; Torres, A.; Knass, P.; Dalcero, A.; Chulze, S. Natural occurrence of ochratoxin A in musts, wines and grape vine fruits from grapes harvested in argentina. *Toxins* **2010**, *2*, 1984–1996. [CrossRef] [PubMed]
4. IARC. *Some Naturally Occuring Substances: Food Items and Constituents, Heterocyclic Aromatic Amines and Mycotoxins*; IARC: Lyon, France, 1993; Volume 56, pp. 489–521.
5. Abouzied, M.; Horvath, A.; Podlesny, P.; Regina, N.; Metodiev, V.; Kamenova-Tozeva, R.; Niagolova, N.; Stein, A.; Petropoulos, E.; Ganev, V. Ochratoxin A concentrations in food and feed from a region with balkan endemic nephropathy. *Food Addit. Contam.* **2002**, *19*, 755–764. [CrossRef] [PubMed]
6. Mantle, P.G.; Faucet-Marquis, V.; Manderville, R.A.; Squillaci, B.; Pfohl-Leszkowicz, A. Structures of covalent adducts between DNA and ochratoxin A: A new factor in debate about genotoxicity and human risk assessment. *Chem. Res. Toxicol.* **2009**, *23*, 89–98. [CrossRef] [PubMed]
7. Abrunhosa, L.; Serra, R.; Venâncio, A. Biodegradation of ochratoxin A by fungi isolated from grapes. *J. Agric. Food Chem.* **2002**, *50*, 7493–7496. [CrossRef] [PubMed]
8. Soliman, K.; Badeaa, R. Effect of oil extracted from some medicinal plants on different mycotoxigenic fungi. *Food Chem. Toxicol.* **2002**, *40*, 1669–1675. [CrossRef]

9. Basílico, M.Z.; Basílico, J.C. Inhibitory effects of some spice essential oils on *Aspergillus ochraceus* nrrl 3174 growth and ochratoxin A production. *Lett. Appl. Microbiol.* **1999**, *29*, 238–241. [CrossRef] [PubMed]
10. Gallo, A.; Perrone, G.; Solfrizzo, M.; Epifani, F.; Abbas, A.; Dobson, A.D.; Mule, G. Characterisation of a pks gene which is expressed during ochratoxin A production by *Aspergillus carbonarius*. *Int. J. Food Microbiol.* **2009**, *129*, 8–15. [CrossRef] [PubMed]
11. Gallo, A.; Knox, B.P.; Bruno, K.S.; Solfrizzo, M.; Baker, S.E.; Perrone, G. Identification and characterization of the polyketide synthase involved in ochratoxin A biosynthesis in *Aspergillus carbonarius*. *Int. J. Food Microbiol.* **2014**, *179*, 10–17. [CrossRef] [PubMed]
12. Gallo, A.; Bruno, K.S.; Solfrizzo, M.; Perrone, G.; Mule, G.; Visconti, A.; Baker, S.E. New insight into the ochratoxin A biosynthetic pathway through deletion of a nonribosomal peptide synthetase gene in *Aspergillus carbonarius*. *Appl. Environ. Microbiol.* **2012**, *78*, 8208–8218. [CrossRef] [PubMed]
13. Crespo-Sempere, A.; Marin, S.; Sanchis, V.; Ramos, A.J. Vea and laea transcriptional factors regulate ochratoxin A biosynthesis in *Aspergillus carbonarius*. *Int. J. Food Microbiol.* **2013**, *166*, 479–486. [CrossRef] [PubMed]
14. Hua, H.; Xing, F.; Selvaraj, J.N.; Wang, Y.; Zhao, Y.; Zhou, L.; Liu, X.; Liu, Y. Inhibitory effect of essential oils on *Aspergillus ochraceus* growth and ochratoxin A production. *PLoS ONE* **2014**, *9*. [CrossRef] [PubMed]
15. Ryu, D.; Holt, D.L. Growth inhibition of *Penicillium expansum* by several commonly used food ingredients. *J. Food Prot.* **1993**, *56*, 862–867.
16. Sinha, K.; Sinha, A.; Prasad, G. The effect of clove and cinnamon oils on growth of and aflatoxin production by *Aspergillus flavus*. *Lett. Appl. Microbiol.* **1993**, *16*, 114–117. [CrossRef]
17. Patkar, K.; Usha, C.; Shetty, H.S.; Paster, N.; Lacey, J. Effect of spice essential oils on growth and aflatoxin b1 production by *Aspergillus flavus*. *Lett. Appl. Microbiol.* **1993**, *17*, 49–51. [CrossRef]
18. Daouk, R.K.; Dagher, S.M.; Sattout, E.J. Antifungal activity of the essential oil of *Origanum syriacum* l. *J. Food Prot.* **1995**, *58*, 1147–1149.
19. Lukas, B.; Schmiderer, C.; Franz, C.; Novak, J. Composition of essential oil compounds from different syrian populations of *Origanum syriacum* L. (lamiaceae). *J. Agric. Food Chem.* **2009**, *57*, 1362–1365. [CrossRef] [PubMed]
20. Soković, M.; Tzakou, O.; Pitarokili, D.; Couladis, M. Antifungal activities of selected aromatic plants growing wild in greece. *Food/Nahrung* **2002**, *46*, 317–320. [CrossRef]
21. da Cruz Cabral, L.; Fernandez Pinto, V.; Patriarca, A. Application of plant derived compounds to control fungal spoilage and mycotoxin production in foods. *Int. J. Food Microbiol.* **2013**, *166*, 1–14. [CrossRef] [PubMed]
22. Omidbeygi, M.; Barzegar, M.; Hamidi, Z.; Naghdibadi, H. Antifungal activity of thyme, summer savory and clove essential oils against *Aspergillus flavus* in liquid medium and tomato paste. *Food Control* **2007**, *18*, 1518–1523. [CrossRef]
23. Rasooli, I.; Abyaneh, M.R. Inhibitory effects of thyme oils on growth and aflatoxin production by *Aspergillus parasiticus*. *Food Control* **2004**, *15*, 479–483. [CrossRef]
24. Dambolena, J.S.; Lopez, A.G.; Canepa, M.C.; Theumer, M.G.; Zygadlo, J.A.; Rubinstein, H.R. Inhibitory effect of cyclic terpenes (limonene, menthol, menthone and thymol) on *Fusarium verticillioides* mrc 826 growth and fumonisin b1 biosynthesis. *Toxicon Off. J. Int. Soc. Toxinol.* **2008**, *51*, 37–44. [CrossRef] [PubMed]
25. Bansod, S.; Rai, M. Antifungal activity of essential oils from indian medicinal plants against human pathogenic *Aspergillus fumigatus* and *A. Niger*. *World J. Med. Sci.* **2008**, *3*, 81–88.
26. Marín, S.; Ramos, A.J.; Sanchis, V. Comparison of methods for the assessment of growth of food spoilage moulds in solid substrates. *Int. J. Food Microbiol.* **2005**, *99*, 329–341. [CrossRef] [PubMed]
27. Baydar, H.; Sağdiç, O.; Özkan, G.; Karadoğan, T. Antibacterial activity and composition of essential oils from origanum, thymbra and satureja species with commercial importance in turkey. *Food Control* **2004**, *15*, 169–172. [CrossRef]
28. Pawar, V.; Thaker, V. In vitro efficacy of 75 essential oils against *Aspergillus niger*. *Mycoses* **2006**, *49*, 316–323. [CrossRef] [PubMed]
29. Helal, G.; Sarhan, M.; Abu Shahla, A.; Abou El-Khair, E. Effects of cymbopogon citratus l. Essential oil on the growth, morphogenesis and aflatoxin production of *Aspergillus flavus* mL2-strain. *J. Basic Microbiol.* **2007**, *47*, 5–15. [CrossRef] [PubMed]

30. Lambert, R.; Skandamis, P.N.; Coote, P.J.; Nychas, G.J. A study of the minimum inhibitory concentration and mode of action of oregano essential oil, thymol and carvacrol. *J. Appl. Microbiol.* **2001**, *91*, 453–462. [CrossRef] [PubMed]
31. Rasooli, I.; Owlia, P. Chemoprevention by thyme oils of *Aspergillus parasiticus* growth and aflatoxin production. *Phytochemistry* **2005**, *66*, 2851–2856. [CrossRef] [PubMed]
32. Esper, R.H.; Gonçalez, E.; Felicio, R.C.; Felicio, J.D. Fungicidal activity and constituents of *Ageratum conyzoides* essential oil from three regions in são paulo state, brazil. *Arq. Inst. Biol.* **2015**, *82*, 1–4. [CrossRef]
33. Murthy, P.S.; Borse, B.B.; Khanum, H.; Srinivas, P. Inhibitory effects of ajowan (trachyspermum ammi) ethanolic extract on *A. ochraceus* growth and ochratoxin production. *Turk. J. Biol.* **2009**, *33*, 211–217.
34. Caceres, I.; El Khoury, R.; Medina, A.; Lippi, Y.; Naylies, C.; Atoui, A.; El Khoury, A.; Oswald, I.P.; Bailly, J.D.; Puel, O. Deciphering the anti-aflatoxinogenic properties of eugenol using a large-scale q-PCR approach. *Toxins* **2016**, *8*. [CrossRef] [PubMed]
35. Ferrara, M.; Gallo, A.; Scalzo, R.L.; Haidukowski, M.; Picchi, V.; Perrone, G. Inhibition of ochratoxin a production in *Aspergillus carbonarius* by hydroxycinnamic acids from grapes. *World Mycotoxin J.* **2015**, *8*, 283–289. [CrossRef]
36. Boutigny, A.L.; Barreau, C.; Atanasova-Penichon, V.; Verdal-Bonnin, M.N.; Pinson-Gadais, L.; Richard-Forget, F. Ferulic acid, an efficient inhibitor of type b trichothecene biosynthesis and *Tri* gene expression in *Fusarium* liquid cultures. *Mycol. Res.* **2009**, *113*, 746–753. [CrossRef] [PubMed]
37. Kim, D.; Yukl, E.T.; Moënne-Loccoz, P.; Ortiz de Montellano, P.R. Fungal heme oxygenases: Functional expression and characterization of Hmx1 from *Saccharomyces cerevisiae* and Cahmx1 from candida albicans. *Biochemistry* **2006**, *45*, 14772–14780. [CrossRef] [PubMed]
38. Yoshinari, T.; Yaguchi, A.; Takahashi-Ando, N.; Kimura, M.; Takahashi, H.; Nakajima, T.; Sugita-Konishi, Y.; Nagasawa, H.; Sakuda, S. Spiroethers of german chamomile inhibit production of aflatoxin G1 and trichothecene mycotoxin by inhibiting cytochrome p450 monooxygenases involved in their biosynthesis. *FEMS Microbiol. Lett.* **2008**, *284*, 184–190. [CrossRef] [PubMed]
39. Carmo, E.S.; Lima, E.D.O.; Souza, E.L.D.; Sousa, F.B.D. Effect of *Cinnamomum zeylanicum* blume essential oil on the rowth and morphogenesis of some potentially pathogenic *Aspergillus* species. *Braz. J. Microbiol.* **2008**, *39*, 91–97. [CrossRef] [PubMed]
40. Sharma, N.; Tripathi, A. Effects of citrus sinensis (L.) osbeck epicarp essential oil on growth and morphogenesis of *Aspergillus niger* (L.) van tieghem. *Microbiol. Res.* **2008**, *163*, 337–344. [CrossRef] [PubMed]
41. Bejaoui, H.; Mathieu, F.; Taillandier, P.; Lebrihi, A. Ochratoxin A removal in synthetic and natural grape juices by selected oenological saccharomyces strains. *J. Appl. Microbiol.* **2004**, *97*, 1038–1044. [CrossRef] [PubMed]
42. Hellemans, J.; Mortier, G.; De Paepe, A.; Speleman, F.; Vandesompele, J. Qbase relative quantification framework and software for management and automated analysis of real-time quantitative PCR data. *Genome Biol.* **2007**, *8*. [CrossRef] [PubMed]
43. Cabañes, F.J.; Bragulat, M.R.; Castellá, G. Characterization of nonochratoxigenic strains of *Aspergillus carbonarius* from grapes. *Food Microbiol.* **2013**, *36*, 135–141. [CrossRef] [PubMed]
44. The National Center for Biotechnology Information advances science and health by providing access to biomedical and genomic information. Available online: http://www.ncbi.nlm.nih.gov/ (accessed on 30 January 2014).
45. Verheecke, C.; Liboz, T.; Anson, P.; Diaz, R.; Mathieu, F. Reduction of aflatoxin production by *Aspergillus flavus* and *Aspergillus parasiticus* in interaction with streptomyces. *Microbiology* **2015**, *161*, 967–972. [CrossRef] [PubMed]

© 2016 by the authors. Licensee MDPI, Basel, Switzerland. This article is an open access article distributed under the terms and conditions of the Creative Commons Attribution (CC BY) license (http://creativecommons.org/licenses/by/4.0/).

Meeting Report

Report from the 5th International Symposium on Mycotoxins and Toxigenic Moulds: Challenges and Perspectives (MYTOX) Held in Ghent, Belgium, May 2016

Sarah De Saeger [1,*], Kris Audenaert [2] and Siska Croubels [3]

1. Department of Bioanalysis, Laboratory of Food Analysis, Faculty of Pharmaceutical Sciences, Ghent University, 9000 Ghent, Belgium
2. Department of Applied Biosciences, Faculty of Bioscience Engineering, Ghent University, 9000 Ghent, Belgium; kris.audenaert@ugent.be
3. Department of Pharmacology, Toxicology and Biochemistry, Laboratory of Pharmacology and Toxicology, Faculty of Veterinary Medicine, Ghent University, 9820 Merelbeke, Belgium; siska.croubels@ugent.be

* Correspondence: sarah.desaeger@ugent.be; Tel.: +32-9-2648137; Fax: +32-9-2648199

Academic Editor: Vernon L. Tesh
Received: 19 April 2016; Accepted: 6 May 2016; Published: 12 May 2016

---

1. Preface

The association research platform MYTOX "Mycotoxins and Toxigenic Moulds" held the 5th meeting of its International Symposium in Ghent, Belgium on 11 May 2016. The Symposium welcomed over 100 scientists, researchers and representatives from industry and government as well as academia to discuss all aspects of mycotoxin research including production, occurrence and detection, the impact on human and animal health, reduction and prevention, toxicology and other topics.

Mycotoxins—toxic fungal secondary metabolites—play a significant role in food and feed safety, as well as in medical and environmental microbiology. Indeed, mycotoxins have been shown to be the number one threat amongst food and feed contaminants regarding chronic toxicity. Economic losses are due to effects on livestock productivity and direct losses in crop yield and stored agricultural products. Legislative limits for a range of mycotoxins continue to develop worldwide, resulting in an increased number of official controls deriving from national food safety plans and for food trade purposes. Furthermore, environmental mycotoxins are a continuous threat to human and animal health.

The challenges presented to those working in mycotoxin research are enormous due to the frequency, the complexity and variability in occurrence. Several aspects make the pre- and post-harvest control of mycotoxins difficult, such as:

- Different fungal species produce mycotoxins;
- Most of the mycotoxin producing fungi are able to produce more than one mycotoxin;
- Mycotoxin levels are influenced by environmental conditions during growth and storage;
- The presence of modified mycotoxins;
- The highly complex influence of environmental factors on the biosynthesis of mycotoxins by fungi.

Other aspects related to human and animal health also contribute to the complexity in mycotoxin research, e.g.,

- The lack of suitable biomarkers to assess exposure of humans and animals;
- The need for guidance levels of mycotoxins in animal body fluids;
- The efficacy and safety testing of mycotoxin detoxifiers;

- Knowledge about toxicokinetics in men and animals.

New developments in mycotoxin analysis focus on faster, multi-mycotoxin, environmentally friendly, cost-effective and fit-for-purpose methods in food, feed, biological tissue and body fluids. A trend towards untargeted metabolic profiling has been noticed.

Mycotoxigenic fungi, mycotoxins and food and feed safety will continue to be a critical interest to researchers for years to come. Likewise, the risks for human and animal health by indoor air contamination and other sources of environmental mycotoxin exposition are still a wide field of research. Innovations take place at a rapid pace. The investigational area is broad (phytopathology, analytical methods, risk management, toxicology, biomarkers for exposure, occupational health risks), but necessary to ensure progress into improvement of public health and animal health, in particular a safe food and feed supply.

Only through multidisciplinary efforts and concerted actions can further progress and solutions be expected for the mycotoxin problem.

The association research platform MYTOX "Mycotoxins and Toxigenic Moulds" was established in 2007 and consists of more than 50 researchers from 12 research laboratories in the Ghent University Association. MYTOX deals with mycotoxin research in a multi-disciplinary way, based on four main units: (1) mycotoxins; (2) toxigenic fungi; (3) mycotoxins and animal health; and (4) mycotoxins and human health. In this way, MYTOX tackles the mycotoxin issue along the production chain from the field to the end consumer, within the 'One Health' concept.

The scientific program included oral and poster presentations related to fungal-related disease monitoring; mycotoxin analysis in food and feed, as well as in animal and human biological samples; prevention on the field; management strategies during food and feed storage and processing.

2. Keynote Lectures

2.1. *Determination of Mycotoxins in Food and Feed—Quo Vadis?*

Stroka, J.

Over the last decade, the number of mycotoxins that are regulated in Europe has steadily increased with the aim of protecting consumers as well as recognizing animal welfare. In parallel scientific progress in many fields such as bio-chemical analysis as well as technical-instrumental progress increased the possibilities to determine mycotoxins with completely new defined scopes, such as shorter time of analysis, parallel determination of mycotoxins, or improved detection capability (LOD/LOQ), while the most important aspect of every chemical determination remained the reliability of the generated value.

The European Union Reference Laboratory (EURL) for mycotoxins has, together with its partner from the national reference laboratories, not only monitored but also evaluated the performance of analytical methods with the aim ensure a reliable measurement capacity in Europe.

This presentation gives an overview of the most important achievements of the work of this network, demonstrated in data generated in the network and gives an outlook how the performance of analytical methods might need to be viewed taking the analytical progress into account and where routes for future improvements might be found.

2.2. *Integrated Management of Mycotoxins in Pre- and Post-Harvest and MycoKey Solutions*

Logrieco, A.F.

Among the emerging issues in food safety, the increase of plant diseases associated with the occurrence of mycotoxigenic fungal species is of major importance. As a result of their secondary metabolism, these fungi can produce mycotoxins, which are low-molecular-weight toxic compounds that represent a serious risk for human and animal health worldwide. The management of good agricultural practices in the pre-harvest is a key issue for minimizing the risk of mycotoxin accumulation in the crops before the harvest. Such practices can involve crop rotation, tillage, proper fertilization and fungicide or biological control distribution, variety selection, timely planting and harvests and the control of the insects which often act as vectors of toxigenic fungi spores. On the other hand, the reduction of mycotoxins along the agro-food chains is also highly dependent on a correct post-harvest management that must aim firstly at the separation of the infected crop products from the healthy material. Therefore, the use of different tools such as manual sorting or optical sensors is also a crucial point for reducing the level of mycotoxin contamination of a given crop. Moreover, it is extremely important to prevent post-harvest contamination during the storage by obtaining low temperature and humidity conditions, in order to limit the development of toxigenic fungal genera. An updated review of an integrated management of pre- and post-harvest practices aiming at minimizing the risk of mycotoxin contamination of the main crops of agro-food importance and main effective solutions proposed by EU project MycoKey will be provided in the presentation.

Acknowledgements: this presentation has been supported by the EU Project MycoKey N. 678781.

2.3. *Analytical Management Strategies to Reduce Mycotoxin Contamination: A View from the Food Industry*

Suman, M.

Mycotoxin contamination in various crops is of major concern since it has significant implications for food and feed safety, food security and international trade. Grain and foods based on grains (e.g., pasta, bread, bakery products) account for the largest contribution to mycotoxin exposure in all age classes. It has been estimated by the Food and Agricultural Organization (FAO) that, worldwide, approximately 25% of the crops are contaminated by molds: the incidence can vary considerably depending on many factors, such as weather conditions, agricultural practices, packaging, transport and storage. Results using the latest state-of-the-art multi-analyte methods show that almost 100% of these crops is contaminated with one or more mycotoxins.

Furthermore, in the next decades, industry will have to face a new "mycotoxins scenario": the so-called "masked/bounded forms". In fact, even if most mycotoxins demonstrate stability at room temperature and under neutral conditions, many factors (e.g., heat, pressure, pH, enzymatic activities, food constituents) must be considered during processing, since the release of native forms as well as formation of conjugated ones by reactions with macromolecular components (such as sugars, proteins or lipids) may be induced. Therefore, nowadays, the food industry clearly demonstrates the need for both rapid screening techniques, which could be also used outside the laboratory environment, and high sensitivity-precision methods for confirmatory purposes. Rapid and user-friendly Lateral Flow Devices (LFD) have been recently made available on the market: sensitivity and selectivity are provided by specific antibodies; generally, they do not require expensive instrumentation and additional chemicals or handling steps, providing in a few minutes a quantitative result for a target mycotoxin. LFD is in fact becoming progressively competitive with the common ELISA methods (Enzyme-Linked Immunosorbent Assay) which still combine good throughput performances for large numbers of samples and simple extraction-detection procedures. In the field of immunochemical methods, there is also a growing interest towards homogeneous approaches like the Fluorescence Polarization Immunoassay (FPIA) and optical-based detection provided by Surface Plasmon Resonance Sensors (SPRS).

Other emerging strategies are related for instance to near-infrared (NIR) spectroscopy and Electronic Noses. An Electronic Nose consists of an array of non-specific chemical detectors that interact with different volatile compounds and provide signals that can be utilised effectively as a fingerprint of the volatile molecules rising from the analysed samples.

Higher accuracy and repeatability can still be guaranteed by chromatographic based techniques like HPLC (High Performance Liquid Chromatography) with ultraviolet (UV) or fluorescence detection

(FLD): pre-concentration and clean-up steps allow the achievement of good performances in very complex matrixes.

Liquid Chromatography coupled with Mass Spectrometry (LC-MS) is actually the most flexible and effective (high sensitivity and selectivity) technique used in order to determine mycotoxins in many different matrixes. At the same time, the natural coexistence of several mycotoxins imposes the need to optimize the global cost of these analyses through the development of multiresidual analytical strategies; in this contest, the new frontier is the High Resolution Mass Spectrometry (HRMS), in particular exploiting high performance Time of Flight (TOF) and Orbitrap^{TM} technology instruments, coupled to Ultra High Performance Liquid Chromatography (UHPLC): in this way, a door is opened to the simultaneous quantification of different categories of chemical contaminants.

2.4. Stepping out of the Dark; How Genomes and Genomics Shed Light on Biosynthetic Pathways

Van der Lee, T.; Hoogendoorn, K.; Medema, M.; Waalwijk, C.

New sequence technologies radically changed sequencing costs, sequencing quality and the time required for the completion of sequencing projects. Recently, high quality genome assemblies and gene annotations were generated for several *Fusarium* species (*F. graminearum*, *F. poae*, *F. culmorum*, *F. pseudograminearum*, *F. langsethiae* and *F. fujikuroi*). This provides unique opportunities to understand the function, organization and dynamics of the genes and chromosomes in these genomes. Our research focusses on genes that encode proteins involved in the production of fungal natural products such as mycotoxins. Although fungal natural products, sometimes referred to as 'biosynthetic dark matter', show an impressive variation in chemical structures and biological activities, their biosynthetic pathways share a number of key characteristics. First, genes encoding successive steps of a biosynthetic pathway tend to be located adjacently on the chromosome in biosynthetic gene clusters (BGCs). Second, these BGCs are often are located on specific regions of the genome and show a discontinuous distribution among evolutionarily related species and isolates. Third, the same enzyme (super)families are often involved in the production of widely different compounds. Fourth, genes that function in the same pathway are often co-regulated, and therefore co-expressed across various growth conditions. The availability of high quality genome assemblies where chromosomes are assembled from telomere to telomere, allow the precise localization and organization of genes that was often lacking from the previously available draft assemblies. The use of the millions of reads generated by RNA-seq data can result in improved annotation of genes that are lowly expressed as is typical for genes involved in the production of secondary metabolites analysis. Computational tools are indispensable to the discovery process, as they are required for the effective exploitation of genomic data to identify potential new secondary metabolites or variants of known natural products. They also can direct functional characterization of the biosynthetic routes. A key step in natural product bioinformatics is the identification of BGCs that encode the enzymatic pathways for compound production, as well as the prediction of their products. Frequently used computational tools for fungal BGC identification, such as antiSMASH, SMIPS/CASSIS and SMURF have largely focused on known types of BGCs, encoding the enzymatic pathways to produce compounds such as polyketides, nonribosomal peptides (NRPs) and terpenes. The intrinsic disadvantage of these programs is that they cannot detect types of BGCs that they were not trained to predict. Recently, algorithms have emerged that also allow the identification of additional, more obscure, BGCs. We will describe the identification and bioinformatic characterization of BGCs in these newly sequenced *Fusarium* genomes with the help of systematized information on BGCs that have been experimentally characterized in the past.

3. Presentations

3.1. Carry-Over of Aflatoxin B_1 from Dairy Cows' Feed to Milk

Van der Fels-Klerx, H.J.; Camenzuli, L.

Ingredients used in the cows' diet, such as maize, can sometimes be heavily contaminated with aflatoxin B_1. This mycotoxin was present in high amounts in maize grown in Italy in 2003 and 2008 and in maize grown in the Balkan area in 2013. Aflatoxin B_1 is metabolized by dairy cows into aflatoxin M_1 appearing in the cows' milk. Aflatoxin B_1 and aflatoxin M_1 are carcinogenic to animals and humans, hence, within the European Community (EC) the presence of both these toxins is regulated. Nowadays, milk production of dairy cows is increasing, as is the rate of maize used in compound feed production for dairy cows. High producing cows were indicated to have a higher transfer rate of the mycotoxin and, consequently, this may affect the appearance of aflatoxin M_1 in the cows' milk.

This study aimed to estimate concentrations of aflatoxin M_1 in dairy cows' milk, given contamination of compound feed with aflatoxin B_1 and feeding consumption regimes. Monte Carlo simulation modeling was applied, with 1000 iterations for each run. The model simulated two types of typical Dutch dairy herds composed of 69 cows. In the first type, all cows started their lactation at the same time, while in the other type, the start date of lactation was spread over the year, with up to two new cows starting lactation each week. The composition of a typical compound feed for dairy cows was used. National monitoring data were used to determine the concentration range of aflatoxin B_1 in each of the ingredients. In addition to this 'standard' contamination, we also used the concentrations of the toxin in maize as reported from the Balkan incident in 2013. Six different equations from literature were used to assess the transfer of aflatoxin B_1 in the cow's body from feed to milk.

Modeling results showed that, in all scenarios considered, the maximum average aflatoxin M_1 concentration in milk produced at the farm (all 69 cows) is below the EC legal limit of 0.05 μg/kg. In some scenarios, however, this maximum was close to this limit (up to 0.04 μg/kg). Apparently, the higher amount of milk produced also dilutes the aflatoxin M_1 in the milk.

3.2. *Biodegradability of Mycotoxins during Anaerobic Digestion*

De Gelder, L.; Audenaert, K.; Willems, B.; Schelfhout, K.; De Saeger, S.; De Boevre, M.

When exceeding maximum limits or guidance values, highly-mycotoxin contaminated food and feed batches are excluded from the Flemish market. To date, it is mandatory by The Public Waste Agency of Flanders (OVAM), which dictates the fate of all waste streams and their appropriate removal, that these contaminated batches should be destroyed by combustion. In light of their policy in achieving maximum valorization of waste streams, OVAM wishes to explore the remediation of mycotoxin contaminated food and feed batches through anaerobic digestion.

Anaerobic digestion, mediated by a consortium of micro-organisms resulting in the production of biogas, is widely employed to treat organic-biological waste (OBW) and manure. In Flanders, 40 biogas plants are in operation or under construction with a total processing capacity of 2.234 million tons per year, of which 9 are thermophilic and 31 are mesophilic installations. OBW comprises 60% of the input in anaerobic digesters but still specific Flemish legislation concerning the presence of certain contaminants in the waste should be respected. Specifically, for mycotoxins, OVAM rejects the processing of OBW in anaerobic digesters due to insufficient guarantees regarding (i) the effect of the mycotoxins on the anaerobic digestion process itself; and (ii) the residual concentrations of mycotoxins which impairs the subsequent usability of the digestate, such as fertilizer or soil improver in agriculture.

In order to address these concerns, lab scale degradation tests were conducted to assess the biodegradability of aflatoxin B1, deoxynivalenol, zearalenone, fumonisin B1, ochratoxin A, T-2 toxin and ergot alkaloids during thermophilic and mesophilic digestion. In 20-day batch tests, where digestate was fed with fructose and spiked with each mycotoxin separately, more than 90% degradation was obtained for aflatoxin B1, deoxynivalenol, zearalenone and ochratoxin A. Remarkably, ergot alkaloids were only degraded for *ca.* 60% under mesophilic conditions but for more than 97% under thermophilic conditions. The presence of the mycotoxins did not result in a significant decrease in methane production. Subsequently, a naturally highly contaminated maize batch (4413 μg/kg deoxynivalenol, 1052 μg/kg zearalenone, 170 μg/kg fumonisin B1+B2) was fed to mesophilic and thermophilic semi-continuous reactors on pilot scale. After one hydraulic retention time (30 days), no mycotoxins were detected except for a small residue of 90 μg/kg deoxynivalenol in the thermophilic reactor. Also, reactor performance and methane production were not hampered by the prolonged feeding of the highly contaminated maize.

These results indicate that anaerobic digestion is generally a safe strategy for the remediation of mycotoxin contaminated food and feed batches.

3.3. Thermal Degradation of T-2 and HT-2 Toxins during Extrusion Cooking in Laboratory and Industrial Scale

Schmidt, H.S.; Becker, S.; Cramer, B.; Humpf, H.-U.

Trichothecene mycotoxins are naturally and worldwide occurring food contaminants, exposing humans and animals to adverse health effects. T-2 and HT-2 toxins are the most prominent members of Type A trichothecenes and are produced by several species of the *Fusarium* genus and are frequently found in oats, which is an increasing problem in times of a growing interest on oat-derived products for a healthy and balanced diet. Concentrations of the sum of both toxins in products for human consumption may exceed the EU-guidance levels of 200 μg/kg, which is regarded critically, due to the high toxic potential of both secondary metabolites.

There are several methods reported to reduce the mycotoxin content in food, ranging from agricultural strategies to prevent the fungal infestation to food processing steps that either remove the toxins or destroy the mycotoxins' molecular structure. Food extrusion is a high-temperature short-time (HTST) process and was shown to efficiently lower the mycotoxin content due to a combination of high temperatures and pressure and severe shear forces.

We therefore investigated the stability of T-2 and HT-2 during food extrusion of toxin-spiked oat flour on a laboratory food extruder with samples differing in water content, die-temperature, pressure and shear rates applied by the screw. The results were compared to an extruder working on industrial scale with naturally contaminated oatmeal, also regarding the formation of degradation products.

3.4. Influence of Maturity Stages, Processing of Maize on Aflatoxins and Fumonisins Levels and Dietary Intake by Some Populations of Cameroon

Nguegwouo, E.; Njumbe Ediage, E.; Njobeh, P.B.; Medoua, G.N.; Ngoko, Z.; Fotso, M.; De Saeger, S.; Fokou, E.; Etoa, F.-X.

Food safety is a call for concern nowadays with mycotoxin contamination particulary aflatoxins and fumonisins as a contributing factor. The aim of this research was to investigate the effect of maturity stage of maize at harvest, post-harvest handling and processing techniques on total aflatoxins (AFT): AFB_1, AFB_1, AFG_1, AFG_1 and total fumonisins (FT): FB_1, FB_2, AFB_3 as well as the dietary intake by some rural Cameroonian populations. Three stages of maturity as currently practiced (80, 85 and 90 days after seed), two drying processes after harvest (sun or barns drying up to two weeks), three periods of storage duration after drying (one, two and three months) and 11 types of maize products were considered during sampling. The analyses of AFT and FT were carried out using quantitative ELISA and those samples with level of mycotoxins higher than regulated level in the European Commission were analyzed by LC-MS/MS to know the sub-types of toxins present. To assess the dietary intake of those mycotoxins, a food consumption survey was conducted on 366 individuals that consisted of 108 children (4–8 years), 102 adolescents (9–14 years) and 156 adults (15 years and over). The results have shown that all analyzed samples were contaminated with AFT (range: 0.8 to 20.0 μg/kg) and FT (range: 100 to 5990 μg/kg). Sun or barn drying for only one week followed by one month usual storage resulted in significant FT contamination, emphasizing the need of at least two weeks of drying. The processing techniques reduced the levels of FT up to 94.74%, mainly in maize dishes that have a screening phase. The estimated intake of AFT and FT respectively showed that children are most at risk (mean: 43.8 × 10^{-3} and 13.2 μg/kg bw/day) followed by adolescents (mean: 31.9 × 10^{-3} and 9.0 μg/kg bw/day), and finally, adults (mean: 27.4 × 10^{-3} and 4.0 μg/kg bw/day).

3.5. *Multi-Mycotoxins Occurrence in Maize and Animal Feed from Assiut Governorate, Egypt*

Abdallah, M.F.; Kilicarslan, B.; Girgin, G.; Baydar, T.

Mycotoxins are secondary metabolites produced by different species of fungi such as *Aspergillus*, *Penicillium* and *Fusarium* with a broad range of toxic effects on human and animals. Over the last decade, several analytical methods have been developed for simultaneous detection of different mycotoxins in maize and feed in one run. Surveys of different countries around the world have shown the existence of the aflatoxins, ochratoxin A, and zearalenone in maize and animal feed. Relatively less information is available about the co-occurrence of different mycotoxins in Upper Egypt. Moreover, the current regulations in Egypt include only total aflatoxins and aflatoxin B1, 20 μg/kg and 10 μg/kg respectively, for both maize and animal feed. We therefore performed a survey of six mycotoxins, aflatoxins B1, B2, G1, and G2, ochratoxin A and zearalenone, in maize (n = 61) and animal feed (n = 17). Multi-mycotoxins immunoaffinity columns were used for clean-up and HPLC-FLD with on-line post-column photochemical derivatization was used for quantification of the mentioned mycotoxins.

A solid–liquid extraction has been done before passing the samples through the immunoaffinity columns. The chromatographic separation is based on the publication of Ofitserova *et al.* (2009). Limits of detection (LOD) were 0.92 μg/kg for ZEA, 0.02 μg/kg for OTA and varied from 0.004 to 0.12 μg/kg for aflatoxins. Limit of quantification (LOQ) were 2.8 μg/kg for ZEA, 0.06 μg/kg for OTA and varied from 0.013 to 0.3 μg/kg for aflatoxins. The mean recovery values were 77%–110% for different concentrations of AFs, OTA and ZEA in spiked maize and feed samples.

OTA, AFG1, and AFG2 were under the limit of detections. AFB1 was detected in both maize (n = 15) and feed (n = 8) with only one maize sample above the maximum permissible level set by Egyptian authorities. AFB2 was detected in six maize samples and in one feed with a maximum value of 0.05 μg/kg. ZEA was detected only in feed samples (n = 4) with a maximum value of 3.5 μg/kg. In conclusion, further surveys are highly recommended in order to establish database for mycotoxins occurrence in Egypt to minimize the possible health risks in animals and human.

Acknowledgements: this work is financially supported by Hacettepe University Scientific Research Projects Coordination Unit (#014 D06 301 002-620).

Ofitserova, M.; Nerkar, S.; Pickering, M.; Torma, L.; Thiex, N. Multiresidue mycotoxin analysis in corn grain by column high-performance liquid chromatography with post column photochemical and chemical derivatization: Single-laboratory validation. *J. AOAC Int.* **2009**, *92*, 15–25.

3.6. *Effects of Climate Change on the Presence of Aflatoxin Producing Aspergillus Species in Central Europe*

Baranyi, N.; Kocsubé, S.; Szekeres, A.; Bencsik, O.; Vágvölgyi, C.; Varga, J.

Global warming can affect the presence of thermotolerant fungi with potential mycotoxin producing abilities in our foods and feeds. In the past few years, this phenomenon was observed in several European countries which did not face this problem before. Furthermore, consequent aflatoxin contamination was observed in agricultural commodities including maize and milk in these countries. Aflatoxins are among the economically most important mycotoxins, they can cause serious yield loss in several agricultural products including cereals, cotton, maize and peanut. Aflatoxins are produced by various *Aspergillus* species mainly belonging to *Aspergillus* section *Flavi*. The economically most important producers of aflatoxins are *A. flavus*, *A. parasiticus* and *A. nomius*. While *A. flavus* produces B-type aflatoxins, *A. nomius* and *A. parasiticus* are also able to produce G-type aflatoxins. Other species able to produce G-type aflatoxins include, e.g., *A. bombycis*, *A. pseudonomius*, *A. parvisclerotigenus*,

A. minisclerotigenes and *A. pseudocaelatus*. Although aflatoxin producers prefer tropical and subtropical climates, these species can occur in temperate climate in increasing quantities. These observations led us to examine the occurrence of mycotoxin producing Aspergilli in agricultural products in Hungary and its neighboring countries. Several potential aflatoxin producing isolates were found, *A. flavus* was the dominant. In this study we examined the aflatoxin producing abilities of *A. flavus*, *A. nomius*, *A. parasiticus* and *A. pseudonomius* isolates that came from cheese (Hungary), indoor air (Croatia), maize (Serbia, Hungary), and wheat (Hungary), respectively. We have found some differences in the ratio and amount of produced B- and G-type aflatoxins under different cultivation conditions in case of *A. nomius*, *A. pseudonomius* and *A. parasiticus* isolates.

Acknowledgements: present work was supported by OTKA grants No. K115690, K84122 and K84077. N.B was also supported by EMET grant No. NTP-EFÖ-P-15-0486 providing infrastructure.

3.7. Mycotoxin Exposure through Biomarker Analysis Using LC-MS/MS

Huybrechts, B.; Heyndrickx, E.; Sioen, I.; De Saeger, S.; Callebaut, A.

The aim of the presented study was to assess human mycotoxin exposure based on the direct measurement of urinary biomarkers via LC-MS/MS in samples of the Belgian population. Urine as target matrix has the advantage of being non-invasive while LC-MS opens the possibility of multi-analyte measurements. These biomarkers are often present in very low concentrations, so many methods published up till now rely on some sample treatment. Morning urine of 155 children (3–12 years) and 239 adults (19–65 years) was collected according to a standardised study protocol. These urine samples were analysed for the presence of 33 urinary mycotoxins and their metabolites. Nine out of 33 potential biomarkers were detected whereby deoxynivalenol (DON), DON-glucuronides (DON-GlcAs), deepoxy-deoxynivalenol-glucuronide (DOMGlcA), ochratoxin A, citrinin and dihydrocitrinone were the most frequently detected. DON15GlcA was the main urinary metabolite found in 100% of the samples and for the first time DOMGlcA was detected in urine of children. Screening a small number of samples via a more sensitive method using immuno-affinity column (IAC) based clean-up revealed the presence of AfM1 at $pg \cdot mL^{-1}$ levels in every urine. Fumonisins were not detected in the Belgian samples.

3.8. Effect of Thymol, 4-Hydroxybenzaldehyde, and Fluconazole on the Aflatoxin B1 Biosynthesis

Dzhavakhiya, V.G.; Voinova, T.M.; Statsyuk, N.V.; Shcherbakova, L.A.

In the course of evolution, plants affected by toxigenic fungi could develop some specific compounds to block the biosynthesis of fungal toxins. This supposition is confirmed by the phenomenon of the aflatoxin B1 (AFB1) biosynthesis inhibition by some terpenoids of plant origin. This group of biologically active substances includes thymol, which antimicrobial properties are widely used in human and veterinary medicine and plant growing. However, thymol has never been considered as a possible inhibitor of the AFB1 biosynthesis.

Our earlier studies showed that thymol and 4-hydroxybenzaldehyde (4-HBA), another compound of plant origin, represent highly efficient sensitizers of plant pathogenic fungi to fungicides. The revealed chemosensitizing effect is similar to the action of antioxidants and is connected with the inhibition of oxidative stress that increases the sensitivity of plant pathogenic fungi to some fungicides. In this study we investigated the ability of thymol and 4-HBA to inhibit AFB1 biosynthesis in *Aspergillus flavus*. In addition, the effect of fluconazole, a synthetic fungicide of a triazole group able to block melaninogenesis in *A. flavus*, on the AFB1 biosynthesis was studied; according to our hypothesis, since both melanin and AFB1 are synthesized by the polyketide pathway, this fungicide was probably able to block the toxigenesis.

According to the obtained data, the thymol addition to a liquid nutrient medium at concentrations of 0.05 and 0.075 μg/mL resulted in a significant reduction of the AFB1 production (79% and 71%, respectively) with the simultaneous blocking of melaninogenesis in fungal mycelium. The addition of 4-HBA (0.5 and 0.75 μg/mL) also inhibited the AFB1 production (90% and 92%, respectively) and caused the discoloration of fungal colonies. Unlike these two compounds, the addition of fluconazole at concentrations of 0.1 and 0.125 μg/mL increased the level of the AFB1 production by 93% and 179%, respectively, and, at the same time, blocked melaninogenesis.

It is known that AFB1 and melanin biosynthetic pathways have common initial stages and then diverge. Based on the obtained results, we can conclude that thymol and 4-HBA block those stages of the polyketide pathway, which are located prior the branch point, whereas fluconazole blocks the melanin biosynthesis stage located after the branch point that probably causes an increased accumulation of AFB1 precursors and the corresponding increase in the AFB1 production level.

This study was financially supported by the Russian Science Foundation (project No. 14-16-00150).

3.9. *Microbial Degradation of Deoxynivalenol*

Vanhoutte, I.; Audenaert, K.; De Saeger, S.; De Gelder, L.

Mycotoxin contamination of food and feed poses major risks for human or animal health and leads to economic losses. Prevention and intervention measures are very well described on the field, but still contaminated batches remain a reality in practice. In order to salvage these resources, feed remediation based on mycotoxin adsorption is already applied through the use of binders. However, adsorption is reversible, pH-dependent and non-specific. Moreover, these binders negatively influence the transfer of medication to the bloodstream. Therefore, there is need to develop more reliable detoxification strategies. This research focuses on the microbial degradation of mycotoxins, in particularly deoxynivalenol (DON) which frequently occurs in crops in Belgium. Several microbial communities, with a possible exposure history to mycotoxins or other complex molecules, are screened for the presence of DON degrading microorganisms. Enrichment cultures of soil and activated sludge showed degradation of DON after two weeks, as analyzed with ELISA, whereas detoxification of DON was confirmed with a bio-assay using *Lemna minor*. Subsequently, several strains, among which one promising isolate was related to *Streptomyces* sp. (derived from soil), have been purified and characterization of their degrading capacities is ongoing.

3.10. *Quantitative UHPLC-MS/MS Method for Citrinin and Ochratoxin A: Prevalence in Food, Feed and Red Yeast Rice Food Supplements*

Kiebooms, J.A.L.; Huybrechts, B.; Thiry, C.; Tangni, E.K.; Callebaut, A.

Mycotoxins have been reported to cause deleterious effects in animals and humans (a.o.: nephrogenic, hepatogenic, carcinogenic, teratogenic, and neurogenic). Some mycotoxins (e.g., aflatoxins and ochratoxin A) have already been well studied, but for citrinin, a nephrotoxic mycotoxin, this has not yet been the case. According to a recent European Food Safety Agency report, occurrence data are lacking for a correct risk assessment of citrinin. Besides, Belgian and German scientific reports have shown that citrinin or its metabolite, dihydrocitrinone are widely (in up to 90% of samples) present in human urine, which might imply chronic exposure. Recently, a maximum limit for citrinin was set in food supplements comprising red yeast (*Monascus purpureus*) fermented rice (RYR), which contains monacolin K, an active component against cholesterolemia. During fermentation the fungus can also produce citrinin. Consequently, the present work aimed to develop a robust and routinely applicable ultrahigh performance liquid chromatography-tandem mass spectrometry (UHPLC-MS/MS) method for the analysis of two incidentally co-occurring nephrotoxic mycotoxins, citrinin and ochratoxin A, in food, feed and in RYR food supplements. The method was successfully validated in RYR food supplements and wheat flour, achieving respective limits of quantification for citrinin of 0.6 μg/kg and 0.6 μg/kg and for ochratoxin A of 10 μg/kg and 0.6 μg/kg. The recoveries varied between 75% and 102%. Furthermore, a preliminary occurrence study in 135 different RYR, food and feed products was executed, proving the potential of this method for future data acquisition within a risk assessment framework regarding citrinin and ochratoxin A (co-)occurrence in food and feed matrices.

3.11. Degradation and Epimerization of Wheat Ergot Alkaloids during French Baking Test

Meleard, B.

With increasing reports of ergot sclerotia detection on cereal grains in EU, regulation (EC) No 1881/2006 has recently been amended to set a maximum level of ergot sclerotia in unprocessed cereals at 0.5 g/kg for human consumption. The wheat ergot sclerotia produce six major alkaloids with variable toxicological effects: ergotamine, ergosine, ergocristine, ergometrine, ergocornine, and ergocryptine. Due to epimerization each (*R*)-configured ergot alkaloid is associated to a (*S*)-configured epimeric form. The two epimers could differ regarding their toxicity with high toxicity of the (*R*)- form whereas the (*S*)- form could be biologically inactive. For this reason, the sum of 12 ergot alkaloids must be considered and the analysis of each specific epimer must be performed.

For our study, a sclerotia grinding containing 3.0 mg/g alkaloids was added to a commercial wheat flour ("*Corde Noire*"). Seven mixes were produced to obtain contaminated flour in a range of concentration for alkaloid from 60 μg/kg to 15,000 μg/kg.

A French baking test was performed at ARVALIS laboratory according to the standard method NF V03-716. The hydration rate is about 60% and cooking temperature is 250 °C. Alkaloid content was determined by means of LC-MS/MS at UGent Laboratory of Food Analysis in dough and bread after cooking and separately in crumb and crust.

For all flours the baking process led to a decrease of alkaloid content and to a conversion from the toxicologically relevant (*R*)-epimer to the biologically inactive (*S*)-epimer.

The degradation rate is variable according to the ergot alkaloid. In this experimentation ergotamine and ergosine were more lowered whereas ergometrine was more stable. The epimerization rate differs for each ergot alkaloid too; ergometrine and ergocornine were more sensitive to this phenomenon in our study. The epimerization increased with the temperature as shown by the decrease of the (*R*)- to (*S*)-epimer ratio from dough to crust.

As a conclusion, flour could be detoxified by both epimerization and degradation of ergot alkaloids. The alkaloid content in bread was reduced by 60% from the flour content whatever the alkaloid concentration was.

3.12. Identification and Classification of Agronomic Factors Involved in Ergot Contents and Its Alkaloids in Small Grain Cereals

Orlando, B.; Maumené, C.; Valade, R.; Maunas, L.; Robin, N.; Bonin, L.

The European Commission Regulation (EU) 2015/1940 of 28 October 2015 amending Regulation (EC) No 1881/2006 as regards maximum levels of sclerotia of ergot in certain unprocessed cereals has established maximum limits for ergot contents (*Claviceps purpurea*) in small grain cereals for human consumption.

Occurrence studies have shown that all the small grains cereals are affected by this disease. Triticale and rye are the most sensitive. Since 2012, ARVALIS—Institut du Végétal has studied more than 2000 farm fields of soft wheat, durum wheat, barley, rye and triticale according to the same methodology, in collaboration with partners.

This work allowed, on the one hand, the determination of the relationship between crop contamination with ergot and the associated production of alkaloids. The total alkaloid content is, on average, 0.32% with variation in μg of alkaloid/g of ergot from 57 to 36,385; that means from 0.006% to 3.6% of ergot bodies weight. The ratio between -ine and -inine forms is 2.6:1. Ergotamine, ergocristine, ergosine and their corresponding epimers represent 74% of total alkaloids. On average the content of total alkaloids of *C. purpurea* is 3103 μg/g. The host plant and the harvest year did not have influence on the alkaloid contents. The ergot contents explain 79% of the variability of alkaloid contents.

On the other hand, the survey allowed the identification and prioritization of some agronomic levers to limit the risk at crop level. A statistical analysis was applied on ln(total alkaloid content). The variance analysis applied to agronomic factors showed that the host plant, the previous crop, grassweeds presence and tillage system, significantly influence alkaloids contents. The relevance of these levers was confirmed by separate analytical works.

This work is a multifactor field prevention tool capable to limit and to manage the ergot and alkaloids risk. Nevertheless, we know that the main factor involved in ergot infection is the weather conditions.

3.13. *Development of a Quantitative PCR Method Specific to Claviceps Purpurea, Ergot of Cereals: Applications to Study Its Dispersion and Evaluation of Its Toxicity*

Dauthieux, F.; Vitry, C.; Ducerf, R.; Leclere, A.; Maumené, C.; Orlando, B.; Valade, R.

The resurgence of ergot of cereals in France is an important health issue due to the production of alkaloids in the sclerotia of *Claviceps purpurea*. A method for quantifying G1 group of *C. purpurea* (pathogenic subspecies of cereals and grasses) by real-time PCR was developed to provide an innovative and powerful tool in order to manage the risk associated with the ergot. The validated method was used to study dispersion of primary and secondary inoculum and to characterize flours for their ergot concentration in relation with their alkaloid content.

Insects were identified as potential vectors of ergot and we confirmed the low dispersion capacity of primary inoculum by wind. The advantage of the qPCR was also demonstrated in order to study the occurrence of the pathogen in flours and to link fungal biomass with the alkaloid content.

3.14. *Effect of Fungicide seed Treatments on Germination of Claviceps Purpurea Sclerotia*

Maunas, L.; Robin, N.; Maumené, C.; Janson, J.P.

Ergot of cereals (*Claviceps purpurea*) may be introduced into healthy parcel through seeds containing ergot sclerotia. These sclerotia are able to germinate in spring and release ascospores in the air. To reduce this source of pollution, a fungicide seed treatment could be a useful complementary tool to seed cleaning, by reducing the viability of the remaining sclerotia. Previous tests in controlled conditions identified potentially effective treatments. This study aims to measure the efficacy of these treatments in the field conditions. The results confirm the interest of two seed treatments, one containing prochloraz and triticonazole, and the other containing carboxin and thiram. They allow a high reduction of the stroma production and, potentially, of the spore production. Their efficacy is not ideal but it permits the possibility of a better control of the disease dispersion due to contaminated seeds.

3.15. *Tillage, an Efficient Lever to Limit Ergot in Cereals*

Maumené, C.; Orlando, B.; Labreuche, J.; Leclere, A.; Maunas, L.

Ergot of cereals (*Claviceps purpurea*) is conserved in the soil as sclerotia. After vernalization, these conservation bodies are able to germinate in the spring and to produce stroma, sorts of head with perithecia supported by a stalk. Buried deep these sclerotia are able to germinate, but fail to emerge and to release ascospores in the air. This work attempts to describe the depth distribution of sclerotia, artificially dispersed on the soil surface, under the influence of different tillage systems (plowing, shallow cultivation, combination of both) over a two-year sequence. It follows, depending on the depth of burial, an estimate of the risks and benefits associated with each single practice or combination. Following results are discussed and shown in tab:toxins-08-00146-t001.

Table 1. Number and distribution (%) of sclerotia in soil profile after one tillage (L: plow + rotative harrow or WS: Simplified tillage, Lemken Smaragd 9 + harrow) or two tillages in sequence L/L, L/WS, or WS/L. The depth of tillage with the cultivator Lemken Smaragd 9 is set to 10 cm, and 20 cm with the plough. The sclerotia were not searched for below 10 cm after simplified tillage.

Number of sclerotia/m^2 and distribution in (%) per layer		0–5 cm	5–10 cm	10–15 cm	15–20 cm	Total
	After tillage: July 2013					
July 2013	L	15.3 (7%)	19.0 (9%)	75.3 (36%)	100.3 (48%)	210.0 (100%)
	WS	103.3 (65%)	56.3 (35%)	0.0 (0%)	0.0 (0%)	159.7 (100%)
	After tillage: October 2014					
October 2014	L/L	17.3 (30%)	16.7 (29%)	10.7 (19%)	12.3 (22%)	57.0 (100%)
	L/WS	0.7 (1%)	8.7 (18%)	18.0 (36%)	22.0 (45%)	49.3 (100%)
	WS/L	1.3 (7%)	5.7 (29%)	5.7 (29%)	7.0 (35%)	19.7 (100%)

After two tillages and almost 10 months later, 14% of the sclerotia were found. The double plowing leads to a relatively homogeneous distribution of sclerotia in the layers observed. It brings back to the surface (0–10 cm), up to 60% of recovered sclerotia. The realization of a single ploughing led to concentrate sclerotia below 10 cm depth. The efficacy (burying more than 10 cm) after two years is respectively to the sequences, L/WS and WS/L, of 81% and 64%. In both cases, less than 10% of sclerotia have been found in the layer 0–5 cm.

3.16. Modulation of Ochratoxin A Genes' Expression in Aspergillus Carbonarius by the Use of Some Essential Oils

El Khoury, R.; Atoui, A.; el Khoury, A.; Cverheec, C.; Maroun, R.; Mathieu, F.

Ochratoxin A is a mycotoxin, mainly produced on grapes by *Aspergillus carbonarius*, that causes massive health problems for humans; therefore, its presence in food and feed is highly controlled. This study aims to reduce the occurrence of this mycotoxin using the 10 following essential oils (E.O): fennel, cardamom, anise, chamomile, celery, cinnamon, thyme, taramira, oregano and rosemary, by evaluating, on Synthetic Grape Medium (SGM), for 4 days, their effects on the growth of *A. carbonarius* S402 cultures and on its ability to produce OTA. Results showed that *A. carbonarius'* growth was reduced up to 100%, when cultured with the E.Os of cinnamon, taramira, oregano and thyme. As for the other six essential oils, their effect on *A. carbonarius'* growth was insignificant, but highly important on the OTA production. In fact, the fennel's oil at 5000 ppm reduced the OTA production up to 88.9% compared to the control, with only 13.8% of growth reduction. These results led us to further investigate the effect of these E.Os on the expression levels of the genes responsible for the OTA production: *acpks*, *acOTApks*, *acOTAnrps* and the two regulation genes *laeA* and *vea*, using the qRT-PCR method. The results revealed that these six E.Os reduced the expression of the five studied genes: the *ackps* was the most downregulated, reaching 99.2% of inhibition with 5000 ppm of fennel's E.O, while the other genes' inhibition levels ranged between 10% and 96% depending on the nature of the essential oil and its concentration in the medium.

3.17. Fast and Sensitive Total Aflatoxins ELISA Development, Validation and Application for Food and Feed Analysis

Oplatowska-Stachowiak, M.; Sajic, N.; Xu, Y.; Mooney, M.H.; Gong, Y.Y.; Verheijen, R.; Elliott, C.T.

Aflatoxins (aflatoxin B_1, B_2, G_1 and G_2) produced by toxigenic fungi can contaminate different agricultural commodities such as corn and peanuts. Due to their toxic effects in humans and animals fast and validated methods for the detection of aflatoxins in food and feed are required for the identification of the contaminated batches before they are processed into final products and placed on the market. In order to address the need for improved detection methods seven monoclonal antibodies for aflatoxins with a good compromise between sensitivity and cross-reactivity were produced. Antibody showing IC_{50} of 0.031 ng/mL for AFB_1 was applied in simple and fast direct competitive ELISA test for the detection of total aflatoxins. The developed ELISA kit was validated for peanut matrix. The detection capability was 0.3 μg/kg for aflatoxin B_1, which is one of the lowest reported values. Critical assessment of the performance of the total aflatoxins ELISA kit for the detection of aflatoxins B_2, G_1, and G_2 was also performed. The kit was used to analyze 32 peanut and peanut butter samples purchased locally and two samples containing small but detectable amounts of aflatoxin B_1 were identified, which was confirmed by LC-MS/MS analysis. The test was further used to analyze 25 feed ingredients samples and 20 maize samples and the results were correlated with these obtained by LC-MS/MS method. The total aflatoxins ELISA kit was also tested in three proficiency schemes and it was demonstrated to have high accuracy. The developed assay has been transformed into commercial product for fast and easy detection of aflatoxins in food and feed.

3.18. Human Biomonitoring and Its Application to Mycotoxin Exposure Assessment: Revealing the Toxicokinetics of Deoxynivalenol in Humans

Heyndrickx, E.; Mengelers, M.; De Saeger, S.

Recently, a study by Heyndrickx *et al.* (2015) demonstrated the presence of several mycotoxins in urine samples (n = 394) of the Belgian population. Deoxynivalenol (DON) and especially its glucuronides were most prevalent, being present in 100% of the urine samples. A risk assessment was performed by deriving estimated dietary intakes from the urinary concentrations. The estimated intake for DON varied between 0.11–19.57 and 0.03–10.08 μg/kg bw/day for children and adults respectively. This could imply a health risk as 56%–69% of children and 16%–29% of the adults were estimated to exceed the tolerable daily intake for DON (1 μg/kg bw/day) depending on the approach applied. However, it has to be highlighted that there are still a lot of uncertainties when estimating the DON intake using urinary biomarkers due to the lack of toxicokinetic data of DON in humans.

For this reason, an intervention study with 14 adults was designed in order to obtain tentative information about the toxicokinetics of DON in humans. The design was partly based on a theoretical kinetic model developed for DON in humans. Prior to the start of this intervention study, each volunteer had to follow a DON restricted diet for 2 days. Additionally, each volunteer received a bolus of DON through a naturally contaminated breakfast while remaining on the DON restricted diet for the rest of the day. Urine samples were collected at different time points in the following 24 h. The urine samples were analysed for the presence of DON as well as its major metabolites de-epoxy-deoxynivalenol, deoxynivalenol-3-glucuronide and -15-glucuronide, using a targeted LC-MS/MS method. As the presence of unknown metabolites can lead to an underestimation of the exposure, additionally an untargeted screening was performed using HR-MS.

Toxicokinetic parameters such as the excretion pattern of DON and metabolites throughout a day and absorption and elimination rates will be calculated using the concentration-time curves in urine. Furthermore, this study could give a decisive answer about the use of morning or spot urine compared to 24-h urine and for the first time natural inter-individual variations will be determined within this group of volunteers. The obtained knowledge will serve to develop a standardised method to estimate DON intake by means of biomarker analysis.

Acknowledgements: this research was financially supported by the Research Foundation Flanders (grant number G.0D4615.N) as part of 'The Food Biomarkers Alliance project (FOODBALL)' within the 'EU Joint Programming Initiative: A Healthy Diet for a Healthy Life: Biomarkers for Nutrition and Health'.

Heyndrickx, E.; Sioen, I.; Huybrechts, B.; Callebaut, A.; De Henauw, S.; De Saeger, S. Human biomonitoring of multiple mycotoxins in the Belgian population: Results of the BIOMYCO study. *Environ. Inter.* **2015**, *84*, 82–89.

3.19. Determination of Blood and Immune Parameters in Broilers Exposed to Fumonisin

Koppenol, A.; Junior, P.M.; Dhaouadi, S.; Caron, L.F.

Fumonisins (FUM) are mycotoxins commonly produced by molds belonging to the *Fusarium* genus. They are usually identified in poultry feed, because these feeds are mainly based on corn. One of the strategies currently available to mitigate the effect of mycotoxins in livestock is the inclusion of anti-mycotoxin additives, with specific adsorbent and/or detoxifying properties against those compounds.

The aim of the present trial was to assess the effects of feeding broiler chickens with FUM naturally contaminated feed and its impact on the poultry's immune response and blood variables as well as to evaluate the protective effect of adding an anti-mycotoxin additive to this feed. In total, 96 male Cobb 500 day-old-chicks were divided over three isolators of 32 chicks each. Animals were fed one out of the three dietary treatments *ad libitum* until 28 days of age; A corn-soybean meal based control diet, a FUM contaminated diet (17 ppm) or a FUM contaminated diet supplemented with an anti-mycotoxin additive. The contaminated diet was formulated by replacing control corn by a naturally contaminated corn with *Fusarium* mycotoxins. The anti-mycotoxin additive supplemented diet was prepared using 0.2% of a commercial product (Elitox®).

Blood samples were taken from eight animals per treatment at 3, 7 and 14 days of age for evaluating circulating lymphocytes, as well as at 14 and 28 days of age for the determination of plasma proteins, albumin and globulin and serum levels of sphingosine (SO) and sphinganine (SA). On day 7, 14 and 28, samples of jejunum were collected from six animals per treatment for intestinal mucosa cell measurement.

Results showed that FUM had detrimental effects in broilers, resulting in decreased hematocrit and increased albumin, albumin/globulin and SA:SO values in the blood as well as decreased T-helper lymphocyte, activated T-cytotoxic lymphocytes and monocytes concentrations. The addition of the anti-mycotoxin additive was shown to ameliorate most of the negative effects of FUM.

3.20. Development of an LC-MS/MS Method for Simultaneous Determination of Beauvericin, Enniatins (A, A1, B, B1) and Cereulide in Cereal and Cereal-Based Food Matrices

Decleer, M.; Rajkovic, A.; Sas, B.; Madder, A.; De Saeger, S.

Beauvericin and the related enniatins are mycotoxins, secondary metabolites mainly produced by different *Fusarium* species that occur naturally on cereal and cereal-based foods and feeds. The emetic toxin cereulide is produced by the specific strains of *Bacillus cereus*. Structurally, these toxins are cyclic depsipeptides with ionophoric properties. Their toxicity is mediated by the ability to initiate cation transport across the cell membrane, disrupting normal intracellular cation levels, leading to apoptosis which is accompanied by DNA fragmentation. They are highly resistant to heat, acidification and proteolytic enzymes. Although these emerging foodborne toxins have different microbial origin, the striking structural and functional similarities enable and provide rationale for their concurrent detection in food matrices. Due to their prevalence in food and feed, and their toxicity it became an imperative to create a new tool in microbial food diagnostics for their reliable detection and quantification. The use of (tandem) mass spectrometry (MS), enabled the sensitive detection and quantification in order to better assess the co-occurrence of toxins. To the best of our knowledge this is a first report of a validated UPLC-MS/MS method for the simultaneous determination of beauvericin and the related enniatins, together with cereulide, in cereal and cereal-based food matrices such as wheat, maize, rice and pasta. A Waters Acquity UPLC system coupled to a Waters Quattro Premier XE™ Mass Spectrometer operating in ESI+ mode was employed. Sample pretreatment involved simple liquid extraction of the target toxins without any further clean-up step. The simple extraction, together with a fast chromatographic separation of only 7 min allowed substantial saving costs and time. The validation of the developed method was performed based on Commission Regulation No. 401/2006/EC and included determination of selectivity, repeatability, limit of detection (LOD), limit of quantification (LOQ), recovery and linearity. The obtained LODs ranged from 0.62 to 3.91 μg/kg and the LOQs from 1.24 to 7.83 μg/kg. The obtained RSD for repeatability was within 5 and 12% and the obtained RSD for intermediate precision was within 5 and 8%. The apparent recovery varied from 89 to 110%. For all compounds the extraction recovery varied between 85% and 105% in the different matrices. The highly sensitive and repeatable validated method was applied to a number of naturally contaminated samples allowing detection of sub-clinical doses of the toxins. Consequently, the influence of matrix (maize, wheat, rice and pasta) on the thermal stability of the target toxins under different conditions was investigated using the developed LC-MS/MS method.

Acknowledgements: this work was financially supported by the BOF Special Research Fund from Ghent University, GOA project No. 01G02213 and FWO mandate of Prof.dr. Andreja Rajkovic.

3.21. *Development of a Multiplex Lateral Flow Immunoassay with Quantum Dots as Innovative Label*

Foubert, A.; Beloglazova, N.; Rajkovic, A.; Sas, B.; Madder, A.; De Saeger, S.

Today, with the increased regulatory requirements in food safety the demand for rapid, sensitive and accurate methods to detect biological and chemical contaminants has increased. In particular, tests that can be completed within minutes would enable processors to take quick corrective actions when contaminants are detected, which is also the case for mycotoxins. Hence, rapid methods like the lateral flow immunoassay (LFIA) are rapid, user-friendly and sensitive on-site tests suitable for this purpose. Here, we present the development of a multi-mycotoxin LFIA system by using quantum dots (QD) as an innovative label. QDs, small semiconductor nanoparticles, are one of the most promising labels due to their unique spectral properties. They are characterized by a high fluorescence quantum yield, stability against photobleaching, and size-tunable absorption and emission bands. They allow simultaneous use of multiple QDs with different spectral characteristics (multiplexing). The stable photoluminescence makes QDs ideal nanoprobes for chemical, biomedical and therapeutic labeling and imaging.

In this work a multi-mycotoxin LFIA based on the use of green, red and orange-emitted QDs was developed and is able to detect four mycotoxins, *i.e.*, deoxynivalenol (DON), zearalenone (ZEN) and T2/HT2 in different matrices (barley and wheat). The test is based on an indirect competitive approach. First, the QDs were solubilized by coating them with polymer or silica, which also made bioconjugation with antibodies (Abs) possible. Next, the specific Abs were immobilized on the QDs by carbodiimide chemistry. The mycotoxin ovalbumin (OVA) conjugates (DON-OVA, ZEN-OVA and T2-OVA) were synthesized and immobilized as three test lines on the membrane. The test is completed within 15 min and there is no need for any mathematical or statistical processing of the obtained results. This detection method is a user-friendly and sensitive detection method with cut-offs (DON: 1000 μg/kg, ZEN: 80 μg/kg, T2/HT2: 80 μg/kg) according to EU legislation. Afterwards, the QD-based LFIA was also compared with a LFIA based on another, often used label, *i.e.*, gold nanoparticles. This showed that the QD-based LFIA was more sensitive and resulted in less use of antibody and antigen.

Acknowledgements: this work is financial supported by the BOF Special Research Fund from Ghent University, GOA project No. 01G02213.

3.22. Dried Blood Spots as a Powerful Tool for Individual Mycotoxin Exposure Analysis in Consumer's Blood

Osteresch, B.; Cramer, B.; Humpf, H.-U.

Mycotoxins are toxic secondary metabolites of moulds that frequently occur in food. The intake varies depending on food contamination and consumption habits. Therefore, individual exposure to mycotoxins and their metabolites is difficult to evaluate when mean food contamination is used as calculation basis. However physiological samples like urine or blood can be analyzed to determine the exposition for each test person individually. In recent years dried blood spots (DBS) came more and more into focus for medical applications and sample collection in comparison to conventional vein puncture. Here, the DBS sample technique in combination with HPLC-MS/MS detection is introduced for the quantitative assessment of mycotoxin exposure in humans.

For sample preparation, blood is collected on commercially available filter paper cards (Whatman 903 protein saver cards™), dried and extracted by an aqueous organic solvent. Then, an aliquot of the extraction solution is evaporated, reconstituted in mobile phase and injected into HPLC-MS/MS. The method has been validated concerning recovery, reproducibility, limit of detection and stability tests. Furthermore, effects on quantification like changes in hematocrit value and spotted blood volume have been investigated. For this, spiked blood samples were spotted and the ratio of detected concentrations in punched DBS were compared to whole DBS of known volume. As a result, altering hematocrit values and spotted blood volumes do not affect the quantification. In the same way the location of the sampling site for blood donation (fingertip or vein) caused no considerable issue.

The developed method provides sufficient mean recovery rates of 80%–120% and, for the majority of the included mycotoxins and metabolites, limits of quantitation in the lower pg/mL range. Storage tests show high stability for ochratoxin A for months. On the contrary compounds such as aflatoxins reveal strong degradation rates within weeks when stored in the darkness at room temperature. In a first study with coffee/noncoffee drinkers (n = 50) ochratoxin A could be found in all samples. In addition, both cohorts could be distinguished by the detection of the thermal degradation product 2'R-ochratoxin A which is formed under coffee roasting conditions.

In conclusion, the developed HPLC-MS/MS method is a new, simple and sensitive way to estimate the mycotoxin exposition in consumer's blood based on DBS.

3.23. Interaction between Fusarium Mycotoxins and Cytochrome P450 drug Metabolizing Enzymes/ABC Drug Transporters in a Porcine Animal Model

Schelstraete, W.; Devreese, M.; Van Bocxlaer, J.; Croubels, S.

The cytochrome P450 (CYP450) enzymes and ABC drug transporters in the intestine and liver play a major role in the pharmacokinetics of drugs and other xenobiotics. When drugs and food or feed are co-administered, this can lead to pharmacokinetic interactions due to inhibition or induction of CYP450 and ABC transporters by feed components. Consequently, a change in absorption, distribution, metabolism and/or excretion (ADME) of substrate drugs may take place. This has led to a number of clinically relevant changes in the pharmacodynamics of drugs. However, little is known about pharmacokinetic interactions between drug and food or feed contaminants. Mycotoxins are such contaminants produced by fungi and are the number one threat regarding chronic toxicity. *Fusarium* is one of the most important mycotoxigenic fungi genera and, consequently, this study will focus on mycotoxins produced by those species, namely deoxynivalenol, T-2 toxin, fumonisin B1 and zearalenone. For instance, our group has demonstrated an inhibition of hepatic CYP3A activity in pigs and the down-regulation of hepatic *CYP3A37*, *CYP1A4*, *CYP1A5* and *MRP2* in broiler chickens after exposure to T-2 toxin (Goossens *et al.* 2013; Osselaere *et al.* 2013). Supported by these previous results, it is the aim of our research to investigate the influence of oral exposure to *Fusarium* mycotoxins on the pharmacokinetic behavior of selected drugs in a porcine animal model.

To tackle this, *in vitro* and *in vivo* studies will be set up. A first goal is to geno- and phenotype key intestinal and hepatic CYP450 and drug transporters in pigs, in order to gain basic insight in the expression and activity of these proteins. Next, *in vitro* modulatory effects of *Fusarium* mycotoxins on these enzymes and transporters will be assessed, as well as the impact on mRNA expression and functional level in *in vivo* trials. These data will finally serve as decision criteria to select those *Fusarium* toxins with a significant impact, in order to conduct an *in vivo* trial to determine the effect of the mycotoxin(s) on the pharmacokinetics of substrate drugs.

Acknowledgements: this research is supported by the Special Research Fund (BOF DOC.2015.0075) from Ghent University.

Goossens, J.; De Bock, L.; Osselaere, A.; Verbrugghe, E.; Devreese, M.; Boussery, K.; Van Bocxlaer, J.; De Backer, P., Croubels, S. The mycotoxin T-2 inhibits hepatic cytochrome P4503A activity in pigs. *Food Chem. Toxicol.* **2013**, *57*, 54–56.

Osselaere, A.; Li, S.J.; De Bock, L.; Devreese, M.; Goossens, J.; Vandenbroucke, V.; Van Bocxlaer, J.; Boussery, K.; Pasmans, F.; Martel, A.; *et al.* Toxic effects of dietary exposure to T-2 toxin on intestinal and hepatic biotransformation enzymes and drug transporter systems in broiler chickens. *Food Chem. Toxicol.* **2013**, *55*, 150–155.

3.24. Presence of Deoxynivalenol in Wheat- and Corn-Based Products Produced in Serbia

Farkas, H.; Marosanovic, B.; Jaksic, M.; Cujic, S.; Marinkovic, D.

Deoxynivalenol (DON) is a natural-occurring mycotoxin mainly produced by *Fusarium graminearum*. It is also known as vomitoxin due to his strong emetic effects. Most often the compound is found in corn, but it is also found in other important crops such as wheat, oats, barley, rice, and other grains. Generally, the *Fusarium* species grow in moist, cool conditions and similarly invade crops under these more favorable conditions. The objective of this study was to investigate the presence of DON in corn- and wheat-based products produced in Serbia. Commercial samples were collected between January 2013 and December 2015 and analyzed by high performance liquid chromatography coupled with fluorescence detection (HPLC-FLD). Samples were prepared by immuno-affinity clean-up column. The method used for detection of the mycotoxins in wheat and corn - based products obtained a recovery rate from 92% to 99%, and limit of quantification was 60 μg/kg. During named period, we have analyzed 3765 samples for presence of DON. In 2013, 0.86% of analyzed wheat- and corn-based products were contaminated by DON, while in 2014 it was 2.63% and in 2015 23.5%. Concentration ranges were, retrospectively 75–498 μg/kg for 2013, 67–434 μg/kg for 2014 and 63–493 μg/kg for 2015. Maximum level for DON according to Commission Regulation (EC) No 1881/2006 in cereals intended for direct human consumption, cereal flour, bran and germ as end product marketed for direct human consumption is 750 μg/kg and for bread (including small bakery wares), pastries, biscuits, cereal snacks and breakfast cereals is 500 μg/kg, which has not been exceeded, but we noticed increased presence of DON in 2015. Despite the low contamination observed in wheat- and corn-based products, monitoring the presence of mycotoxins in foods is important to ensure safety.

3.25. Review of Principal Mycotoxins Occurrence in Food and Feed in Serbia during 2004–2014 Period

Udovicki, B.; Rajkovic, A.

This paper is a comprehensive review of relevant published investigations during a 10-year period (2004–2014) aimed at providing the necessary overview of occurrence of principal mycotoxins in food and feed in Serbia. This work has been done in the framework of the Horizon2020 project "MycoKey". Mycotoxins, as secondary metabolites produced by several genera of fungi, are usually found in agricultural commodities and subsequently in animal products from animals that consumed such commodities as feed. Serbia is an agricultural country and grain production, especially wheat and maize, is an important economic factor as Serbia represents one of the largest maize producers and exporters in Europe. Cereals are an essential food for population in Serbia, having high social and nutritional relevance and to represent a food group with high risk for acute and chronic exposure to mycotoxins. In February–March 2013 several European countries, including Serbia, Croatia and Romania reported nationwide contamination of milk for human consumption (and possibly of derivative products) with aflatoxins. It was reported in March of the same year that feed originating from Serbia and imported in the Netherlands and Germany was contaminated. It was also reported in March that tests revealed contamination in milk produced by two Dutch farms. This large scale incident inspired many research efforts in the subsequent period. Out of 2595 samples tested in the period 2004–2014 75.7% were positive for aflatoxins, with 34.9% over the EU limits. It must be stated that most of positive samples are after year 2012 (after heavy drought in 2012). In that period 760 samples are tested with 70% of positive samples with 43.6% over EU limits. Out of 921 samples tested for DON 44.8% were positive, with 3.4% over the EU limits. Out of 490 samples tested for ZEA 34.7% were positive, with 5.7% over the EU limits. Out of 360 samples tested for FUMs 67.5% were positive, with only 0.3% over EU limits. Out of 363 samples tested for T-2/HT2 48.2 were positive, with 12.7% over EU regulation. Out of 483 samples tested for ochratoxin A 29.4% were positive, with 3.7% over the EU limits. Apart from the increase in the number of positive samples during years with favorable conditions for fungi growth and toxins production, in recent years that increase may also be attributed to more sensitive immunosorbent assays and HPLC/MS methods.

3.26. Analysis of 648 Feed Samples Sourced in Europe in 2015 for More than 380 Mycotoxins and Secondary Metabolites

Muccio, M.; Naehrer, K.; Kovalsky, P.; Krska, R.; Sulyok, M.

A total of 648 feed samples such as corn, wheat, barley, silage, as well as finished feed and others were screened for more than 380 mycotoxins and other secondary metabolites. The feed samples were collected in Europe in 2015 and analyzed with a multi-mycotoxin LC-MS/MS method at IFA-Tulln according to Vishwanath *et al.* (2009). The analytical method was transferred to a more sensitive mass spectrometer (QTrap® 5500) and extended to cover more than 380 metabolites (Malachova *et al.* 2014; Streit *et al.* 2013). The accuracy of the method is monitored by regular participation in proficiency tests, which includes a separate testing scheme on "animal feed" (BIPEA, Gennevilliers, France).

On average 31 different metabolites were detected per sample. Seventy-seven percent of the samples tested positive for deoxynivalenol, 67% for nivalenol, 66% for zearalenone and 62% for deoxynivalenol-3-glucoside (average of positives 681, 71, 207, 108 μg/kg; max. 34,861, 5771, 6239, 2741 μg/kg). Thirty-eight percent of the samples were contaminated with fumonisins, 38% with A-trichothecenes and 81% with B trichothecenes (average of positives 241, 10, 689 μg/kg; max. 18,411; 557; 39,158 μg/kg).

The "emerging mycotoxins" emodin, enniatin B1 and beauvericin were found in over 81%; 79% and 56% of the samples analyzed respectively (average of positives 102, 45, 41 μg/kg; max. 2667; 662; 1601 μg/kg).

As the sensitivity of LC-MS/MS increased by 200-fold in the last 10 years, more mycotoxins and other secondary metabolites are detected per sample nowadays. In consequence, further data on the metabolic fate, mode of action, toxicity and interactions of mycotoxins are required to interpret the health risk. Nevertheless, *Fusarium* mycotoxins like deoxynivalenol, zearalenone and fumonisins are still among the most frequently occurring agriculturally relevant mycotoxins.

Malachova, A.; Sulyok, M.; Beltran, E.; Berthiller, F.; Krska, R. Optimization and validation of a quantitative liquid chromatography-tandem mass spectrometric method covering 295 bacterial and fungal metabolites including all regulated mycotoxins in four model food matrices. *J. Chromatogr. A* **2014**, *1362*, 145–156.

Streit, E.; Schwab, C.; Sulyok, M.; Naehrer, K.; Krska, R.; Schatzmayr, G. Multi-mycotoxin screening reveals the occurrence of 139 different secondary metabolites in feed and feed ingredients. *Toxins* **2013**, *5*, 504–523.

Vishwanath, V.; Sulyok, M.; Labuda, R.; Bicker, W.; Krska, R. Simultaneous determination of 186 fungal and bacterial metabolites in indoor matrices by liquid chromatography/tandem mass spectrometry. *Anal. Bioanal. Chem.* **2009**, *395*, 1355–1372.

3.27. Exploring the Effects of Gaseous Ozone (O_3) and 1-Methylcyclopropene (1-MCP) Treatments on the Development of Penicillium Expansum and Patulin Production on Apple Fruits (cv. Granny Smith)

Testempasis, S.; Myresiotis, C.; Tanou, G.; Molassiotis, A.; Karaoglanidis, G.S.

"Blue mold" caused by *Penicillium expansum* is considered to be one of the most destructive postharvest diseases of apple fruit. The pathogen may lead to great quantitative losses while contributing to qualitative deterioration of apple products due to the production of a great variety of mycotoxins, such as patulin and citrinin that impose a risk for human health. Control of the disease is achieved by fungicide treatments. However, development of fungicide resistance and social concerns regarding pesticide residues, necessitate research for alternative control methods. The aim of this study was to evaluate the effect of 1-MCP (0.5 μL/L, 24 h, 0 °C) and ozone treatments (0.3 μL/L) on blue mold severity and patulin production on artificially inoculated apple fruits (cv. Granny Smith). In detail, artificially inoculated apple fruit, treated or not with 1-MCP, subjected for 2 and 4 months to cold storage (0 °C, RH > 95%) either in an O_3 enriched atmosphere or in a conventional cold chamber. Results showed that disease severity was higher in both O_3 and/or 1-MCP treated fruit, compared to the non-treated control fruit. To elucidate whether the increased susceptibility of apple fruit to ozone treatments was mediated by a predisposition of apple due to ozone applications, an additional experiment was conducted. 1-MCP treated apple fruit were exposed to ozone for 4 and 8 days prior the inoculation and incubated for 15 days in room temperature (20 °C, RH > 95%). Measurements of disease severity showed that the 1-MCP treated fruit, which had/had not previously been exposed for 4 and 8 days to O_3, were more susceptible than the untreated fruit. In both experiments, patulin production was measured with High Performance Liquid Chromatography-Diode array detector (HPLC-DAD). 1-MCP-treated fruit that had been exposed to O_3 for 8 days, showed 35 times higher patulin production than the untreated fruits. Similarly, 1-MCP treated fruit, stored in ozone chambers for 2 and 4 months, revealed higher patulin concentration than the untreated fruit. Such results emphasize that 1-MCP and O_3 treatments not only do not contribute to the control of the disease but in addition, appear to be directly related to an increased patulin production. Further research is required to explore the molecular basis of this increase in blue mold incidence and patulin production in 1-MCP and O_3-treated apple fruit.

3.28. Biomarkers for Exposure of Mycotoxins in Pigs and Broiler Chickens

Lauwers, M.; De Baere, S.; Letor, B.; De Saeger, S.; Croubels, S.; Devreese, M.

Poultry and pigs are highly exposed to mycotoxins due to their cereal based diet. Acute and chronic exposure to mycotoxin contaminated feed can cause deleterious effects on the performance and wellbeing of the animal and leads to economic losses. Therefore, it is of great importance to assess mycotoxin exposure in livestock and to correlate this with the health status of the flock or herd. This can be done by using biomarkers as indicator of exposure. Typical biomarkers for exposure are parent compounds and their phase I and II metabolites, which can be detected in biological matrices, such as plasma and excreta.

Besides exposure assessment, alleviating the effects of mycotoxins on animals is of critical importance. Mycotoxin detoxifying agents, or mycotoxin detoxifiers, are feed additives which aim to bind (mycotoxin binders) or to alter the chemical structure (mycotoxin modifier) of the mycotoxin in the gastro-intestinal tract of the animal, thereby diminishing the exposure of animals to mycotoxins. The European Food Safety Authority (EFSA) has stated that the efficacy of these compounds should be evaluated *in vivo* by analysing the concentration of a suited biomarker, *i.e.*, the parent compound and/or the metabolites, in biological matrices.

The first objective of this PhD research is to develop an ultra-high performance liquid chromatography coupled to HR-MS (type Synapt G2-Si HDMS) screening method to determine the mycotoxins aflatoxins, deoxynivalenol, T-2 toxin, fumonisins, zearalenone and enniatins and their major phase I and II metabolites in plasma, urine and faeces of pigs and plasma and excreta of poultry. Optimization of the LC parameters (column, gradient, solvents, temperature, *etc.*) led to the successful separation and detection of 24 mycotoxin analytical standards. Secondly, feed and selected feedstuffs will be screened for mycotoxin contamination. In the future, toxicokinetic studies of the most prevalent mycotoxins will be performed in pigs and broiler chickens to evaluate species dependent differences in absorption, distribution, metabolisation and excretion (ADME) processes of these mycotoxins in both animal species. Furthermore, the most appropriate biomarkers of exposure will be identified for each mycotoxin in both animal species.

Next, the effect of candidate mycotoxin detoxifiers on the toxicokinetic behavior and selected biomarkers of the mycotoxins will be evaluated.

Finally, the most potent mycotoxin detoxifier(s) will be retained and assessed in long-term feeding trials where a mycotoxin contaminated diet is fed to the animals with and without detoxifier.

3.29. MIP Loaded Porous Scaffolds as SPE Sorbent for Ergot Alkaloid Analysis

De Middeleer, G.; Dubruel, P.; De Saeger, S.

Mycotoxins are naturally occurring contaminants in food and feed which are produced by various fungal species. Although these secondary metabolites are only present in ppb-ppt concentrations, they can be toxic to humans and animals. Within the group of mycotoxins, ergot alkaloids are produced by *Claviceps* species and they occur mostly in cereal based products. Since infected crops can be processed for consumption, the economic consequences of such contaminations should not be underestimated. Therefore, rapid, sensitive and accurate analysis is obligatory to control and secure food and feed safety. In general, mycotoxin analysis includes rapid screening and confirmatory methods, mostly focussing on a multi-analyte approach. Therefore, selective recognition elements are required such as antibodies or molecularly imprinted polymers (MIP) which can bind with different target analytes. Since antibodies suffer from several disadvantages such as low stability and high production costs, this research aims to use MIP for the development of a solid phase phase extraction (SPE) application prior to ergot alkaloid LC-MS/MS analysis. This application needs to cover the extraction of the six major ergot alkaloids and their corresponding epimers.

Sub-micrometer sized MIP for ergot alkaloids analysis have been produced by precipitation polymerization. Equilibrium experiments indicated that MIP particles bind higher amounts of metergoline template molecules compared to non-imprinted particles. In addition, different conditions were tested to evaluate binding characteristics of the MIP more specifically, to study the binding of a mixture of the six major ergot alkaloids and their corresponding epimers through recovery experiments. Since the final application implies MIP to be immobilized onto poly-ε-caprolactone (PCL) structures, the particles were deposed by means of Pluronic® F127 bismethacrylate hydrogel building blocks. The optimization of the immobilization protocol and selection of the optimal hydrogel concentration were first examined on 2D PCL-spincoated glass plates. These immobilization experiments and sol-gel tests have shown that 7.5% and 10% of hydrogel resulted in successful particle immobilization and respectively an 87.4% and 83.3% gel-fraction of the corresponding hydrogel network. Second, MIP particles need to be immobilized on 3D structures to meet the intended SPE application. Therefore, the Bioplotter™ technology was used to produce 3D PCL scaffolds which are characterized by micrometer sized interconnective pores. The immobilization of MIP onto these scaffolds was successful as shown by SEM analysis. In a next step, the functionality of the MIP particles onto the 3D structures was investigated to examine whether the MIP binding capacity is still sufficient after immobilization on 3D scaffolds.

Acknowledgements: the authors thank the agency for Innovation by Science and Technology (IWT) for the financial support.

3.30. Elucidation of Bikaverin Biosynthesis in Fusarium Fujikuroi by Genetic Engineering, High Resolution Mass Spectrometry and NMR

Arndt, B.; Studt, L.; Wiemann, P.; Tudzynski, B.; Köhler, J.; Krug, I.; Humpf, H.-U.

In this study, we have identified an unknown metabolite in *Fusarium fujikuroi*, an ascomycetous fungus normally infecting rice plants resulting in yield losses. This fungus gained attention due to produced phytohormones, gibberellines, which constitute a virulence factor of the fungus but are nowadays also used as plant growth regulators in agriculture. Since the genome of *F. fujikuroi* is fully sequenced, the identified metabolite could be assigned to the corresponding gene cluster through genomic engineering, showing that the substance is dependent on the bikaverin gene cluster. Although all genes for the biosynthesis of the PKS-derived pigment are known, only two bikaverin precursors, nor-bikaverin and pre-bikaverin, are established so far.

With the help of genomic engineering and high-performance liquid chromatography (HPLC) coupled to high resolution mass spectrometry (HRMS) followed by isolation and detailed structure elucidation, the new substance could be designated as an unknown bikaverin precursor, missing two methyl- and one hydroxy group, hence named oxo-pre-bikaverin. To decipher the whole bikaverin biosynthetic pathway and to overcome negative regulation circuits, the structural cluster genes *BIK2* and *BIK3* were overexpressed independently in the ΔΔ*bik2/bik3* + OE::*BIK1* mutant background by using strong constitutive promoters. With the help of the software MZmine 2, the metabolite spectra of the created mutants were compared, revealing further possible intermediates.

To analyze the potential cytotoxic properties of this new compound, we compared the effects of bikaverin and the new intermediate on Hep G2 cells using a cell proliferation and cytotoxicity assay (CCK-8). Oxo-pre-bikaverin (1 nM–100 μM) showed no cytotoxic effect but a concentration dependent increase of cell viability by up to 62%. Treatment with bikaverin showed no cytotoxic effects.

Acknowledgements: financial support by the DFG and the NRW Graduate School of Chemistry is gratefully acknowledged.

3.31. Occurrence and Quantitation of the Non-Cytotoxic Tenuazonic Acid Isomer Allo-Tenuazonic Acid in Tomato Products

Hickert, S.; Krug, I.; Cramer, B.; Humpf, H.-U.

Tenuazonic acid (TeA) is a mycotoxin mostly produced by fungi of the genus *Alternaria* on various foodstuff. *Allo*-tenuazonic acid (*allo*-TeA) is a known acid and base catalysed degradation product of TeA. Routine methods described in literature for the quantitative determination of TeA in food samples do not distinguish between TeA and *allo*-TeA as the chromatographic separation of both diastereomers is challenging. The separation of both compounds using inorganic salts as mobile phase additives is described in literature—but these chromatographic approaches are not compatible to HPLC-MS/MS as these additives are not volatile. The scope of this work was to develop a chromatographic method for the separate detection of both diastereomers using an HPLC-MS/MS compatible solvent system and its application to food samples. Furthermore, *allo*TeA should be isolated and tested against TeA for its cytotoxicity in cell culture.

Allo-TeA and TeA could be separated on a preparative scale using RP-amide material. TeA and *allo*-TeA were obtained in isomeric purities >98%. Application of *allo*-TeA to HT-29 cells revealed no cytotoxicity (10–800 μM) while TeA showed toxic effects starting at 250 μM. A chromatographic method separating both isomers on an analytical scale using hypercarb material and a binary gradient consisting of methanol and water, both containing 1% FA and 10 mM NH_4OAc, was successfully developed. Using a previously synthesized $^{13}C_2$-labeled standard of TeA and *allo*TeA [4], 20 tomato samples were analyzed for their TeA and *allo*TeA levels, showing that all 20 tomato products contained both epimers. TeA was found in concentrations from 5.3–540 μg/kg (average: 110 μg/kg) and *allo*-TeA in a range of 1.4–270 μg/kg. On average, *allo*-TeA represents 27% of the sum of both epimers. *Allo*-TeA can be found in small amounts (<5%) when *Alternaria alternata* is cultivated on tomato puree.

As *allo*-TeA shows differences in toxicity compared to TeA and constitutes a major portion of the sum of both epimers, approaches quantifying both as a sum parameter might overestimate the risk to consumers. The co-occurrence of both isomers may have implications on future legal limits in the European Union.

3.32. *Piperazine and Benzodiazepine-Derived Metabolites from Penicillium Aurantiogriseum*

Kalinina, S.; Hickert, S.; Cramer, B.; Humpf, H.-U.

Fungi are considered as major plant and insect pathogens as well as important agents of disease in vertebrates. They produce a multitude of low-molecular-weight compounds known as mycotoxins, which are toxic in low concentrations, causing acute and chronic diseases. Additionally, around 25% of crops worldwide are contaminated by molds and affected by mycotoxins, and the estimated loss extends to billions of dollars. Among others *Penicillium* species represent important toxigenic fungi, which are often found in food products and the range of mycotoxin classes produced by these fungi is much broader than that of any other genus. *Penicillium* toxins often affect liver and kidney function of vertebrates and are represented mainly by structurally diverse compounds such as penitrems, roquefortines, patulin, citrinin, griseofulvin, auranthine, PR-toxin and others. Nevertheless, some species of these genera are not well characterized and can produce several unknown toxic food contaminants. In the course of our study of toxins from *Penicillium* species in food, piperazine- and benzodiazepine-derived toxins from *P. aurantiogriseum* were successfully isolated. For that purpose, fungal cultures were cultivated on Czapek Dox Agar in the dark at 22 °C. After three weeks of cultivation fungal cultures were extracted with ethyl acetate followed by separation of compounds using flash chromatography (reversed-phase column SNAP KP-C18-HS) with an acetonitrile/water gradient. Final purification of piperazine and benzodiazepine-derived toxins was carried out with normal-phase flash chromatography using a gradient of cyclohexane/ethyl acetate. The structure of the isolated compounds was elucidated by NMR and HPLC-MS experiments. Substantial toxicity of the isolated compounds in cell culture experiment encourages further method development for their detection and quantification in food.

3.33. *Monitoring Chemical Contaminants and Residues in Insects: Focus on Mycotoxins*

Kowalski, E.; Wauters, J.; Croubels, S.; Claes, J.; Vanhaecke, L.

A growing world population together with ecological as well as economic concerns related to livestock industry enhances the quest towards alternative protein sources of which insects are acknowledged to have great potential. The new European Novel Food legislation (Regulation 2015/2283) requires that, before 1 January 2018 for all (products of) insects for human consumption, an application for authorization must be submitted for authorizing the placing on the market of these products. Until that date, the current Belgian tolerance of 10 insect species continues to apply. In this Novel Food regulation scientific evidence of the safety for human health must be demonstrated.

In the advice of the Federal Agency for the Safety of the Food chain (FASFC) and Superior Health Council (SHC), the potential microbiological, chemical, and physical hazards related to the consumption of edible insects, is questioned. In an attempt to guarantee health-safe end products, we aim to map and monitor the relevant (organic) chemical contaminants and residues. To achieve the latter purpose, the development of a broad-range analysis method, specific for insect tissues as well as their feed/substrates is mandatory.

The compounds of interest include at least 25 pesticides (herbicides and insecticides), 29 relevant veterinary drugs and coccidiostats, a bacterial toxin and a total of 25 mycotoxins. The mycotoxins originate from different fungal genera, *i.e.*, *Aspergillus, Penicillium, Fusarium* and *Alternaria.* For separation and detection, ultra-high performance liquid chromatography coupled to quadrupole Orbitrap high-resolution mass spectrometry (UHPLC-Q-Orbitrap^{TM}-HRMS) is the platform of choice. Optimization of the parameters of both the LC (gradient, column, solvents, temperature, *etc.*) and MS (gases, temperature, voltage, *etc.*) resulted in a short method (10 min) that proved successful for the separation and detection of the 80 analytical standards. Assisted by a fractional factorial design, generic extraction protocols for several insect species (including *Hermetica illucens, Tenebrio mollitor, Locusta migratoria* and *Acheta domesticus)* and their potential substrates (including wheat-bran, grass, waste products (supermarket), dry cat food, chicken blood and faeces) are currently under development. Next, the analytical methods will be validated and applied for the targeted and untargeted analysis of the specified chemical contaminants and residues in edible insects and their substrates.

3.34. *Comparative Analysis of Fusariotoxins Occurrence in Wheat, Barley and Corn Grain*

Laurain, J.; Nacer Khodja, E.; Marengue, E.

Feed raw materials show different fusariotoxins occurrence depending on the type of culture, as shown by the Scientific Cooperation (SCOOP) survey in 2003. The aim of this study is to identify the occurrence of fusariotoxins in the three main cereals used by the feed industry: wheat, barley and corn grain. This study uses the 'Laboratoire Public Conseil, Expertise et Analyse' (LABOCEA) database composed of chromatography analyses run with LC-MS/MS from 2013 to 2015. Twenty-four fusariotoxins are tested in each analysis. The percentage of positive samples (>LOQ) and the median level of contamination (ppb) per mycotoxin are the two main criteria used. In order to avoid any geographical interaction, samples from one restricted area only (France) are considered: wheat (n = 274), barley (n = 104) and corn grain (n = 336). Data show that corn grain is more poly-contaminated than wheat and barley (per sample, on average seven fusariotoxins are positive for corn grain, two for wheat and three for barley). As a consequence, the percentage of positive samples per mycotoxin is more important in corn grain than for other cereals. Deoxynivalenol (DON) is the most frequent fusariotoxin (>90% of positive) in all cereals, but its median level of contamination is far higher for corn grain (740 ppb) than for wheat (215 ppb) and for barley (75 ppb). The levels of 15-*O*-acetyl DON, zearalenone (ZEA) and fumonisins are also significantly higher for corn grain (153; 135 and 345 ppb, respectively) than for wheat (15, 25 and 50 ppb, respectively) and barley (20, 25 and 30 ppb, respectively). Focusing only on straw cereals, wheat shows higher median contamination in DON, T-2 toxin and fumonisins (215, 35 and 50 ppb respectively) than barley (75, 10 and 30 ppb respectively), whereas barley is more often contaminated (% of positive samples) in DON acetylated forms (40.4% in 15-*O*-acetyl and 17.3% in 3-*O*-acetyl DON) than wheat (5.8% in 15-*O*-acetyl and 16.4% in 3-*O*-acetyl DON). The different cropping parameters (time of harvest, use of fungicide, *etc.*) of corn could explain the important differences in fusariotoxins occurrence compared to straw cereals. *Fusarium* strains have different developments between wheat and barley, which may explain the variable fusariotoxins occurrence.

3.35. *Comparative Analysis of Fusariotoxins Occurrence in Different Types of Corn Materials*

Laurain, J.; Nacer Khodja, E.; Marengue, E.

Feed raw materials show different profiles of mycotoxin contamination depending on the type of culture and storage conditions. Corn is known as a major source of mycotoxins and particularly of fusariotoxins as described in 2003 by SCOOP survey. The aim of the present study is to identify the occurrence of fusariotoxins in different types of corn materials: dry corn grain, humid corn grain (silage corn grain) and corn silage (fermented full corn plant). This study uses the LABOCEA database composed of chromatography analyses run with LC-MS/MS from 2013 to 2015. Twenty-four fusariotoxins are tested in each analysis. The percentage of positive samples (>LOQ) and the median level of contamination (ppb) per mycotoxin are the two main criteria used. In order to avoid any geographical interaction, samples from one restricted area only (France) are considered: dry corn grain (n = 337), humid corn grain (n = 119) and corn silage (n = 557). Data show that all types of corn materials are poly-contaminated with fusariotoxins (per sample, on average seven fusariotoxins are positive) and that deoxynivalenol (DON) is the most frequent (>90% of positive) and the most present (median value > 600 ppb) fusariotoxin. The profiles of fusariotoxins (% of positive samples) are very similar between corn raw materials whereas the level of median contamination depends on the type of corn material. Corn silages have higher DON median level of contamination (1090 ppb) than humid corn grains (980 ppb) than dry corn grains (720 ppb). Deoxynivalenol acetylated forms (15-*O*-acetyl and 3-*O*-acetyl DON) contaminations are similar for all types of corn materials. Nevertheless, nivalenol (NIV) median level in corn silage is 4 times higher than in corn grain (290 ppb *vs.* 68 ppb, respectively). On the contrary dry corn grains have higher median sum of fumonisins (320 ppb) than corn silage (40 ppb) and humid corn (68 ppb). Zearalenone (ZEA) occurrence is similar in all types of corn materials with contamination levels far lower than for DON (median level: <200 ppb for ZEA; >700 ppb for DON). Regarding type-A trichothecenes the profiles of contamination are equivalent for all types of corn materials apart for one metabolite (monoacetoxyscirpenol, MAS) which is more often present (% positive samples) in corn silage (66.4%) than in humid corn grain (32.8%) and in dry corn grain (11%). Some parameters like time of harvest and type of preservation may explain the variable profile of fusariotoxins among corn materials.

3.36. *Interactions between Mycotoxins and the Rumen, Their Possible Toxicological Effects on the Gastrointestinal Tract and Their Intestinal Absorption in Dairy Cattle: An in Vitro Approach*

Debevere, S.; Devreese, M.; De Baere, S.; De Saeger, S.; Haesaert, G.; Fievez, V.; Croubels, S.

Mycotoxins are more and more associated with subclinical health problems for high productive dairy cows reflected by vague and non-specific symptoms and periodic decrease in milk production. Indeed, the risk of mycotoxin contamination of dairy diets is high since the main components such as corn, grass silage and small grain cereals are susceptible to infection with toxigenic fungi. Moreover, considering the wide diversity of toxigenic fungal species on crops and the ability of several fungi to produce more than one mycotoxin, a multiple contamination can be expected. Hence, dairy cows are exposed to various mycotoxins which may lead to depletion of the detoxifying capacity of the microbiota in the rumen. To date more and more dairy farmers and veterinarians are concerned about the impact of mycotoxin on the health and performance of dairy cows. Therefore, research about this topic is needed.

This multidisciplinary doctoral research will elucidate the degradation of individual mycotoxins and relevant mycotoxin combinations in the rumen, the interactions between mycotoxins and the rumen function as well as the possible toxicological effects of mycotoxins on the gastrointestinal tract and their intestinal absorption. Mycotoxins that contaminate corn silage, the most important part of the diets of dairy cattle, will be studied by means of *in vitro* rumen simulations (static and continuous fermentation system). Also varying rumen conditions will be included (e.g., varying dietary ratios roughage/concentrate up to circumstances of subacute acidosis) as they may influence mycotoxin degradation. Ultra-high performance liquid chromatography coupled to high-resolution mass spectrometry (UHPLC-HR-MS, type Synapt G2-Si HDMS) will be used to identify and quantify mycotoxins and their metabolites in the rumen contents. Intestinal cytotoxicity and absorption of mycotoxins will be determined by means of relevant cell cultures. In addition, the efficacy of mycotoxin binders will be tested at the level of rumen metabolism, cytotoxicity and intestinal absorption.

Acknowledgements: this project is co-financed by Flanders Innovation & Entrepreneurship (VLAIO).

3.37. Beyond Aflatoxins: MS-Based Metabolomics Approaches to Investigate Known and Novel Fungal Secondary Metabolites in Aspergillus Flavus

Diana Di Mavungu, J.; Arroyo-Manzanares, N.; Uka, V.; Zheng, H.; Cary, J.W.; Ehrlich, K.C.; Bhatnagar, D.; Calvo, A.; Vanhaecke, L.; De Saeger, S.

The *Aspergillus flavus* metabolome is of considerable interest because *A. flavus* produces the secondary metabolites aflatoxins, the most toxic and carcinogenic compounds known from fungi that occur world-wide as contaminants of foods and animal feeds. Examination of the fungal genome revealed 55 gene clusters predicted to be involved in secondary metabolite biosynthesis, suggesting that *A. flavus* is capable of producing many more potentially significant metabolites than have thus far been reported.

To identify the metabolites produced by these yet unexplored gene clusters and the enzymatic steps involved in their biosynthetic pathways, we established a strategy based on the metabolic profiling of extracts of *A. flavus* wild-type and mutant strains, followed by the structural elucidation of differentially expressed metabolites. To date, we have closely examined metabolite production by wild type *A. flavus* and mutants in biosynthetic genes in clusters 23, 27 and 39. Our results show that cluster 27, with close similarity to the aflatoxin cluster, encodes a polyketide synthase (PKS) producing four anthraquinone metabolites including asparasone A, a compound previously identified in *A. parasiticus*. The hybrid PKS-nonribosomal peptide synthetase (NRPS) of cluster 23 is responsible for the production of a class of 2-pyridone compounds, the leporins, while the PKS of cluster 39 produces 4,7 didesmethyl-siderin, a precursor that is converted to a set of methylated and hydroxylated coumarin derivatives involved in the formation of aflavarin and other previously unidentified C3-C8′ linked bicoumarins.

In this presentation, the potential of high resolution mass spectrometry in both the structural confirmation and *de novo* identification of fungal secondary metabolites will be demonstrated using data from the above mentioned *A. flavus* gene clusters studies. In particular, the opportunity offered by new developments in mass spectrometry instrumentation and data processing tools for improving the current state of metabolite identification strategies will be discussed.

3.38. Development of an Immunochemical Test for Mycotoxin Detection Using Luminescent Nanobiolabel

Goftman, V.; Goryacheva, I.; De Saeger, S.

For the successful development of sensitive and reproducible immunoassay methods, the researcher has to settle two tasks: exploiting highly efficient signal-transduction labels and adopting simple, sensitive signal-transduction methods. In this regard, luminescent sensing provides fast, sensitive, reliable and reproducible detection of the target molecules. The qualitative characteristics of luminescent methods strongly depend on the label used. Quantum dots (QDs) have shown great potential as luminescent nanobiolabels because of their unique properties: nanoscale size, broad excitation spectra for multicolor imaging, robust, narrow-band emission and versatility in surface modification. First used for bioimaging, QDs have later become a useful tool for immunoassay in traditional microtiter-plate format (fluorescent-linked immunosorbent assay, FLISA), sensors and rapid tests.

Our research is focused on the development of a stable luminescent nanobiolabel based on QDs for the sensitive detection of mycotoxins, which are toxic secondary metabolites produced by molds. To this end, first, an efficient synthesis of QDs with different emission colors was carried out. Second, silanization was applied as a modification technique to obtain functional biolabels. In particular, the nanolabel surface was modified with diverse functional groups (amino-, carboxyl-, epoxy-, mercapto-), as such controlling its charge. High buffer stability and low non-specific sorption was achieved by using PEG-derivate silanes.

Third, bioconjugation with secondary rabbit anti-mouse antibodies (IgG) was performed in order to prepare the fluorescent label for indirect immunoassay. An indirect format of immunoassay was chosen in order to increase the sensitivity of the developed method. Moreover, the synthesized conjugate of QDs and IgG is a universal reagent for all monoclonal mouse antibodies, which opens an opportunity to use it for the detection of different analytes, including multiplex detection of different antigens.

Finally, the synthesized luminescent labels were successfully applied in FLISA for the determination of deoxynivalenol (DON) mycotoxin. To decrease time of analysis and to obtain a rapid system for on-site detection, a novel polyethylene frit-based immunoassay for DON determination was also developed. Both methods showed good precision, but the frit-based immunoassay allowed a decrease in IC_{50} by a factor of 20 (IC_{50} FLISA was 473 ng/mL, while IC_{50} for frit-based immunoassay was 20 ng/mL). The total assay time for FLISA was 16 h (using laboratory equipment at each stage), while for frit-based rapid test—only 1 h. Additional advantage of frit-based assay is the possibility to work in a non-laboratory environment (in the case of availability of a handheld reader).

3.39. Impact of Chronic Multi-Mycotoxin Exposure in Europe on Cancer Incidence: a Basis to Develop Future Public Health Strategies

De Ruyck, K.; Heyndrickx, E.; De Boevre, M.; Huybrechts, I.; De Saeger, S.

Mycotoxins are fungal toxins, estimated by the Food and Agricultural Organization (FAO) to contaminate 25% of the world's most frequently consumed foods and feeds. Several fungi may co-occur on crops, resulting in co-occurrence of multiple mycotoxins. Given the ubiquity of many fungi worldwide, an urgent need exists for a coordinated international response to the problem of dietary mycotoxins.

In terms of chronic toxicity, mycotoxins are estimated to be the most hazardous food contaminants. The International Agency for Research on Cancer (IARC) identifies aflatoxins B1, G1, and M1 as sufficiently evident carcinogens, while other mycotoxins are possibly or probably carcinogenic (e.g., ochratoxin A and fumonisins). Consumed with food, some mycotoxins may be absorbed by the blood stream to most commonly affect the liver, where they are metabolized, though not always inactivated. Further down the gastro-intestinal tract, other less bio-available mycotoxins may cause significant interactions with colon cancer cells.

Though the current 'gold standard' for estimating dietary mycotoxin exposure consists of food-based mycotoxin occurrence data cross-referenced against food consumption data, this estimate may be considered the 'external exposure' quantity. Alternatively, measurements made through physiological tissues may be considered representative of an individual's 'internal exposure' to mycotoxins.

A previously published UHPLC-MS/MS method for simultaneous determination of multiple mycotoxins has recently been augmented with methods for the determination of several major mycotoxin metabolites. Particularly, glucuronides of deoxynivalenol, HT-2 toxin, and zearalenone may be present in the blood or urine. Measurement of these metabolites may provide useful insights toward characterizing dietary mycotoxin exposure.

Preliminary analyses of combined data from EFSA and the European Food Consumption Validation (EFCOVAL) Project reported unexpectedly high external exposures, some above upper tolerance levels (unpublished data). This dataset will have been validated through UHPLC-MS/MS quantification of mycotoxin 'exposure biomarkers' in serum and urine, by the time of this conference.

This is the first large-scale cohort study investigating the effect of multi-mycotoxin intakes on incidence of specific forms of cancer. The resulting mycotoxin databases will be instrumental for further characterizing the health effects of real exposure.

3.40. *Identification of Mercapturic Acid Conjugates of Deoxynivalenol in Cereals*

Stanic, A.; Uhlig, S.; Rise, F.; Hofgaard, I.S.; Miles, C.O.

We recently synthesized and characterized the L-cysteine and glutathione (GSH) conjugates of 4-deoxynivalenol (DON). The amino acid and the tripeptide were shown to be linked to DON via conjugation of the 9,10-double bond or the 12,13-epoxide. We also reacted DON with γ-glutamyl-cysteine, cysteinyl-glycine and *N*-acetyl cysteine ("mercapturic acid") and characterized the product mixtures using liquid chromatography coupled to low- and high-resolution mass spectrometry. Extracts from cereal samples ($n = 40$, oats and spring wheat), which were naturally contaminated with DON at a concentration of 1.1 to 11 mg/kg, were screened for the presence of DON-GSH conjugates and their breakdown products. Both C-10 and C-13 linked DON-GSH conjugates were detected in two oats samples. Furthermore, two *N*-acetyl cysteine conjugated DON isomers were detected in 15 samples (both grain species). The LCMS-data (retention time, HRMS^2) suggested that the *N*-acetyl cysteine moiety in the two isomers was likewise linked to DON via C-10 and C-13 conjugation. Our data: (1) confirms that GSH-conjugation of DON occurs in plants; (2) shows that conjugation occurs both at C-10 and C-13, and; (3) suggests that DON-mercapturate could be a major breakdown product of such DON-GSH conjugates in planta.

3.41. *Occurrence of Fusarium Mycotoxins and Their Modified Forms in Major Cereal Crops and Their Processed Product (ogi) from Nigeria*

Chilaka, C.A.; De Boevre, M.; Atanda, O.; De Saeger, S.

In Nigeria, maize, sorghum and millet rank as the most important crops. They are prepared in several forms and consumed on daily basis. These crops are prone to fungi infestation, and as a result may be contaminated with toxic secondary metabolites. Despite the health hazards caused by these mycotoxins, Nigeria has devoted minimal attention to these arrays of toxins especially those produced by *Fusarium* fungi. *Fusarium* mycotoxins such as fumonisins, zearalenone and trichothecenes are the most common mycotoxins contaminating a wide range of crops, especially cereals and cereal-based foods worldwide. They have been reported to cause a variety of toxic effects in animals as well as in humans.

A great concern regarding *Fusarium* mycotoxin contamination of foods is the error of underestimation of mycotoxin levels as a result of modification of these mycotoxins. Modification can be conjugated by plants, animal or fungi, matrix-related or occurring during food processing. They easily escape routine mycotoxin detection methods, and when ingested, may subsequently hydrolyse to free toxins in the digestive tract of organisms. To date, a detailed study on the occurrence of *Fusarium* mycotoxins and their modified forms remains a knowledge gap in Nigeria. We aim to survey and investigate the occurrence of *Fusarium* mycotoxins and their modified forms in Nigerian maize, sorghum, millet and their processed product (*ogi*) available in the market from four agro-ecological zones in the country.

A total of 363 samples (maize $n = 136$, sorghum $n = 110$, millet $n = 87$ and *ogi* $n = 30$) were sampled between September 2015 and October 2015. Analytical methods using LC-MS/MS have been developed and validated for the different matrices in this study. The expected results will contribute to the formulation of a national food safety action plan.

Acknowledgements: authors would like to acknowledge Ghent University Special Research Fund (BOF-01W01014) for funding first author's doctoral studies.

3.42. *Unraveling the Detoxification Mechanism of Deoxynivalenol in Aphids Using Targeted and Untargeted Analysis*

Arroyo-Manzanares, N.; De Boevre, M.; de Zutter, N.; Smagghe, G.; Haesaert, G.; Audenaert, K.; De Saeger, S.

Mycotoxins are toxic, low-molecular-weight, secondary metabolites produced by fungi. Although they are produced by several fungal genera their function often remains elusive. One exception is the trichothecene deoxynivalenol (DON) which is a well-known virulence factor during the infection process of *Fusarium graminearum* helping the fungus to colonize wheat ears. *F. graminearum* is not the only pathogen present on wheat ears: grain aphids (*Sitobion avenae*) co-occur with *F. graminearum* and the aphids come in contact with DON by ingesting plant phloem of *F. graminearum* infected ears. This tripartite interaction between grain aphids, *Fusarium* and wheat was investigated in current study.

In vitro studies revealed that grain aphids can thrive well on artificial diets amended with high concentrations of DON (up to 100 ppm) while pea aphids (*Acyrthosiphon pisum*), which are insects infecting vegetables and never encounter DON, are susceptible to DON. This result suggests that grain aphids have adapted to living in the presence of *F. graminearum* and its mycotoxin DON. We are the first research group providing evidence for a detoxification mechanism in grain aphids enabling them to detoxify DON.

The presented study reports the targeted and untargeted analysis of *Sitobion avenae* to unravel the detoxification mechanism of DON in these insects. Via extensive analyses the authors were able to state that *grain aphids* are more efficient in converting DON into glycosylated forms than pea aphids. For the first time, DON, deoxynivalenol-3-glucoside (DON3-G) and three isomers of DON-diglucoside were identified in *Sitobion avenae.* DON and DON3-G were quantified, and more DON-3G was present in *Sitobion avenae* compared to DON ($p < 0.001$). In addition, DON-diglucosides were identified for the first time, but could not be quantified as no reference standards were available.

This report points out the analytical methods used and the identification of the modified DON-forms.

3.43. *Multidetection of Urinary Ochratoxin A, Deoxynivalenol and Its Metabolites: A Pilot Time-Course Study and Risk Assessment in Catalonia (Spain)*

Vidal, A.; Cano-Sancho, G.; Marín, S.; Ramos, A.J.; Sanchis, V.

The prevalence of two main mycotoxins, ochratoxin A (OTA) and deoxynivalenol (DON), is widespread in cereal-based foodstuffs marketed in Europe. The objective of this study was to assess the urinary and plasma concentrations of OTA, ochratoxin α (OTα), DON, deoxynivalenol-3-glucoside (DON-3-glucoside), deoxynivalenol-3-glucuronide (DON-3-glucuronide), 3-acetyldeoxynivalenol (3-ADON) and de-epoxy-deoxynivalenol (DOM-1) in a preliminary follow-up trial in Catalonia (Spain). Urine and plasma mycotoxin levels and food dietary intake were prospectively monitored in a group of volunteers throughout a restriction period followed by a free-diet period. The results showed that urinary OTA, DON and its metabolites were detected in most of samples, displaying moderate reductions after the restriction period, and subsequently recovering the background levels. Despite the restriction period, some DON metabolites, such as 3-ADON or DOM-1, were found in most of urine samples, placing other alternative sources of exposure under suspicion. DON and DON-3-glucuronide were significantly associated with consumption of bread ($r = 0.362$, $p < 0.001$) and pastries ($r = 0.239$, $p < 0.01$), while OTA was only associated with consumption of wine and breakfast cereals. The urinary levels of OTA were significantly correlated with plasmatic levels of OTA and OTα, supporting the results from urine that allows the simultaneous determination of OTA and DON forms. Also it is more convenient to enroll donors in large-scale studies. The results also showed that the high exposure to DON could be held throughout the time by the same person, exceeding the total daily intakes (TDIs) systematically instead of eventually. The estimates of OTA exposure through urine were largely higher than those obtained with the dietary approach. The background levels found in urine revealed that the exposure to DON and OTA could be of concern for the Catalonian population, thus, further studies applying this biomonitoring methodology in a larger sample of population are needed to accurately characterize the human health risks at population level.

3.44. Mycotoxin Contamination in Sugar Cane Grass and Juice: First Report on Multi-Toxins Detection and Exposure Assessment in Humans

Abdallah, M.F.; Krska, R.; Sulyok, M.

Sugar cane, *Saccharum officinarum*, is a tropical tall perennial grass used to produce raw sugar, molasses (brownish black viscous syrup), and ethanol in addition to the grass left over (bagasse) which is used as animal feed. In Africa, sugar cane is the second most cultivated crop after cassava. In Egypt, around 97% of the total sugar cane production, 15.8 million tons in 2013, is cultivated in the upper part of the country. Sugar cane juice is considered the most popular fresh juice in Egypt where cane juice shops are spreading through all the Egyptian cities. Few previous studies from African and Asian countries discussed the isolation of different fungal species from the plant. No reports have been published for the natural mycotoxin occurrence in cane grass and juice. Moreover, no regulations in Egypt or other countries for this commodity have been set.

Therefore, the aim of the present study was to screen for toxic fungal and bacterial metabolites in sugar cane grass (n = 21) and juice (n = 40) sold in Assiut city, Egypt. Quantification of the target analytes has been done using liquid chromatography-tandem mass spectrometry (LC-MS/MS).

Overall, 33 different metabolites in juice and 29 in cane grass have been quantified. The contamination with major mycotoxins such as aflatoxin B1 and aflatoxin G1 in juice was 23 (58%) and 7 (18%) and in grass was 10 (48%) and 2 (10%), respectively. The other most prevalent metabolites in juice were asperphenamate, tryptophol, emodin, citreorosein, ilicicolin B, versicolorin C, averufin, and iso-rhodoptilometrin while in cane grass, asperphenamate, emodin, tryptophol, citreorosein, iso-rhodoptilometrin, n-benzoyl-phenylalanine and kojic acid were the commonly detected ones. For exposure assessment, online and paper based questionnaire forms are being distributed to be filled in by Assiut city inhabitants (current sample size n = 66). The participants were of different ages and levels of education of both sexes. The preliminary results showed that more sugar cane juice consumption and consequently mycotoxins exposure occur in summer season and male inhabitants consume more than female.

In conclusion, follow-up detection studies in summer and in other cities of the country will provide better insights regarding the number of contaminating mycotoxins and their quantities in both sugar cane grass and juice.

3.45. Detoxification Efficacy of FreeTox towards DON and Its Masked Metabolites in Piglets

Goderis, A.; Van de Mierop, K.; Jin, L.; Michiels, J.

Mycotoxins are toxic fungal metabolites that can contaminate a wide array of cereals. One of the most prevalent mycotoxins is the trichothecene deoxynivalenol (DON). The toxic effects induced by DON are well characterized in all animal species, with pigs being the most susceptible. Occasionally, symptoms due to intake of mycotoxin-contaminated feed are more serious than what would be expected based on the feed contamination level. This observation has led to the discovery of masked mycotoxins which escape detection by routine analytical methods. Consequently, this would imply an underestimation of the degree of contamination upon analysis. Masked metabolites of DON include 3-acetyl-DON (3ADON), 15-acetyl DON (15ADON) and deoxynivalenol-3-β-D-glucoside (DON3G). These may have a direct toxic effect, or may be hydrolyzed to DON in the digestive tract of animals, resulting in higher exposure levels to DON. Another point of concern is the potential synergistic effect when DON and its metabolites are simultaneously present. An *in vivo* trial was set up to investigate the detoxification capacity of the mycotoxin binder FreeTox when pigs were fed a diet contaminated with a mixture of DON, 3ADON and 15ADON. It was hypothesized that FreeTox, which contains specific clay minerals and yeast derived cell walls, could reduce DON-induced growth depression. A total of 120 piglets, weaned at 3.5 weeks old, were assigned to 20 pens. From day 1 until 14 of the experiment, the diet was contaminated with 3 ppm of a mixture of DON (2.6 ppm), 3ADON (0.1 ppm) and 15ADON (0.3 ppm), after which the contamination level was reduced to 1 ppm from day 15 until 37. The piglets were assigned to one out of four dietary treatments: (T1) control uncontaminated diet, (T2) uncontaminated diet + 1 g FreeTox per kg feed, (T3) diet contaminated with DON mixture and (T4) diet contaminated with DON mixture + 1 g FreeTox per kg feed. During the first 14 days of the experiment, the pigs that received the DON-contaminated feed without FreeTox (T3) significantly consumed less feed when compared to pigs that received the same DON-contaminated feed with FreeTox (T4) (227 *vs.* 272 g/d; $p = 0.039$) which resulted in significant growth depression (159 *vs.* 205 g/d; $p = 0.024$). From day 15 until 37 and over the total experimental period, no significant differences were seen between the different treatments. However, numerically the same trend was seen when compared to the first 14 days. In conclusion, FreeTox alleviated the negative effects induced by DON and DON-metabolites in piglets.

3.46. Silanized Liposomes Loaded with Quantum Dots as Label for the Detection of Mycotoxins

Goryacheva, O.A.; Sobolev, A.M.; Foubert, A.; De Saeger, S.; Goryacheva, I.Y.; Beloglazova, N.V.

Nowadays different labels are developed for application in immunochemical assays in order to reach the highest sensitivity. Quantum dots (QDs) are one of the most promising labels because of their unique spectral properties. Their stable photoluminescence makes QDs ideal nanoprobes for chemical, biomedical and therapeutic labeling and imaging. However, when QDs are synthesized they are hydrophobic, only soluble in organic solutions and as such not applicable as label in assays. To solve this QDs can be loaded into liposomes.

The spherical double-layer structure of liposomes allows encapsulation of both hydrophobic and hydrophilic compounds and makes them biocompatible. Until now, a major drawback is the lack of stability in experimental conditions. This could result in leakage of compounds out of the liposomes.

In this research liposomes, loaded with QDs, were covered with a silica coating in order to increase the stability of the liposomes during storage and application and to decrease non-specific interaction. The silica coating also made bioconjugation with antibodies possible. Here, the hydrophilized QDs were coupled with antibodies against two mycotoxins (aflatoxin B1 (AFB1) and zearalenone (ZEN)). This allowed the performance of simultaneous detection of AFB1 and ZEN in cereals by FLISA. The silanized liposomes loaded with QDs allowed a significant increase of label stability, decrease of non-specific interaction and an increase of the assay sensitivity.

3.47. A Decision Support System to Control Mycotoxin Contamination in Maize Silages

Vandicke, J.; Debevere, S.; Fievez, V.; Croubels, S.; De Saeger, S.; Audenaert, K.; Haesaert, G.

Mycotoxins are toxic secondary metabolites produced by a variety of fungal species, such as *Fusarium*, *Penicillium* or *Aspergillus*, among others. Contamination of feed with mycotoxins can cause severe health problems in dairy cattle. Especially high yielding dairy cows with a high feed uptake and rapid ruminal flow are susceptible to gastroenteritis, reduced reproduction and reduced milk production, as a result of mycotoxin contamination. Maize silage is one of the main components of dairy feed in the region of Flanders, Belgium, and is therefore one of the main sources for mycotoxin uptake in dairy cows. This research aims towards providing dairy farmers in Flanders with a user-friendly prediction model, able to foresee mycotoxin contamination based on weather, cultivation, harvest and silage conditions. This model will be constructed based on analyses of maize silages across Flanders, and on our research focusing on methods to prevent mycotoxin contamination. One hundred maize silages will be selected based upon geographical spread, cultivation technique and silage conditions. These silages will be sampled once during harvest and 2–3 times during feeding every year for four years, and analyzed for mycotoxin and fungal contamination. Our research will be divided into five separate work packages, with the following topics: biofumigation of the soil using green crop manures, treatment of crop residues with antagonistic microbial populations, impact of harvest date and dry matter content on mycotoxin contamination, microbial detoxification in the silage, and toxicity of mycotoxins in dairy cattle. These results will aid in constructing and validating the prediction model.

Acknowledgements: this project is co-financed by Flanders Innovation & Entrepreneurship (VLAIO).

3.48. Identification and Characterization of Alternaria Species Causing Early Blight on Potato in Belgium

Vandecasteele, M.; Landschoot, S.; Audenaert, K.; Höfte, M.; De Saeger, S.; Haesaert, G.

Alternaria species, including *A. solani* and *A. alternata* are a serious threat for potato cultivation since heavy infections can lead to significant yield and quality losses. Both species cause necrotic symptoms, which cannot be visually distinguished. Over the past years, both pathogens have become increasingly important in NW Europe. Although the exact cause for this emerging problem remains elusive, it might be attributed to the combined effect of climate change, a reduced use of the fungicide mancozeb, the increased specificity of active ingredients to control *Phytophtora infestans* and the production of high-yielding susceptible cultivars. Furthermore, little is known about the Belgian *Alternaria* population and the contribution of both *A. solani* and *A. alternata* to the disease. The main goal of this research is to identify the primary causal agent of potato Early Blight, to determine inter- and intraspecific diversity within the *Alternaria* population in Flanders, and to unravel the complex interaction between stress-related hormones and the *Alternaria* infection.

To achieve these objectives, 124 fields were monitored throughout Flanders from 2013 to 2015. Results of this disease survey unequivocally show that the disease incidence and—pressure for all seasons was low and no significant differences between regions could be found. In a second part, we identified the population structure at the species level on different time points during the growing season. Therefore, leaf samples were collected during all three growing seasons and using a microscopic and molecular approach, we concluded that *A. alternata*, rather than *A. solani*, was the predominant species at the beginning of the growing season, while *A. solani* became apparent at later time points. Additionally, the genetic diversity of a subset of small-spore isolates was explored using a multilocus sequence typing analysis based on three conserved genetic regions. Our results show a high degree of diversity among the population and hints at a complex of different species, including *A. tenuissima* and *A. arborescens*. The same subset of isolates is now being tested for their pathogenicity and toxin fingerprint using high-throughput *in vitro* infection assays and an LC-MS/MS method respectively. Based on the outcome of these analyses, a subset of isolates will be used to investigate the complex interaction between the disease progress and stress-related hormones such as ethylene or auxins. Indeed, previous research shows that ethylene is an important factor in *Alternaria* spore germination and that it is a key component in upstream signaling of Programmed Cell Death induced by host-specific A. *alternata f. sp. lycopersici* (AAL)-toxin.

3.49. Evaluation of Next Generation Liquid Chromatography—Single Quadrupole Mass Spectrometry for Screening and Quantitative Analysis of Multiple Mycotoxins in Foods

Van Hulle, M.; Claereboudt, J.; Lattanzio, V.M.T.; Ciasca, B.; Stead, S.; McCall, E.; Powers, S.; Visconti, A.

Mycotoxin detection is of great importance in regulated environments such as food and animal feed analysis. Maximum permitted levels for the major mycotoxins, namely aflatoxins (AFB_1, AFB_2, AFG_1, AFG_2), ochratoxin A (OTA), fumonisins (FB_1, FB_2), deoxynivalenol (DON), and zearalenone (ZEA) are set by the European legislation (1881/2006/EC, 1126/2007/EC). Indicative maximum levels for the sum of T-2 and HT-2 toxins have been recently issued in the Recommendation 2013/165/EU. Although not regulated, attention is paid to the occurrence of nivalenol (NIV), another *Fusarium* toxin that frequently occurs in cereals also in combination with DON. Analytical methods for determination of major mycotoxins in food matrices need to be sensitive, selective and robust to provide accurate data when applied for monitoring, risk assessment, quality control and research. Most of the reference methods currently used for quality control purposes are based on immunoaffinity columns (IAC). However, due to the specificity of the antibodies used in these columns towards individual mycotoxins, different methods have been developed for a single mycotoxin or for closely related mycotoxins. In the last decade, the commercial availability of multi-mycotoxin IACs has opened new frontiers to the analysis of these contaminants in food control laboratories.

A liquid chromatography/mass spectrometry (LC-MS) method was developed for the simultaneous determination of aflatoxins (B_1, B_2, G_1, G_2), OTA, fumonisins (B_1 and B_2), NIV, DON, ZEA, T-2 and HT-2 in cereal based foods. A double extraction approach, based on high speed blending with water followed by methanol was applied for the effective co-extraction of the 12 mycotoxins under investigation in 4 min. Multi-toxin immunoaffinity columns (Myco6in1+^{TM}, Vicam) were used for cleanup of the extract. The simultaneous detection and quantification of the 12 mycotoxins was performed by LC-MS evaluating performances of a new mass detector based on single quadrupole technology (Acquity QDa Detector, Waters). Furthermore, mycotoxin fragmentation patterns obtained by collision induced dissociation (CID) were investigated to identify two characteristic masses per each mycotoxin. The resulting detection approach enabled the acquisition of quantitative and confirmatory information in a unique chromatographic run. Reliability of this approach was evaluated by establishing the repeatability of ion ratios (quantifier/qualifier ion) by repeated measurements in standard solution and matrix extracts. Method performances such as linearity range, recoveries from spiked samples, quantification limits were evaluated and proved the method to be suitable to assess with a single analysis, compliance of the selected food commodities with the EU maximum permitted or recommended levels for all regulated mycotoxins. The developed LC-MS detection approach could represent, in routine mycotoxin analysis, a cost-effective alternative tool to more sophisticated LC-MS/MS equipments.

3.50. Using Ion Mobility Mass Spectrometry and Collision Cross Section Areas to Elucidate the α and β Epimeric Forms of Glycosylated T-2 Toxin

Claereboudt, J.; Stead, S.; McCullagh, M.; Busman, M.; McCormick, S.; Crich, D.; Kato, T.; Lattanzio, V.; Maragos, C.

Toxigenic fungi often grow on edible plants, thus contaminating food and feed with fungal metabolites. Plants can alter the chemical structure of mycotoxins as part of their defence against xenobiotics. The extractable conjugated or non-extractable bound mycotoxins formed remain present in the plant tissue but are currently neither routinely screened for in food nor regulated by legislation, thus they may be considered "masked". *Fusarium* species mycotoxins are prone to metabolism or binding by plants. Toxicological data are scarce, but several studies highlight the potential threat to consumer safety from these substances. In particular, the possible hydrolysis of masked mycotoxins back to their toxic parents during mammalian digestion raises concerns. Masked mycotoxins may also elude conventional analysis because of modified physicochemical properties. All of these effects may lead to a potential underestimation or overestimation of the total mycotoxin content of the sample.

In this study we report the use of High Definition Mass Spectrometry (HDMS) as a powerful tool for the separation and characterisation of α and β epimeric forms of glycoslated T-2 and related toxins. The α-glycosylated T-2 standard was isolated from *Blastobotrys muscicola* cultures following exposure to T-2 and the β form produced *via* chemical synthesis. HDMS is a combination of high resolution mass spectrometry and high efficiency ion mobility separation. Ion mobility spectrometry (IMS) is a rapid orthogonal gas separation phase technique which allows another dimension of separation to be obtained within an UPLC timeframe. Compounds can be differentiated based on their size, shape and charge. In addition, both precursor ion and fragment ion information can be simultaneously acquired in a single injection in an HDMS experiment, referred to as $HDMS^E$. $HDMS^E$ data not only provides additional peak capacity but also insights into the molecular characteristics of the analytes for example, the elucidation of different isomeric species and intra-molecular sites of protonation.

The ion mobility drift time data acquired under $HDMS^e$ mode was used to calculate the collision cross section area (CSS) values for both precursor ion and fragments within the data processing software of 244.85 and 251.33 Angstroms for the α and β T-2 glycosides, respectively. The combination of CCS, retention time, exact mass and fragmentation information provides a unique characteristic signature for the compounds. The individual CCS values derived for the α and β epimers can be used to determine which epimeric form of the toxin is present in the sample and can serve as a valuable tool during toxicological and profiling studies.

4. Author Affiliations

- Abdallah, M.F., Assiut University, Assiut, Egypt
- Arndt, B., Westfälische Wilhelms-Universität Münster, Münster, Germany
- Arroyo-Manzanares, N., Ghent University, Ghent, Belgium
- Atanda, O., McPherson University Ogun State, Nigeria
- Atoui, A., Lebanese Atomic Energy Commission, Beirut, Lebanon
- Audenaert, K., Ghent University, Ghent, Belgium
- Baranyi, N., University of Szeged, Szeged, Hungary
- Baydar, T., Hacettepe University, Ankara, Turkey
- Becker, S., Westfälische Wilhelms-Universität Münster, Münster, Germany
- Beloglazova, N., Ghent University, Ghent, Belgium
- Bencsik, O., University of Szeged, Szeged, Hungary
- Bhatnagar, D., USDA, New Orleans, LA, USA
- Bonin, L., Arvalis, Boigneville, France
- Busman, M., USDA-ARS National Center for Agricultural Utilization Research Peoria, Illinois, USA
- Callebaut, A., CODA-CERVA (Veterinary and Agrochemical Research Center), Tervuren, Belgium
- Calvo, A., Northern Illinois University, Dekalb, IL, USA
- Camenzuli, L., RIKILT, Wageningen University and Research centre, Wageningen, the Netherlands
- Cano-Sancho, G., University of California at Davis, Davis, USA
- Caron LF, Universidade Federal do Parana, Curitiba, Brazil
- Decleer, M., Ghent University, Ghent, Belgium
- Cary, J.W., USDA, New Orleans, LA, USA
- Chilaka, C.A., Ghent University, Ghent, Belgium
- Ciasca, B., National Research Council of Italy, Bari, Italy
- Claereboudt, J., Waters NV/SA, Zellik, Belgium
- Claes, J., Faculty of Engineering Technology, KU Leuven, Geel, Belgium

- Cramer, B., Westfälische Wilhelms-Universität Münster, Münster, Germany
- Crich, D., Wayne State University, Detroit, USA
- Croubels, S., Ghent University, Merelbeke, Belgium
- Cujic, S., SP Laboratorija AD, Becej, Serbia
- Cverheec, C., Université de Toulouse, INP-ENSAT, Castanet-Tolosan, France
- Dauthieux, F., Arvalis, Thiveral-Grignon, France
- De Baere, S., Ghent University, Merelbeke, Belgium
- De Boevre, M., Ghent University, Ghent, Belgium
- De Gelder, L., Ghent University, Ghent, Belgium
- De Middeleer, G., Ghent University, Ghent, Belgium
- De Ruyck, K., Ghent University, Ghent, Belgium
- De Saeger, S., Ghent University, Ghent, Belgium
- De Zutter, N., Ghent University, Ghent, Belgium
- Debevere, S., Ghent University, Merelbeke, Belgium
- Devreese, M., Ghent University, Merelbeke, Belgium
- Dhaouadi, S., Impextraco NV, Heist-op-den-Berg, Belgium
- Diana Di Mavungu, J., Ghent University, Ghent, Belgium
- Dubruel, P., Ghent University, Ghent, Belgium
- Ducerf, R., Arvalis, Boigneville, France
- Dzhavakhiya, V.G., All-Russian Research Institute of Phytopathology, Bolshie Vyazemy, Russia
- Ehrlich, K.C., Tulane University, New Orleans, LA, USA
- El Khoury, A., Université Saint-Joseph, Mkalles-Liban
- El Khoury, R., Université Saint-Joseph, Mkalles-Liban
- Elliott, C.T., Queen's University Belfast, Belfast, United Kingdom of Great Britain and Northern Ireland
- Etoa, F.-X., University of Yaoundé I, Cameroon
- Farkas, H., SP Laboratorija AD, Becej, Serbia
- Fievez, V., Ghent University, Ghent, Belgium
- Fokou, E., University of Yaoundé I, Cameroon
- Fotso, M., Centre for Food and Nutrition Research, Yaoundé, Cameroon
- Foubert, A., Ghent University, Ghent, Belgium
- Girgin, G., Hacettepe University, Ankara, Turkey
- Goderis, A., AGRIMEX nv, Lille, Belgium
- Goftman, V., Saratov State University, Saratov, Russia
- Gong, Y.Y., Queen's University Belfast, Belfast, United Kingdom of Great Britain and Northern Ireland
- Goryacheva, I., Saratov State University, Saratov, Russia
- Goryacheva, O.A., Saratov State University, Saratov, Russia
- Haesaert, G., Ghent University, Ghent, Belgium
- Heyndrickx, E., Ghent University, Ghent, Belgium
- Heyndrickx, E., Ghent University, Ghent, Belgium
- Hickert, S., Westfälische Wilhelms-Universität Münster, Münster, Germany
- Hofgaard, I.S., Norwegian Institute of Bioeconomy Research, Ås, Norway
- Höfte, M., Ghent University, Ghent, Belgium
- Hoogendoorn, K., Plant Research International, Wageningen University and Research centre, Wageningen, the Netherlands
- Humpf, H.-U., Westfälische Wilhelms-Universität Münster, Münster, Germany

- Huybrechts, B., CODA-CERVA (Veterinary and Agrochemical Research Center), Tervuren, Belgium
- Huybrechts, I., International Agency for Research on Cancer, Lyon, France
- Jaksic, M., SP Laboratorija AD, Becej, Serbia
- Janson, J.P., FNAMS, Troyes, France
- Jin, L., Ghent University, Ghent, Belgium
- Kalinina, S., Westfälische Wilhelms-Universität Münster, Münster, Germany
- Karaoglanidis, G.S., Aristotle University of Thessaloniki, Thessaloniki, Greece
- Kato, T., Wayne State University, Detroit, USA
- Kiebooms, J.A.L., CODA-CERVA (Veterinary and Agrochemical Research Center), Tervuren, Belgium
- Kilicarslan, B., Hacettepe University, Ankara, Turkey
- Kocsubé, S., University of Szeged, Szeged, Hungary
- Köhler, J., Westfälische Wilhelms-Universität Münster, Münster, Germany
- Koppenol, A., Impextraco NV, Heist-op-den-Berg, Belgium
- Kovalsky, P., BIOMIN Holding GmbH, Herzogenburg, Austria
- Kowalski, E., Ghent University, Merelbeke, Belgium
- Krska, R., University of Natural Resources and Life Sciences, Vienna, Austria
- Krug, I., Westfälische Wilhelms-Universität Münster, Münster, Germany
- Labreuche, J., Arvalis, Boigneville, France
- Landschoot, S., Ghent University, Ghent, Belgium
- Lattanzio, V.M.T., National Research Council of Italy, Bari, Italy
- Laurain, J., OLMIX, Bréhan, France
- Lauwers, M., Ghent University, Merelbeke, Belgium
- Leclere, A., Arvalis, Boigneville, France
- Letor, B., Innovad, Antwerpen-Berchem, Belgium
- Logrieco, A.F., Research National Council, Bari, Italy
- Machado, P., Junior, Impextraco Latin America, Curitiba, Brazil
- Madder, A., Ghent University, Ghent, Belgium
- Maragos, C., USDA-ARS National Center for Agricultural Utilization Research Peoria, Illinois, USA
- Marengue, E., LABOCEA public lab, Ploufragan, France
- Marín, S., University of Lleida, Lleida, Spain
- Marinkovic, D., SP Laboratorija AD, Becej, Serbia
- Marosanovic, B., SP Laboratorija AD, Becej, Serbia
- Maroun, R., Université Saint-Joseph, Mkalles-Liban
- Mathieu, F., Université de Toulouse, INP-ENSAT, Castanet-Tolosan, France
- Maumené, C., Arvalis, Boigneville, France
- Maunas, L., Arvalis, Montardon, France
- McCall, E., Waters Corporation, Manchester, UK
- McCormick, S., USDA-ARS National Center for Agricultural Utilization Research Peoria, Illinois, USA
- McCullagh, M., Waters Corporation, Wilmslow, UK
- Medema, M., Plant Research International, Wageningen University and Research centre, Wageningen, the Netherlands
- Medoua, G.N., Centre for Food and Nutrition Research, Yaoundé, Cameroon
- Meleard, B., Arvalis, Boigneville, France
- Mengelers, M., RIVM, Bilthoven, the Netherlands
- Michiels, J., Ghent University, Ghent, Belgium
- Miles, C.O., Norwegian Veterinary Institute, Oslo, Norway

- Molassiotis, A., Aristotle University of Thessaloniki, Thessaloniki, Greece
- Mooney, M.H., Queen's University Belfast, Belfast, United Kingdom of Great Britain and Northern Ireland
- Muccio, M., BIOMIN Holding GmbH, Herzogenburg, Austria
- Myresiotis, C., Aristotle University of Thessaloniki, Thessaloniki, Greece
- Nacer Khodja, E., OLMIX, Bréhan, France
- Naehrer, K., BIOMIN Holding GmbH, Herzogenburg, Austria
- Ngoko, Z., Catholic University of Cameroon, Bamenda, Cameroon
- Nguegwouo, E., University of Yaoundé I, Cameroon
- Njobeh, P.B., University of Johannesburg, Johannesburg, South Africa
- Njumbe Ediage, E., Ghent University, Ghent, Belgium
- Oplatowska-Stachowiak, M., Queen's University Belfast, Belfast, United Kingdom of Great Britain and Northern Ireland
- Orlando, B., Arvalis, Boigneville, France
- Osteresch B, Westfälische Wilhelms-Universität Münster, Münster, Germany
- Powers S, VICAM, Milford, MA, USA
- Rajkovic A, Ghent University, Ghent, Belgium
- Ramos, A.J., University of Lleida, Lleida, Spain
- Rise, F., University of Oslo, Oslo, Norway
- Robin, N., Arvalis, Montardon, France
- Sajic, N., EuroProxima B.V., Arnhem, Netherlands
- Sanchis, V., University of Lleida, Lleida, Spain
- Sas, B., Ghent University, Ghent, Belgium
- Schelfhout, K., OVAM, Mechelen, Belgium
- Schelstraete, W., Ghent University, Merelbeke, Belgium
- Schmidt, H.S., Westfälische Wilhelms-Universität Münster, Münster, Germany
- Shcherbakova, L.A., All-Russian Research Institute of Phytopathology, Bolshie Vyazemy, Russia
- Sioen, I., Ghent University, Ghent, Belgium
- Smagghe, G., Ghent University, Ghent, Belgium
- Sobolev, A.M., Saratov State University, Saratov, Russia
- Vandicke, J., Ghent University, Ghent, Belgium
- Stanic ,A., Norwegian Veterinary Institute, Oslo, Norway
- Statsyuk, N.V., All-Russian Research Institute of Phytopathology, Bolshie Vyazemy, Russia
- Stead, S., Waters Corporation, Manchester, UK
- Stroka, J., IRMM Geel, Geel, Belgium
- Studt, L., University of Natural Resources and Life Sciences, Vienna, Austria
- Sulyok, M., University of Natural Resources and Life Sciences, Vienna, Austria
- Suman, M., Barilla Advanced Laboratory Research, Parma, Italy
- Szekeres, A., University of Szeged, Szeged, Hungary
- Tangni, E.K., CODA-CERVA (Veterinary and Agrochemical Research Center), Tervuren, Belgium
- Tanou, G., Aristotle University of Thessaloniki, Thessaloniki, Greece
- Testempasis, S., Aristotle University of Thessaloniki, Thessaloniki, Greece
- Thiry, C., CODA-CERVA (Veterinary and Agrochemical Research Center), Tervuren, Belgium
- Tudzynski, B., Westfälische Wilhelms-Universität Münster, Münster, Germany
- Udovicki, B., University of Belgrade, Belgrade, Serbia
- Uhlig, S., Norwegian Veterinary Institute, Oslo, Norway
- Uka, V., Ghent University, Ghent, Belgium

- Vágvölgyi, C., University of Szeged, Szeged, Hungary
- Valade, R., Arvalis, Thiveral-Grignon, France
- Van Bocxlaer, J., Ghent University, Ghent, Belgium
- Van de Mierop, K., AGRIMEX nv, Lille, Belgium
- Van der Fels-Klerx, H.J., RIKILT, Wageningen University and Research centre, Wageningen, the Netherlands
- Van der Lee, T.A.J., Plant Research International, Wageningen University and Research centre, Wageningen, the Netherlands
- Van Hulle, M., Waters NV/SA, Zellik, Belgium
- Vandecasteele, M., Ghent University, Ghent, Belgium
- Vanhaecke, L., Ghent University, Merelbeke, Belgium
- Vanhoutte, I., Ghent University, Ghent, Belgium
- Varga, J., University of Szeged, Szeged, Hungary
- Verheijen, R., EuroProxima B.V., Arnhem, Netherlands
- Vidal, A., University of Lleida, Lleida, Spain
- Visconti, A., National Research Council of Italy, Bari, Italy
- Vitry, C., Arvalis, Thiveral-Grignon, France
- Voinova, T.M., All-Russian Research Institute of Phytopathology, Bolshie Vyazemy, Russia
- Waalwijk, C., Plant Research International, Wageningen University and Research Centre, Wageningen, The Netherlands
- Wauters, J., Ghent University, Merelbeke, Belgium
- Wiemann, P., University of Wisconsin-Madison, Madison, WI, USA
- Willems, B., Innolab CVBA, Ghent, Belgium
- Xu, Y., Queen's University Belfast, Belfast, United Kingdom of Great Britain and Northern Ireland
- Zheng, H., Ghent University, Ghent, Belgium

© 2016 by the authors. Licensee MDPI, Basel, Switzerland. This article is an open access article distributed under the terms and conditions of the Creative Commons Attribution (CC BY) license (http://creativecommons.org/licenses/by/4.0/).

MDPI AG
St. Alban-Anlage 66
4052 Basel, Switzerland
Tel. +41 61 683 77 34
Fax +41 61 302 89 18
http://www.mdpi.com

Toxins Editorial Office
E-mail: toxins@mdpi.com
http://www.mdpi.com/journal/toxins

www.ingramcontent.com/pod-product-compliance
Lightning Source LLC
Chambersburg PA
CBHW041214220326
41597CB00033BA/5881

* 9 7 8 3 0 3 8 4 2 4 8 4 0 *